T0339606

GEOMETRY FOR NAVAL ARCHITECTS

GEOMETRY FOR NAVAL ARCHITECTS

ADRIAN BIRAN
Faculty of Mechanical Engineering
Technion - Israel Institute of Technology
Technion City, Israel

Butterworth-Heinemann
An imprint of Elsevier

Butterworth-Heinemann is an imprint of Elsevier
The Boulevard, Langford Lane, Kidlington, Oxford OX5 1GB, United Kingdom
50 Hampshire Street, 5th Floor, Cambridge, MA 02139, United States

Copyright © 2019 Elsevier Ltd. All rights reserved.

No part of this publication may be reproduced or transmitted in any form or by any means, electronic or mechanical, including photocopying, recording, or any information storage and retrieval system, without permission in writing from the publisher. Details on how to seek permission, further information about the Publisher's permissions policies and our arrangements with organizations such as the Copyright Clearance Center and the Copyright Licensing Agency, can be found at our website: www.elsevier.com/permissions.

This book and the individual contributions contained in it are protected under copyright by the Publisher (other than as may be noted herein).

Notices

Knowledge and best practice in this field are constantly changing. As new research and experience broaden our understanding, changes in research methods, professional practices, or medical treatment may become necessary.

Practitioners and researchers must always rely on their own experience and knowledge in evaluating and using any information, methods, compounds, or experiments described herein. In using such information or methods they should be mindful of their own safety and the safety of others, including parties for whom they have a professional responsibility.

To the fullest extent of the law, neither the Publisher nor the authors, contributors, or editors, assume any liability for any injury and/or damage to persons or property as a matter of products liability, negligence or otherwise, or from any use or operation of any methods, products, instructions, or ideas contained in the material herein.

Library of Congress Cataloging-in-Publication Data
A catalog record for this book is available from the Library of Congress

British Library Cataloguing-in-Publication Data
A catalogue record for this book is available from the British Library

ISBN: 978-0-08-100328-2

For information on all Butterworth-Heinemann publications
visit our website at https://www.elsevier.com/books-and-journals

Working together
to grow libraries in
developing countries

www.elsevier.com • www.bookaid.org

Publisher: Matthew Deans
Acquisition Editor: Carrie Bolger
Editorial Project Manager: Carrie Bolger
Production Project Manager: Anitha Sivaraj
Designer: Matthew Limbert

Typeset by VTeX

DEDICATION

To my beloved wife Suzi

CONTENTS

Part 3.   **Computer Methods**

Part 4.   Applications in Naval Architecture

ABOUT THE AUTHOR

Adrian Biran, born Birbănescu, received a *Cum laude* Diploma of Engineering, in the field of ship engineering, from the Bucharest Polytechnic Institute, and Master of Science and Doctor of Science degrees from the TECHNION — Israel Institute of Technology, the latter two with theses related to software for Marine Engineering and Naval Architecture. After graduating he worked as Design Engineer, Chief of Department and Project Leader at IPRONAV — the Institute for Ship Design, Bucharest. In continuation he managed the design office of the Bucharest Studios and worked as Project Leader at IPA — the Institute for Automation Design, Bucharest. In Israel Adrian Biran worked as Senior Engineer at the Israel Shipyards, Haifa, and as Research Fellow and Research Engineer at the Technion R&D Foundation. In parallel with the above activities Adrian Biran served as project instructor at the Military Technical Academy, Bucharest, and at the Beersheba University. He has been for many years an adjunct teacher at the TECHNION — Israel Institute of Technology, since 1995 as Adjunct Associate Professor. Adrian Biran is the author of several technical papers on various subjects including also Naval Architecture. He wrote in Romanian a book of popular science about ships. With Moshe Breiner as coauthor Adrian wrote a book on *MATLAB for Engineers* that was published in three English, three German, two French, and two Greek editions. Based on the course he has been delivering at the Technion, Adrian Biran published in 2003, with Butterworth-Heinemann, the book *Ship Hydrostatics and Stability* that was immediately translated into Turkish. In 2005 the book was reprinted with contributions by Rúben López-Pulido. A chapter on stability regulations was reproduced in the Maritime Engineering Reference Book edited by Anthony Molland. In 2011 Adrian Biran published with Taylor and Francis the book *What every Engineer should know about MATLAB and Simulink*. In 2014 a second edition of Ship Hydrostatics and Stability was issued with Rúben López-Pulido as coauthor. The book *Geometry for Naval Architects* also started from lectures delivered at the Technion.

PREFACE

Naval Architecture is based on geometry, for theory as well as for practice. Naval Architecture is also the field of technology in which some of the basic ideas of geometric modelling first appeared. Archimedes was the first to study theoretically the properties of floating bodies. He analyzed the stability of floating paraboloids and introduced geometrical notions such as *centre of buoyancy*. In the *Handbook of computer aided geometric design* (Kim et al., 2002), Gerald Farin writes

> *'The earliest recorded use of curves in a manufacturing environment seems to go back to early AD Roman times, for the purpose of shipbuilding. A ship's ribs ... were produced based on templates that could be reused many times. Thus a vessel's basic geometry could be stored ... These techniques were perfected by the Venetians from the 13th to the 16th century. The form of ribs was defined in terms of tangent continuous circular arcs — NURBS in modern parlance...'*

These matters are discussed in Section 2.1 and it is mentioned there that the idea of defining a ship section by 'pieces' with common tangents in the joining points is basic for the mathematical procedures called today 'splines'. Farin mentions the appearance in 1752 of the name *spline* for the wooden instrument used to draw smooth curves, and concludes the first part of his historical introduction with the comment

> *'This "shipbuilding connection," described by H. Nowacki ... was the earliest use of constructive geometry to define free-form shapes...'*

As Nowacki and Ferreiro (2003) point out, the modern study of the stability of ships began with the publication of the *Traité du Navire* by Bouguer, in 1746, and the *Scientia Navalis* by Euler, in 1749. From the beginning this study involved geometric properties of plane figures, such as areas, centroids, first and second moments of areas. Bouguer introduced the basic notions of metacentric radius and the metacentric evolute, two notions belonging to *differential geometry*. In 1814 and 1822 Charles Dupin studied the surface of centres of buoyancy by means of differential geometry. During the last centuries two-dimensional drawings were the main tools for ship design and construction. In 1765 Gaspard Monge introduced his *descriptive geometry*, a fully fledged system for representing in two dimensions a three-dimensional object seen from any desired direction. The method was first considered a military secret and was published only in 1799. Since then engineering drawings are based on Monge's method. The advent of digital computers enabled the development of new methods for geometric modelling, but traditional drawings still remain the main means for presenting and understanding a ship design. Generally, the first drawing of any boat or ship project is the *lines drawing*, a particular, but obvious application of descriptive geometry. The new techniques for modelling and displaying ship hulls and calculating their properties are based on branches of geometry that previously presented more theoretical, rather than practical interest. These branches include *affine geometry*, *projective geometry*, and again *differential geometry*. As Gallier (2011) puts it,

'Thus, there seems to be an interesting turn of events. After being neglected for decades, stimulated by computer science, old-fashioned geometry seems to be making a comeback as a fundamental tool used in manufacturing, computer graphics, computer vision, and motion planning, just to mention some key areas.'

Devlin (2001) too reminds that the teaching of geometry has been neglected for many years, and he points out to its role in the development of reasoning abilities. Let us insert a short quotation.

'Younger people may not have taken a geometry class. The subject was reclassified as optional some years ago in the mistaken belief that it was no longer sufficiently relevant to today's world, a view that demonstrates the ignorance of many of the people who make such decisions. Although it is true that hardly anyone ever makes direct use of geometrical knowledge, it was the only class in the high school curriculum that exposed children to the important concept of formal reasoning and mathematical proof.
Exposure to formal mathematical thinking is important for at least two reasons. First, a citizen in today's mathematically based world should have at least a general sense of one of the major contributors to society. Second, a survey carried out by the United States Department of Education in 1997 (The Riley Report) showed that students who completed high school geometry performed markedly better in gaining entry to college and did better when at college than those students who had not taken such a course, regardless of the subjects studied at college.'

My experience fully corroborates Devlin's comments and I appreciate that the above quotations are particularly valid in modern ship design. In this book proofs for many theorems and procedures are provided. By studying the proof the reader is not only convinced that the proven statement is correct, but also understands its limits of validity. Moreover, this opens the way for extending or modifying the results to other cases. As some authors say, this is the *discovery* propriety of proofs.

Present-day CAD software for naval architectural uses is inconceivable without the use of the branches of geometry mentioned here; anyone who wants to get a good insight in this field and be able to participate in the development of software should acquire the basics of these mathematics. Many books dedicated to the branches of mathematics mentioned above have been available for a long time. Books dedicated to computer-aided applications appeared in the last decades. Their titles include such terms as splines, computer graphics, or geometric modelling. By writing this book, we had no intention to replace any of the books dedicated to one subject. Our goals are:

- to fill gaps in certain scholar programs that ignore one or more of the fields required by Naval Architecture;
- to bring together the basics of these fields and explain them at the level required in practical engineering;
- to show specific applications in Naval Architecture and related fields.

These are starting points and any reader interested in deepening the knowledge in a particular field should refer to more specialized books, some of them referred to in our book. Finally, let us remark that not only some roots of computer-aided geometric design stem from Naval Architecture and ship construction, but also some of its most beautiful fruits rise from these fields. No wonder that some authors chose them to illustrate books on computer-aided geometric modelling. The internal cover of Rogers and Adams (1990) shows the lines of the America's Cup yacht *Stars & Stripes*, and the cover of Mortenson (1997) contains the keel and bulb lines of the yacht *Black Magic*.

THE ORGANIZATION OF THE BOOK

The book is divided into four parts, as detailed below.

Traditional methods — In this part we present an introduction to descriptive geometry as the basis of engineering drawing, the definition of the hull surface by means of the lines drawing, and the calculation of geometric properties of plane figures.

Differential geometry — Here we introduce the parametric representation of curves and surfaces, the fundamental notion of curvature and other related concepts, the curvatures of surfaces. We also deal with ruled and developable surfaces, two subjects of paramount importance in ship design and production.

Computer methods — We start with a chapter on cubic splines. A chapter on geometrical transformations deals with translation, scaling, rotation and reflections in usual coordinates, and continues with their treatment in homogeneous coordinates. As an introduction to the following two chapters, we define affine combinations and show how they preserve collinearity, coplanarity, and invariance under affine transformations. A whole chapter is devoted to Bézier curves and ends with a short treatment of rational Bézier curves. B-splines and NURBS are treated shortly in the last chapter of this part.

Applications in Naval Architecture — The first chapter in this part describes computer methods for transforming ship hulls to achieve desired properties. The second chapter is a short introduction to *conformal mapping*. This application of functions of complex variables has been used for calculating hydrodynamic properties that intervene in seakeeping calculations.

SOFTWARE

The book includes many examples in MATLAB, a versatile computing environment of *Mathworks* that greatly simplifies prototyping by providing many built-in functions and direct graphic procedures accessible at various levels of sophistication. Exercising in MATLAB the reader may understand in detail how the various methods work. There are also examples produced in *MultiSurf*, a product of Aerohydro. This is a user-friendly software remarkable by its visible relationship to theory and that is very well documented.

This book reuses procedures developed in the book referenced as Biran (2011). These procedures are: ArcDim, arrow, pline, point; their sources can be found on the site www.mathworks.com looking for Adrian Biran.

It may be useful to read this book in parallel with the book Biran and López-Pulido (2014).

NOTATION

Teletype characters are used for keyboard keys, for example `Ctrl`, or terms that appear in the Multisurf interface, for example `Insert`, `Point`, `Curve`, `Surface`.

ACKNOWLEDGEMENTS

Reflecting on how I began to understand and love geometry I owe much to my teachers of mathematics at the *Cultura* high school in Bucharest. Artur Holingher opened my eyes to the beauty of geometry and of geometrical proof. Isidor Sagher taught us in detail how to write a proof and how to achieve precision. Rafael Faion insisted on the importance of exercises. I thank the editors of Butterworth-Heinemann, Hayley Gray, Charlie Kent, Chelsea Johnston for initializing the project of this book, Carrie Bolger for realizing it, and Anitha Sivaraj for managing the production process. Larrie Ferreiro helped me to reach interesting historical sources. Reinhard Siegel provided information and valuable comments on the use of the *MultiSurf* software. My discussions with the late Yakov Kohanov of the University of Haifa introduced me to nautical archeology. Our librarians, Shelly Imberg and Vered Sayag helped to obtain part of the literature used in this book. Malin Joakimson of the *Swedish Maritime Museum* informed me that I can insert a reproduction from Chapman's famous book *Architectura Navalis Mercatoria* (1768).

MATLAB is a registered trademark of The Mathworks, Inc. For product information, please contact *The Mathworks, Inc.*, 3 Apple Hill Drive, Natick, MA 01760-2098, USA.

PART 1

Traditional Methods

CHAPTER 1

Elements of Descriptive Geometry

Contents

Geometry for Naval Architects
https://doi.org/10.1016/B978-0-08-100328-2.00010-9

Copyright © 2019 Elsevier Ltd.
All rights reserved.

3

1.1 INTRODUCTION

English	descriptive geometry
French	géométrie descriptive
German	darstellende Geometrie
Italian	geometria descrittiva
Spanish	geometría descriptiva

The graphic definition of the hull surface that will be discussed in Chapter 2 is an application of the descriptive geometry; that is why we present in this chapter the principles of this branch of mathematics. The hull surface of a real ship cannot be defined by simple analytical expressions although, as mentioned in Chapter 2, attempts to do this were made. A graphic definition has been used for centuries and it still is the common way of visualizing ship forms and use them in design. A simple surface, composed of plane faces, can be defined by projections; more complex surfaces, for example those of ship hulls, require also sections. Thus, for detailed viewing of 3D objects we employ 2D representations. The formal basis of the method was developed by Gaspard Monge (French, 1746–1818) and it starts with the fact that two 2D projections are necessary and sufficient to define a point in 3D space.

To understand how we represent an object we must first understand how we see it. The human eye, like any photographic or TV camera, sees the world in *central projection*. Drawing and measuring in central projection requires special techniques. Therefore, in engineering drawing it is necessary to accept certain assumptions that greatly simplify the work. These assumptions are the bases of the parallel, and of the orthogonal projections. The systematic treatment of the latter is achieved in descriptive geometry, a branch of mathematics that allows drawing and measuring by purely graphic means. The aim of descriptive geometry is to define a three-dimensional object by two-dimensional projections so that the exact shape and dimensions can be retrieved. Most engineering drawings are orthogonal projections, each projection showing two coordinates of the drawn points. Therefore, to store three coordinates and completely represent an object, at least two projections are necessary. The largest part of this chapter is devoted to such projections. Taking as examples such simple surfaces as that of a cube, a cylinder or a cone, we show that given two projections it is possible to obtain a third one, as well as plane sections of the surface. In Chapter 2 we apply these methods to the much more complex ship-hull surface as represented in conventional lines drawings.

It is possible to show all three coordinates in one axonometric projection. As this kind of display is included in most CAD packages available today, we insert a brief introduction to this subject.

Descriptive Geometry works with 2D drawings. The drawings of ship lines described in the next chapter are also 2D projections and sections. The CAD software available today allows 3D modelling. However, as pointed out in Anonymous (2014), the documentation submitted for approval by regulating bodies of the shipbuilding industry includes 2D drawings, mainly of ship structures. The same source also mentions that

> *'... to the trained eye, a properly executed set of 2D class approval drawings provide a decent enough 3D picture of the vessel being designed.*
>
> *...*
>
> *Accurately detailing and interpreting 2D approval drawings in the marine industry continues to require extensive specialized knowledge and skill.'*

This specialized knowledge is based in the first place on descriptive geometry, even if the name of this discipline is not mentioned in some recent textbooks of engineering drawing.

When dealing with conic surfaces we show that their plane sections are curves such as the circle, the ellipse, the parabola, or the hyperbola. Their collective name is *conic sections*. We recall their definitions as loci and derive their canonical equations. For more details we refer the reader to books of analytical geometry, such as Fuller and Tarwater (1992), or Bugrov and Nikolsky (1982).

Bodies having certain simple surfaces can be produced by bending a planar cardboard or metal sheet. Therefore, we show how to develop the surfaces of a cube, a cylinder, and a cone. The notion of developed surface is essential in ship production and we'll continue the discussion of the subject in Chapters 2 and 5.

This chapter ends with two appendices that introduce the reader to the implementation of geometric concepts on computers. The first appendix shows that geometric operations can be performed by linear algebra, and, therefore, in MATLAB. The second introduces the reader to MultiSurf.

1.2 NOTATIONS

We summarize here the notations used in this chapter. Alternative notations may be found in the literature of specialty. The indicated colours are for the electronic version of the book.

Object	Notation	Examples
Point	upper-case Latin letters	$A, B, ...P$
Straight line	lower-case Latin letters	l, g
Plane	upper-case Greek letter	Δ, Γ
Projection planes	Greek letter π with subscripts	π_1 for the horizontal plane π_2 for the frontal plane π_3 for the side (profile) plane π_4 for an additional plane
Projected object	prime marks	P' horizontal projection of P P'' frontal projection of P P''' side projection of P P'''' projection on π_4 Similarly for lines
Object line	solid, black line	see Fig. 1.1
Hidden object line	dashed, black line	see Fig. 1.1
Line connecting two projections of the same point (in French *ligne de rappel*)	dashed, blue line	see Fig. 1.1
Axis of symmetry	dash–dot line	see Fig. 1.1
Auxiliary lines	continuous, red line	see Fig. 1.1

The *prime marks* used in this chapter as superscripts for projections should not be confounded with the prime notation of derivatives (see Section 4.5). With our notation we can define, for example, a set of points P_1, P_2, P_3, P_4. Then, the line composed of the segments that connect these points is $P_1P_2P_3P_4$, its horizontal projection is $P_1'P_2'P_3'P_4'$, a.s.o.

1.3 HOW WE SEE — THE CENTRAL PROJECTION

Before learning how to represent an object, we must understand how we see it. In Fig. 1.2 the object is the letter V; it lies in the **object plane**. Rays of light coming from the object are concentrated by the **lens** of the

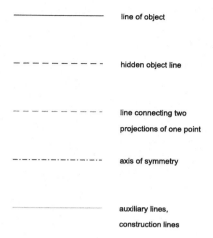

——————————   line of object

— — — — — — — —   hidden object line

— — — — — — — —   line connecting two
projections of one point

—·—·—·—·—·—·—·—·—   axis of symmetry

————————   auxiliary lines,
construction lines

Figure 1.1 Conventions for line notation

How we see the world, central projection

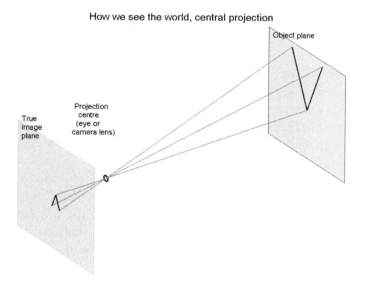

Object plane

Projection
centre
(eye or
camera lens)

True
image
plane

Figure 1.2 How we see the world

eye (*cornea*) and projected on the *retina*, forming there the **image** of the object. This image is *inverted*. The retina is not plane, but, for the small angles of ray bundles involved, we can consider it to be planar.

Images in photographic cameras are formed in the same way as in the eye, but in this case the image is, indeed, plane. In the older, non-digital

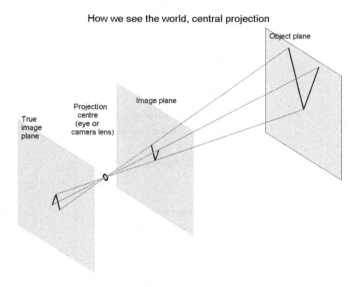

Figure 1.3 How we see the world

photographic cameras the image is formed on films. In digital cameras the image is formed on arrays of electronic sensors. In photographic cameras too the image is inverted, and the image plane is known as the **focal plane**. In the *pinhole*, or *camera obscura*, a device known for at least 2000 years, light is admitted through a narrow hole into a dark chamber. An inverted image is formed on the opposing wall.

To simplify the geometrical reasoning, it is convenient to consider an image plane situated between the lens and the object, at the same distance from the lens as the true image plane. In Fig. 1.3 we see the true image plane and the one introduced here, while in Fig. 1.4 we see only the latter. It is obvious that the size of the image is equal to that on the true image plane, but the position on the new image plane is the same as on the object plane and not inverted. Therefore, from here on we will base our explanations on the latter description.

1.4 CENTRAL PROJECTION

1.4.1 Definition

In the preceding section we considered an object and its image. We say that the image is the **projection** of the object on the plane of the retina, film

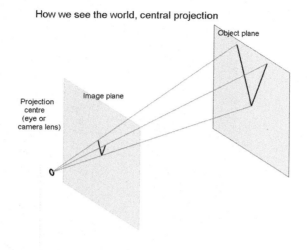

How we see the world, central projection

Figure 1.4 Conventional image plane

or array of electronic sensors. All the lines that connect one point of the object with its projection (that is its image) pass through the same point that we call **projection centre**. This kind of projection is called **central projection**. In French we find the term *perspective centrale* (see, for example, Flocon and Taton, 1984), in German *Zentralprojektion* (see, for example, Reutter, 1972), and in Italian *prospettiva centrale* (see, for example, Chirone and Tornincasa, 2005). Definitions regarding the central projection can be read in ISO 5456-4:1996(E).

1.4.2 Properties

Fig. 1.4 allows us to discover a first property of the central projection. If the object and the image planes are parallel, the image of a plane object is **geometrically similar** to the object.

We examine now other situations. In Fig. 1.5 the point P is the object and E the eye. We consider a system of coordinates in which the xy-plane is horizontal and contains the point P, the yz-plane is the image plane, and the eye lies in the zx-plane. The image, or projection of the point P is the point P' in which the *line of sight PE* pierces the image plane. Theoretically, we can consider object points situated before or behind the image plane.

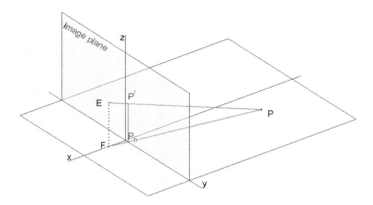

Figure 1.5 A point in central projection

In Fig. 1.5 we use a graphic method to find the projection P'. If the method is not obvious at this stage, we invite the reader to accept it now as such and to return to it after reading the whole chapter. We draw through E a perpendicular on the horizontal, xy-plane and find the point F in which it intersects the x-axis. The line FP intersects the y-axis in P_h. We raise the vertical from P_h and intersect the line EP in P'. By construction the point P' belongs both to the image plane and to the line of sight; it is the central projection of the point P.

The central projection of the projection centre, E, is not defined. Neither can we find the projection of the point F. In general, we cannot construct the projection of a point that lies in the plane that contains the projection centre and is parallel to the image plane. These images are sent to infinity. In practice, this fact may not appear as a disadvantage, but it is a nuisance for mathematicians. To build a more convenient theory it is necessary to include the points at infinity into the *space* under consideration. This is assumed in the branch of mathematics called **projective geometry** where the additional points are called **ideal points** (see, for example, Penna and Patterson, 1986, or Courant and Robbins, 1996).

Let us turn now to Fig. 1.6. The segment $\overline{EE_p}$ is perpendicular to the image plane and E_p belongs to this plane. We cannot find in the plane xy a point whose image is E_p; however, we can think that if we consider a point P on the axis x and its image, when the point P moves to infinity (in either direction) its image approaches E_p. To understand the geometrical significance of the point E_p we must see how straight lines are projected under a central projection.

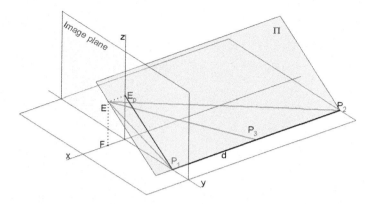

Figure 1.6 Central projection of line perpendicular to the image plane

In Fig. 1.6 we consider the line d belonging to the horizontal plane xy, and perpendicular to the image plane. The point P_1 belongs both to the line d and to the image plane. Therefore, this point coincides with its projection. We can choose two other points, P_2 and P_3, of the line d and look for their projections. The line d and the point E define a plane. Let us call it Π. The plane Π contains all the projection rays of points on d; it is the plane that **projects** d on the image plane, with E as the projection centre. It follows that the projection of d is the intersection of the plane Π with the image plane. Then, the projections of the points P_1, P_2 and P_3, and, in general, the projections of all points of d lie on the line that passes through the points P_1 and E_p. As $\overline{EE_p}$ is perpendicular to the image plane, it is parallel to d and belongs to the plane Π. Let us think now that the point P_2 moves towards infinity. Then, its projection moves towards E_p; this is the image of the ideal point of the line d.

In the reasoning above we made no assumption as to the position of the line d; all we said is that it is perpendicular to the image plane. We can conclude that the projections of all lines perpendicular to the image plane end in the point E_p. In other words, E_p is the projection of the ideal points of all lines perpendicular to the image plane. The reader may generalize the above considerations to lines that are not perpendicular to the image plane, but make any angle other than zero with it. In fact, in *projective geometry* — the branch of geometry that evolved from the study of the central projection — all lines parallel to a given direction have the same ideal point.

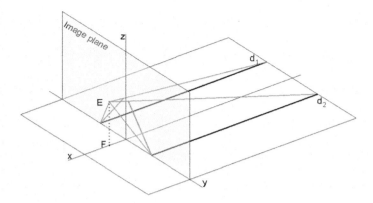

Figure 1.7 Central projection, lines perpendicular to the image plane

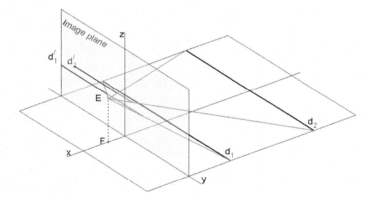

Figure 1.8 Central projection, lines parallel to the image plane

The particular case of lines parallel to the image plane is illustrated in Fig. 1.8 where we consider two lines, d_1 and d_2, parallel to the image plane. Their projections, d_1' and d_2', are parallel to the y-axis and, thus, are parallel one to the other.

The analysis of Figs 1.7 and 1.8 leads us to the conclusion that *in central projection parallel lines are projected as parallel if and only if they are parallel to the image plane.* In mathematical terms we like to say that the parallelism of lines parallel to the image plane is an **invariant** of central projection. In general, however, parallelism is not conserved under central projection.

Fig. 1.8 reveals another property of central projection. The segments of the lines d_1 and d_2 shown there are equal, but the projection of the

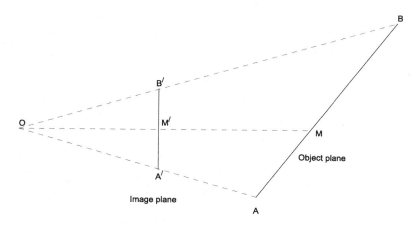

Figure 1.9 Central projection, the simple ratio is not invariant

segment of d_2 is shorter than the projection of the segment of d_1. The farther a segment is from the image plane, the shorter is its projection. This phenomenon is known as **foreshortening** and it contributes to the sensation of depth conveyed by central projection.

There is another property that is not invariant under central projection. In Fig. 1.9 the point O is the centre of projection, $\overline{AB}$ is a line segment in the object plane, and $\overline{A'B'}$, its central projection on the image plane. As one can easily check, M' is the midpoint of the segment $\overline{A'B'}$, while M is not the midpoint of the segment $\overline{AB}$. This simple experiment shows that the central projection does not conserve the **simple ratio** of three collinear points. Formerly we can write that, in general,

$$\frac{\overline{A'M'}}{\overline{M'B'}} \neq \frac{\overline{AM}}{\overline{MB}}$$

An exception is the case in which the object and the image plane are parallel.

The central projection, however, conserves a more complex quantity, the **cross-ratio** of four points. In Fig. 1.10 we consider the projection centre, O, and four collinear points, A, B, C, D, of the object plane. The four points lie on the line d that together with the projection centre defines a projecting plane; let its intersection with the image plane be the line d'. The projection of the point A is A', that of B, B', a.s.o. Following Reutter (1972) we draw through O a line parallel to the object line, d. It intersects d' in F. We also draw through D' a line parallel to d. Its intersections with

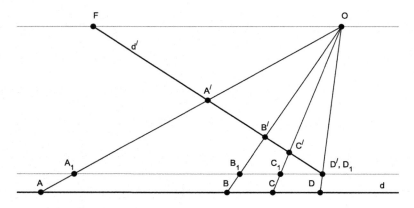

Figure 1.10 The invariance of the cross-ratio

the projection rays are A_1, B_1, C_1 and D_1, the latter being identical, by construction, with D'. Reutter uses the figure to prove that

$$\frac{\overline{AC}}{\overline{BD}} : \frac{\overline{AD}}{\overline{BD}} = \frac{\overline{A'C'}}{\overline{B'C'}} : \frac{\overline{A'D'}}{\overline{B'D'}} \tag{1.1}$$

The quantity at the left of the equal sign is the **cross-ratio** of the points A, B, C, D and is noted by $(ABCD)$. The quantity at the right of the equal sign is the cross-ratio of the points A', B', C', D' and is noted by $(A'B'C'D')$. The *cross-ratio of four collinear points is an invariant of central projection*. This property has important theoretical applications, but it is also used in practice, for instance in the interpretation of aerial photos. In French, the term cross-ratio is translated as *birapport*, in German as *Doppeltverhältnis*, and in Italian as *birapporto*. More about the subject can be read, for example, in Brannan et al. (1999), or in Delachet (1964).

1.4.3 Vanishing Points

Let us list the properties revealed by studying the central projections of various lines.

- Parallel lines parallel to the image plane are projected as parallels.
- Parallel lines not parallel to the image plane are projected as converging lines.
- All lines parallel to a given direction are projected as lines converging to the same point.

The above properties determine the character of images formed under central projection. Thus, Fig. 1.11 shows a room defined by lines parallel

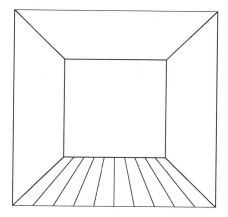

Figure 1.11 One-point perspective

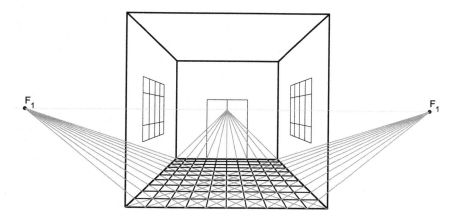

Figure 1.12 The construction of a one-point perspective

or perpendicular to the image plane. While the former are projected as parallel lines, the latter appear as lines converging to one point, as seen in the centre of the figure. Fig. 1.12 shows the construction of the lines belonging to the pavement. The point of convergence is called **vanishing point**. In Figs 1.11 and 1.12 there is only one vanishing point. Therefore, such a projection is known as **one–point perspective**.

In a more general situation, the object can be defined by a set of lines parallel to the image plane, usually vertical, and sets of lines parallel to two directions oblique to the image plane. We obtain then a **two–point perspective**, as in Fig. 1.13.

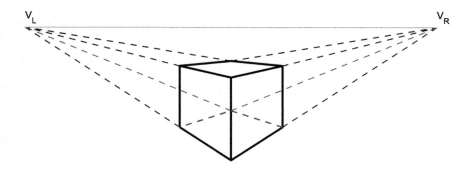

Figure 1.13 Two-point perspective

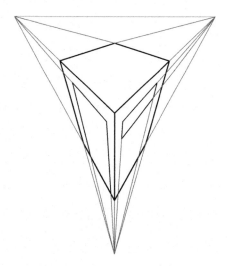

Figure 1.14 The construction of a three-point perspective

In the most general situations there are no lines parallel to the image plane. If we can identify sets of lines parallel to three directions, we can construct a **three-point perspective**, as in Fig. 1.14. Such perspectives are used, for example, to obtain *bird views* of tall buildings.

When Renaissance artists discovered the laws of perspective, they liked to exercise with it and draw spectacular images. One particularly popular item was the *pavement* (in Italian *pavimento*) which, as exemplified in Figs 1.11 and 1.12, greatly contributed to the sensation of space. More recently, the Dutch graphic artist Maurits Cornelis Escher (1898–1972) and the Spanish surrealist painter Salvador Dali (1904–1989) exploited in remarkable manners the visual effects of perspectives.

1.4.4 Conclusions on Central Projection

The central projection produces the **true** perspective, that is images as we see them, and as photographic cameras (film or digital) and TV cameras see. As we saw, the central projection conserves the parallelism of lines parallel to the image plane, the geometric similarity of objects in planes parallel to the image plane, and the cross–ratio. On the other hand, the central projection does not conserve the parallelism of lines that have any direction oblique to the image plane, does not conserve the simple ratio, and the foreshortening effect can lead to ambiguity of size. As a result, to draw in central projection one must use elaborate techniques, and Fig. 1.12 is just a very simple example. Still worse, to pick up measures from a central-projection drawing one has to use special calculations. Architects and graphic artists draw in central projection in order to obtain realistic effects. The quantitative interpretation of central projections, such as aerial photos and images seen by a robot requires the use of elaborate techniques and not-so-simple mathematics. In engineering drawing, however, we are interested in using simpler techniques, both for drawing and for reading a drawing. Therefore, we must accept assumptions that make our work simpler, even if we depart from reality. These matters are described in the next sections. We think, however, that the preceding introduction was worth reading because

1. the user of projections described in continuation must know to what extent engineering drawings depart from reality and why we have to accept this fact;
2. the Naval Architect who wants to retrieve ship dimensions from photographic images must know that this task requires specialized techniques;
3. the student who wants to deepen the knowledge or continue to the study of computer vision, must know in which direction to aim;
4. last, but not least, software packages for Naval Architecture, such as MultiSurf, provide the option of displaying in central projection.

1.5 A NOTE ON STEREOSCOPIC VISION

The projections used in engineering drawing are 2D images. The popular claim that present-day CAD programs yield 3D images is an abuse of language. The model built and stored by the computer may be, indeed, three-dimensional, but the display on a conventional screen is as two-dimensional as the screen itself is. This is not the way a man with healthy vision sees the world. The images formed on the retinas of the two eyes differ slightly one from the other because the two eyes see the object from slightly different positions. The sensorial information obtained by the eyes is transmitted to the brain where it is processed to yield the three-dimensional perception. It follows that for *true* 3D view one needs two different images and two eyes. In professional terminology we say that the two images constitute a *stereo pair*. The observer must view this pair by means of a device or system that enables each eye to see only one of the two images. This is not what the vast majority of the available CAD software and hardware does today. Various methods have been devised and implemented for creating the three-dimensional illusion, but they still are not common tools for engineering design. If some drawings and paintings give us the sensation of depth, this feeling is obtained by other means. For static images these means are the foreshortening of dimensions, the convergence of parallel lines, and shadows. In coloured photos and paintings the more distant an object is, the more its colours are fading and tending to be bluish. Movies provide a further clue for the third dimension. When the camera turns around a set of objects, some distant objects are hidden by closer ones, while other distant objects appear from behind closer ones. Convincing as they may be, such images are not three-dimensional.

This is the place to point out that stereoscopic vision is not a uniformly distributed attribute of mankind. Only a relatively small percentage of the population enjoys full three-dimensional vision, while a similar percentage does not see three-dimensionally at all. The quality of stereo vision of the remaining population ranges between the two extremes.

There is a particularly simple method of achieving the stereoscopic vision of man-made images, namely the *anaglyphs*. In this method the stereoscopic pair consists of two images in different, suitably chosen colours. One frequent choice is red for the left image and cyan for the right one. To view the stereo pair the observer has to wear glasses with the left-hand lens red, and the right-hand one cyan. The observer perceives a black three-dimensional image. Vuibert has used anaglyphs in his book *Les anaglyphes géometriques* published in 1912. More recently, other authors have added anaglyphs to their books on descriptive geometry (see, for example, Schmidt, 1977).

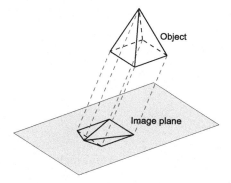

Figure 1.15 Parallel projection of a pyramid

1.6 THE PARALLEL PROJECTION

1.6.1 Definition

To simplify our work, we accept a first, strong assumption: *the projection centre is sent to infinity*. Obviously, all projection lines become parallel. What we obtain is called **parallel projection** (sometimes *oblique projection*). An example is shown in Fig. 1.15. As we shall see in the next section, by letting the projection rays be parallel we add a few properties that greatly simplify the execution and reading of drawings. In real life we cannot think about looking to objects from infinity. The farther we go from an object, the smaller its image is. From a certain distance we practically see nothing. We ignore this fact and, as said, we assume that we do see the world from infinity.

1.6.2 A Few Properties

In Fig. 1.16 the line d belongs to the object plane and the line d' belongs to the image plane. The parallel projection of the point A is A', that of B is B', and that of C is C'. The projection lines, AA', BB', CC', are parallel; they can make any angle with the image plane, except being parallel to it. Let us draw through the point C' the line d_1 parallel to d. The projection lines intersect the line d_1 in the points A_1, B_1, C_1. By construction

$$\overline{A_1 C_1} = \overline{AC}, \ \overline{B_1 C_1} = \overline{BC} \tag{1.2}$$

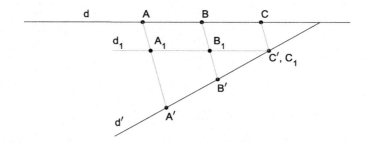

Figure 1.16 Parallel projection — The invariance of the simple ratio

Applying the theorem of Thales in the triangle $A'C'A_1$ yields

$$\frac{\overline{B'C'}}{\overline{A'C'}} = \frac{\overline{B_1C_1}}{\overline{A_1C_1}} \qquad (1.3)$$

Substituting the values given by Eq. (1.2) into Eqs (1.3) we obtain

$$\frac{\overline{BC}}{\overline{AC}} = \frac{\overline{B'C'}}{\overline{A'C'}} \qquad (1.4)$$

We conclude that *the simple ratio of three collinear points is an invariant of parallel projection*. Another property of parallel projection can be deduced from Fig. 1.15: *parallel lines are projected as parallel*. Finally and most important, Fig. 1.17 shows that *when the object plane is parallel to the image plane, distances and sizes of angles are invariant*.

1.6.3 The Concept of Scale

We know now that in parallel projection the length of the projection of a segment parallel to the image plane is equal to the real length of that segment. We say that the projection is in *true length*. This property greatly simplifies the process of drawing. Apparently, it would be sufficient to arrange the object so as to have as many segments parallel to the image plane and then draw their projections in real size. In practice, however, this is not always possible. Many objects, for example ships, are larger than the paper sheets we use. Other objects are too small and drawing them in real size would result in an unintelligible image. To solve the problem we must multiply all dimensions by a number such that the resulting image can be well accommodated in the drawing sheet and can be easily understood.

The international standard ISO 5455–1979 defines **scale** as the '*Ratio of the linear dimension of an element of an object as represented in the original drawing*

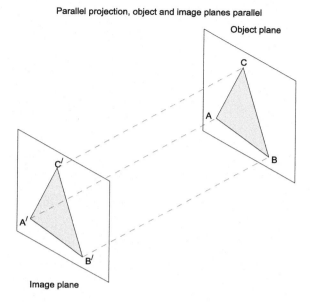

Parallel projection, object and image planes parallel

Figure 1.17 Parallel projection — The invariance of lengths and angles

Table 1.1 The scales recommended by ISO 545-1979

Enlargement	2:1 5:1 10:1 20:1 50:1
Reduction	1:2 1:5 1:10 1:20 1:50 1:100 1:200 1:500 1:1000 1:2000 1:5000 1:10000

to the real linear dimension of the same element of the object itself'. Drawing at **full size** means drawing at the scale 1:1, **enlargement** means a drawing made at a scale larger than 1:1, and **reduction**, one made at a scale smaller than 1:1. For example, if a length is 5 m, that is 5000 mm, when we use the scale 1:20 we draw it as 250 mm. On the other hand, a rectangle 2 mm by 5 mm drawn at the scale 2:1 appears as a rectangle with the sides 4 and 10 mm. Table 1.1 shows the scales recommended by ISO 5455-1979.

1.7 THE ORTHOGONAL PROJECTION

1.7.1 Definition

Parallel projections are used for *oblique axonometric projections* in Engineering Graphics, Architecture, and, to a greater extent, in general graphics. In

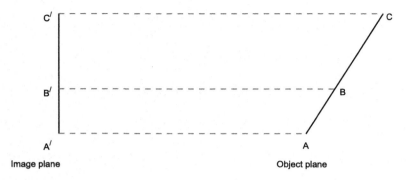

Figure 1.18 Orthogonal projection

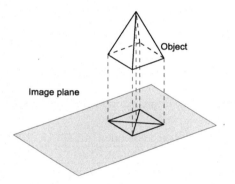

Figure 1.19 The orthogonal projection of a pyramid

Descriptive Geometry, however, we go one step farther and we assume that *the projection rays are perpendicular to the image plane.* One simple example is shown in Fig. 1.18. We obtain thus the **orthogonal projection**. This is a particular case of the parallel projection; therefore, it inherits all the properties of the parallel projection. In addition, the assumption of projection rays perpendicular to the image plane yields a new property that is described in the next section. Fig. 1.19 shows a more interesting example of orthogonal projection; the reader is invited to compare it with Fig. 1.15.

In technical drawing, when using several orthogonal projections to define one object, we talk about **orthographic projections**. Standard definitions related to orthographic representations can be read in ISO 5456-2:1996(E).

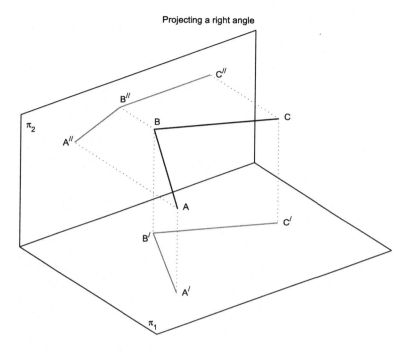

Figure 1.20 Projecting a right angle

1.7.2 The Projection of a Right Angle

Fig. 1.20 shows two line segments, $\overline{BA}$ and $\overline{BC}$, and two image planes, π_1 and π_2, perpendicular one to another. As shown in Section 1.8, such *projection planes* are used in Descriptive Geometry. In Fig. 1.20 the angle $\widehat{ABC}$ is right and the side $\overline{BC}$ is horizontal, i.e. parallel to the projection plane π_1. The figure shows also the *horizontal* and the *frontal* projections of this angle. By **horizontal projection** we mean the projection on the horizontal plane π_1, and by **frontal projection**, the projection on the frontal plane π_2. The following theorem states an important property of orthogonal projections.

If one of the sides of a right angle is parallel to one of the projection planes, the (orthogonal) projection of the angle on that plane is also a right angle.

It is possible to prove this theorem by considering the scalar products of the vectors $\overrightarrow{BA}$ and $\overrightarrow{BC}$ and of their horizontal projections. Let us give, however, an experimental proof of the right-angle theorem. We invite the

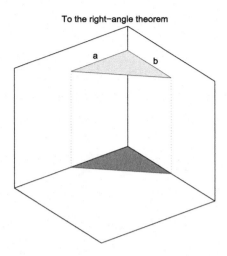

Figure 1.21 Experimental proof of the right-angle theorem

reader to take a cardboard box and a square set (triangle). First, locate the square set in the corner of the box, and orient it parallel to the bottom, as in Fig. 1.21. In the electronic version of book the square set is shown in yellow, and its horizontal projection in a darker, brown hue. This projection is right-angled.

Let now the side b of the square set remain horizontal, while we rotate the side a as in the left-hand side of Fig. 1.22. The square set still fits well in the corner of the box and the horizontal projection remains right-angled. Alternatively, keep the side a horizontal and rotate the side b as in the right-hand side of Fig. 1.22. The square set again fits well in the corner of the box and the horizontal projection remains right-angled. Finally, try to rotate both sides a and b of the square set so that none of them is horizontal, or, in other words, none of them is parallel to the bottom. It is impossible to do this without changing the angle of the box corner. In fact it must be increased. This means that the horizontal projection is no more right-angled.

Is this theorem valid only for orthogonal projections? Doesn't it hold also for an oblique parallel projection? The answer is that in the general case it does not hold. Without loss of generality, we invite the reader to check the case of a parallel, oblique projection in which the projecting rays are parallel to the frontal projection plane, π_2, but make the angle α with the horizontal projection plane, π_1.

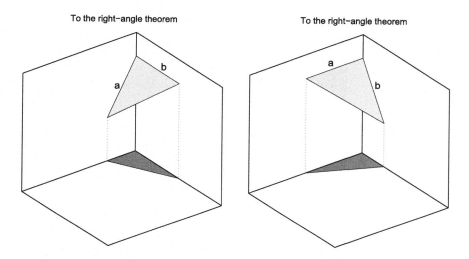

Figure 1.22 Experimental proof of the right-angle theorem

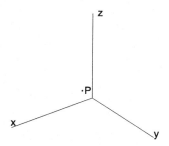

Figure 1.23 A point in a 3D system of coordinates

1.8 THE METHOD OF MONGE

It seems that the great German painter and printmaker Albrecht Dürer (1471–1528) was the first to show how three-dimensional objects can be represented by projections on two or more planes (Coolidge, 1963, pp. 109–12, Bertoline and Wiebe, 2003, pp. 394–5). The French mathematician Gaspard Monge was the first to organize in a coherent system the methods for doing so. Therefore, he is considered to be the founder of descriptive geometry. Monge was one of the scientists who accompanied Bonaparte in the Egypt campaign and one of the founders of the École Polytechnique in Paris.

Fig. 1.23 shows a point *P* and a system of cartesian coordinates. The basic idea of Descriptive Geometry is to use two *projection planes*, a *horizontal*

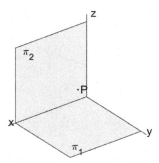

Figure 1.24 The horizontal and frontal projection planes

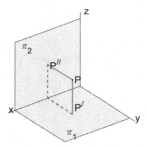

Figure 1.25 The horizontal and frontal projections of point P

one, defined by the axes x and y, and a *frontal* one, defined by the axes x and z. Let us note the horizontal plane by π_1, and the frontal one by π_2. These projection planes are shown in Fig. 1.24.

The next step is to project orthogonally the point P on π_1 and π_2. The horizontal projection, P', is the point in which a ray emitted from P, perpendicularly to π_1, pierces the plane π_1. Similarly, the frontal projection, P'', is the point in which a ray emitted from P, perpendicularly to π_2, pierces the plane π_2. These projections are shown in Fig. 1.25. Finally, as shown in Fig. 1.26, we rotate the plane π_2 until the right angle between it and π_1 becomes $180°$ and π_2 is coplanar with π_1. We obtain thus what we call the **sketch** of the point P. A sketch of a point R_1 is shown in Fig. 1.27. Sometimes it is necessary to add a third projection plane defined by the axes Ox and Oz. We note this plane by π_3 and call the projection on it *side projection*. Fig. 1.29 shows the definition in 3D space and the horizontal,

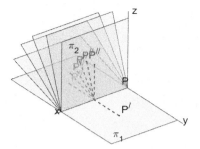

Figure 1.26 Opening the projection diedre

frontal, and side projections of a point with coordinates

$$\mathbf{P} = \begin{bmatrix} 3 \\ 5 \\ 6 \end{bmatrix}$$

We invite the reader to check that any pair of projections contains all three coordinates. The various projections of the same point are connected by dashed lines (blue in the electronic version). These lines are perpendicular to the lines that separate two projection planes. In particular, the line that separates the π_1 and π_2 planes is called *ground line*. Some authors, like Rising and Almfeldt (1964), use the term *hinge lines* for the lines that separate two projection planes.

1.9 POINTS

Points are projected as points. In a sketch, the position of the projections of a point relative to the ground line depends on the position of the point with respect to the projection planes. To explain this we refer to Fig. 1.30. The horizontal projection plane, π_1, and the frontal projection plane, π_2, extend to infinity in all directions; we can draw only small regions of them. The above-mentioned planes divide the space into four *quadrants*. We number the four quadrants going in the direct trigonometric sense. The point $\mathbf{R_1}$ is situated in the first quadrant, that is above π_1 and before π_2. As shown in Fig. 1.27, in the sketch the projection R_1' appears under the ground line, and the projection R_1'', above. We say that the point is drawn in *first angle view*. This has been, and still is the representation preferred in Europe. The point $\mathbf{R_3}$ is situated in the third quadrant, that is under π_1 and behind π_2.

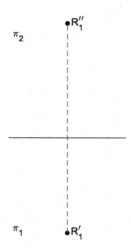

Figure 1.27 The sketch of a point in first angle view

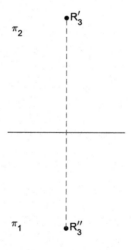

Figure 1.28 The sketch of a point in third-angle view

As shown in Fig. 1.28, in the sketch the projection R'_3 appears above the ground line, and the projection R''_3, under. We say that the point is drawn in *third angle view*. This is the representation used in USA. In these two cases the two projections of a point are well separated and, therefore, easy to distinguish. We invite the reader to check that the two projections of the points $\mathbf{R_2}$ and $\mathbf{R_4}$ appear on the same side of the ground line. As en-

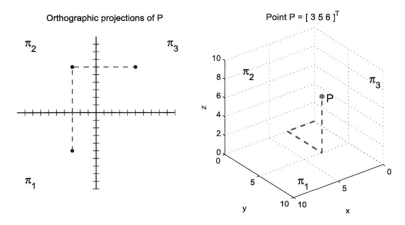

Figure 1.29 The projections of a point with given coordinates

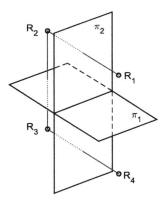

Figure 1.30 Points in four quadrants

gineering drawings contain many points and lines, such projections would be unreadable. Therefore, second and fourth angle views are not used.

1.10 STRAIGHT LINES

1.10.1 The Projections of a Straight Line

Straight lines are generally projected as straight lines, in special cases as points. As an example we show in Fig. 1.31 the sketches of two angles, each one defined by two straight lines. All four straight lines are projected as straight lines. The theorem stated in Section 1.7.2 enables us to know

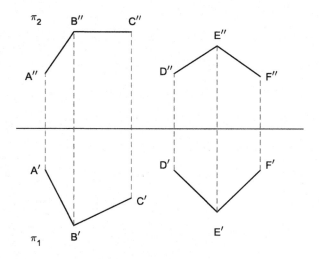

Figure 1.31 The sketches of two angles

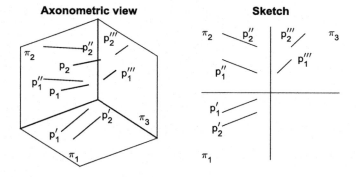

Figure 1.32 The projections of two parallels

that the angle $\widehat{ABC}$ is right, while the angle $\widehat{DEF}$ is not. Further, as the segment $\overline{BC}$ is parallel to the horizontal projection plane, π_1, the length of the projection $B'C'$ is equal to that of the segment $\overline{BC}$. We say that we read in π_1 the *true length* of the segment.

Parallel lines are projected as parallel lines on all planes, except planes perpendicular to the plane defined by the parallel lines. See Fig. 1.32 where the left-hand side is an *axonometric projection*, a term explained in detail in Section 1.17. The corresponding sketch is shown in the right-hand side of the figure.

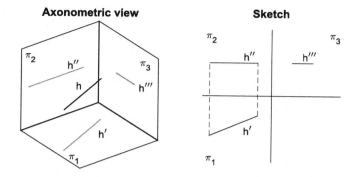

Figure 1.33 A horizontal line

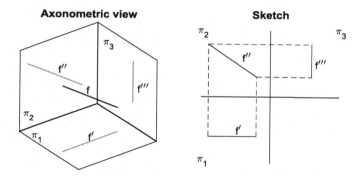

Figure 1.34 A frontal line

Fig. 1.33 describes a horizontal line, h. We identify it as such because its frontal projection, h'', is parallel to the ground line. Fig. 1.34 shows a frontal line, f. In the sketch, the horizontal projection, f', is parallel to the ground line. Fig. 1.35 shows the axonometric projection and the sketch of a line perpendicular to the side projection plane, π_3. This line is both horizontal and frontal; therefore, the projections s' and s'' are parallel to the ground line, while the side projection, s''', is a point. In the drawing of the ship lines the x-axis is such a line.

1.10.2 Intersecting Lines

In the left-hand side of Fig. 1.36 we see the projections of the lines l_1 and l_2. These lines intersect; they have a common point J. The x-coordinate of the point J is the same in the horizontal and the frontal projections. Hence, the line that connects the two projections, J', J'', is perpendicular to the ground

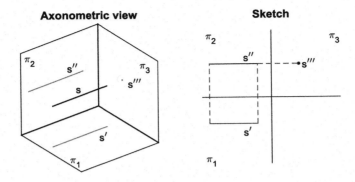

Figure 1.35 A line perpendicular on π_3

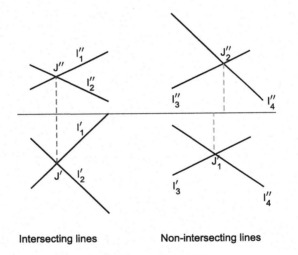

Figure 1.36 Intersecting and non-intersecting lines

line. The right-hand part of Fig. 1.28 shows the projections of the lines l_3 and l_4. These lines do not intersect. Let us draw from the projections J_1' and J_2'' perpendiculars to the ground line. These perpendiculars do not coincide. The points J_1' and J_2'' have different x-coordinates, which means that they are not the projections of the same point.

1.11 PLANES

Two planes that are not parallel intersect along a straight line. We are particularly interested in the intersection of planes with the projection planes

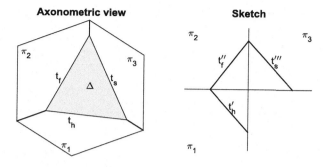

Figure 1.37 The traces of a plane

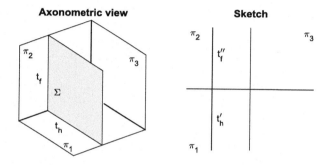

Figure 1.38 The traces of a plane parallel to π_3

and we call such intersections **traces**. To be more specific, we say that the intersection of the plane Δ with the projection plane π_i is the *trace of* Δ *on* π_i. This definition is exemplified in Fig. 1.37. To simplify the drawing, we mark in the sketch only the projections of the trace on the respective projection plane, for example t'_h on π_1. The frontal projection of the trace t_h and the horizontal projection of the trace t_f coincide with the ground line that separates π_1 from π_2. The reader may generalize to other projections of traces. The traces allow us to identify special planes. We are going to show four special planes that are used for cutting the hull surface and generate the sections that compose the drawing of ship lines discussed in Chapter 2.

Fig. 1.38 describes a plane Σ parallel to π_3; it is perpendicular to both the horizontal projection plane, π_1, and to the frontal projection plane, π_2. In the sketch, the horizontal trace, t'_h, and the frontal trace, t''_f, are collinear and perpendicular to the ground line. Such planes are used to cut the hull surface and obtain the transversal sections called *stations*. The set of all projections of these sections on π_3 is called *body plan*.

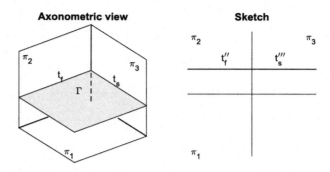

Figure 1.39 The traces of a horizontal plane

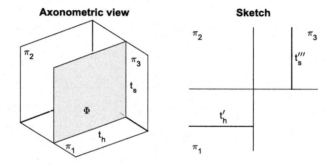

Figure 1.40 The traces of a frontal plane

Fig. 1.39 describes a horizontal plane (i.e. parallel to π_1); it is perpendicular to both the frontal projection plane, π_2, and to the side projection plane, π_3. In the sketch the frontal trace, t_f'', and the side trace, t_s''', are collinear and perpendicular to the line that separates π_2 and π_3. Such planes are used to cut the hull surface and obtain the horizontal sections called *waterlines*.

Fig. 1.40 describes a frontal plane (i.e. parallel to π_2); it is perpendicular to both the horizontal projection plane, π_1, and to the side projection plane, π_3. In the sketch, the horizontal trace, t_h', is perpendicular to the continuation of the line that separates π_2 and π_3. The side trace, t_s''', is perpendicular to the ground line. Such planes are used to cut the hull surface and obtain the longitudinal sections called *buttocks*. The set of all projections of these sections on π_2 is called *sheer plan*. The most important curve in this set is the section by the plane of symmetry of the ship; it is the *outline* of the hull.

Fig. 1.41 describes a plane perpendicular to π_3. In the sketch the horizontal trace, t_h', and the frontal trace, t_f'', are parallel to the ground line. The

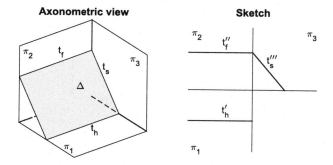

Figure 1.41 The traces of a plane perpendicular to π_3

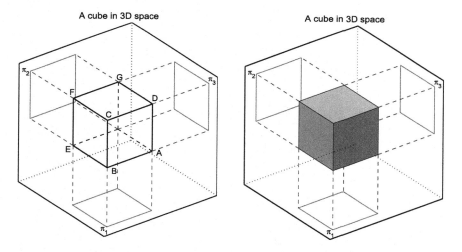

Figure 1.42 The orthographic projections of a cube

side trace, t_s''', is oblique. Such planes are used to cut the hull surface and obtain the sections called *diagonals*.

1.12 AN EXAMPLE OF PLANE-FACETED SOLID — THE CUBE

To show how descriptive geometry enables us to define a body, we begin with the very simple example of a cube. The left-hand side of Fig. 1.42 shows an axonometric view (this subject is treated in Section 1.17) of a cube and its projections on the planes π_1, π_2 and π_3. To make the illustration more 'picturesque', in the right-hand side of the same figure we add shading. The corresponding sketch is shown in Fig. 1.43. We see that some

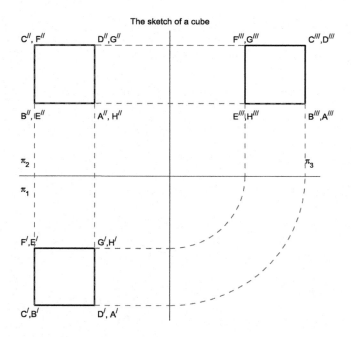

Figure 1.43 The sketch of a cube

projections coincide. For example, in the horizontal projection (projection plane π_1) the projection of the point D is superimposed on that of the point A. Also, the projection of the face $ABCD$ coincides with that of the face $EFGH$. To name the projections of the eight corner points we first write the name of the point that we see first, then a comma, and finally the name of the point that is hidden under, or behind the first point. For example, in the projection plane π_1, as we are looking from above, we see the point D, while the point A is hidden under D. Therefore, we note the point that corresponds to both projections D', A'.

Assuming that we are given the horizontal and frontal projections, the side projection is well-defined and can be drawn. To explain this, let us find the side projection, A''', of the point A whose coordinates are x_A, y_A, z_A. This projection requires two coordinates that we obtain as follows:

- y_A; we retrieve it from the horizontal projection A';
- z_A; we find it from the frontal projection A''.

Graphically, we may start by drawing from A' the dashed line parallel to the ground line. Using the compass we continue with a quarter of circle. From here we raise a vertical line; let us call it l_1. All points on this line have

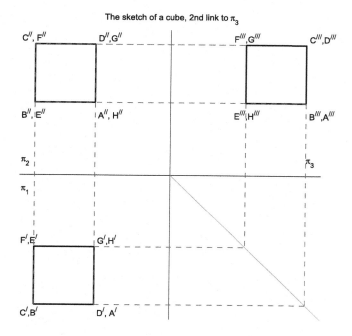

Figure 1.44 Cube sketch — Another way to relate projections

the y-coordinate equal to y_A. Now, we draw from the frontal projection A'' a horizontal line into the side projection; let us call it l_2. All points on this line have the z-coordinate equal to z_A. The projection A''' we look for lies at the intersection of the lines l_1 and l_2. The generalization of this procedure is obvious. A slight variation of the method is shown in Fig. 1.44. Here, instead of using a compass to draw half a circle, the line at 45° helps in retrieving the y-coordinates.

Fig. 1.45 shows a section parallel to the base of the cube. The horizontal projection of the section is a square, while the frontal and side projections are line segments parallel to the ground line.

1.13 A SPACE CURVE — THE HELIX

The *helix* is the line described by a point that rotates with constant radius, r, around a fixed axis and advances at constant speed in parallel with the mentioned axis. The distance, p, travelled in parallel with the axis for one complete rotation is called *pitch*. Fig. 1.46 illustrates the generation of the

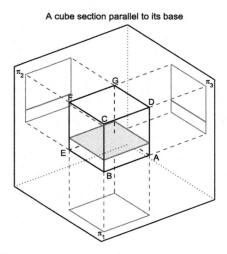

Figure 1.45 Cube section parallel to base

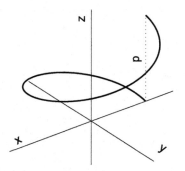

Figure 1.46 A helix

helix, and Fig. 1.47 contains a sketch of the helix. Assuming that the helix is drawn on an opaque cylinder we show its visible part as a solid line, and the hidden part as a dashed line. In Section 4.4.3 we give the parametric equations of this curve. The helix finds and important application in Naval Architecture as the geometry of screw propellers is based on this curve.

1.14 THE CYLINDER

For the general definition of the **cylinder** we show in Fig. 1.48 a curve called *directrix*, and a straight line called *generatrix*. As the generatrix moves along the directrix so that it stays always parallel to a given direction, it

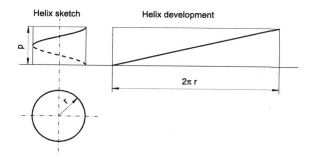

Figure 1.47 The sketch and the development of a helix

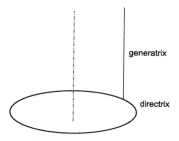

Figure 1.48 Generating a right, circular cylinder

generates the surface of a cylinder. Popularly, the term cylinder is associated with the particular case in which the directrix is a circle, and the generatrix is perpendicular to the plane of the directrix. This case is illustrated in Fig. 1.48 and the corresponding sketch appears in Fig. 1.50. Fig. 1.49 shows a cylinder whose directrix is an ellipse, a curve defined in Section 1.16.3. We can distinguish several lines parallel to the generatrix, and a number of lines parallel to the directrix. As detailed in Chapter 2, the middle body of many ships is a cylinder whose directrix is the midship section.

Cutting a right, circular cylinder with planes perpendicular to its axis yields circles. Alternatively, we may say that all sections parallel to the base are circles. If the cutting plane makes an angle with the axis, other than 0° or 90°, the section is an ellipse. In Fig. 1.51 we show how to find points on such a section. The cutting plane is perpendicular to the frontal projection plane, π_2. Therefore, the trace h' on the horizontal projection plane is perpendicular to the ground line. The trace f'' on the frontal projection plane shows the slope of the cutting plane. Let us choose a point **P** on the section line, for example one on the visible part of the cone surface. We start from

Figure 1.49 An elliptic cylinder

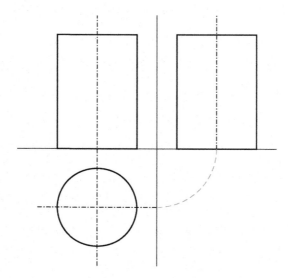

Figure 1.50 The sketch of a right, circular cylinder

the frontal projection P''. We know that the horizontal projection of all the surface is a circle. Therefore, to find the horizontal projection, P', it is sufficient to draw from P'' a perpendicular to the ground line and find the intersection with the circle in the π_1 plane. As we mean a point on the visible side of the cylinder, we are interested in the intersection with the 'forward' half-circle. Thus we obtain the y-coordinate of the point; we transfer it to the side projection plane, π_3, and draw there the dashed, ver-

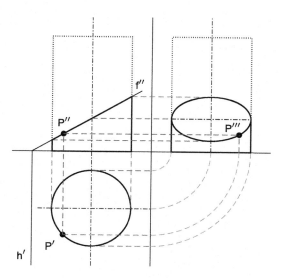

Figure 1.51 Oblique section of a right, circular cylinder

tical line corresponding to this coordinate. From the frontal projection, P'', we draw the dashed horizontal whose points have the z-coordinate of the point **P**. The intersection of the two dotted lines is the side projection P'''. The figure shows also how to obtain the centre and the points corresponding to the extremities of the two axes of the ellipse. Our solution started from the frontal projection, but it is possible to start from the horizontal projection.

1.15 THE CONE

1.15.1 Introduction

The **cone** is another simple surface treated in several branches of geometry. For the general definition we show in Fig. 1.52:

- a plane curve called *directrix*;
- a point, V, called *vertex*, that does not lie in the plane of the directrix;
- a straight line, called *generatrix*, that passes through the point V and touches the directrix.

As the generatrix moves along the directrix so that it always passes through the vertex, it generates the surface of the cone. Popularly, the term 'cone' is associated with the particular case in which the directrix is a circle and the vertex lies on the line that passes through the centre of the circle

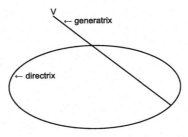

Figure 1.52 Generating a cone

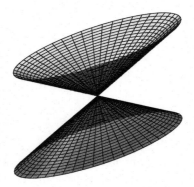

Figure 1.53 An elliptic cone

and is perpendicular to the plane of this curve. This is the *circular, right cone* an example of which can be seen in Fig. 1.54. In mathematics, however, the directrix can be any plane curve and there is no restriction on the position of the vertex. Moreover, it is assumed that the generatrix extends to infinity in both directions from the vertex. Therefore, the generated surface has two *nappes*, as exemplified in Fig. 1.53. In this figure we see several positions of the generatrix and a number of curves that are geometrically similar to the directrix.

1.15.2 Points on the Cone Surface

Fig. 1.54 shows a right, circular cone. Assuming this knowledge, and given a projection of a point belonging to the surface of the cone, all other projections of that point are well defined. We are going to show two ways for finding a second projection. This exercise will help to understand the relationships between the various sections shown in the lines drawing of a vessel.

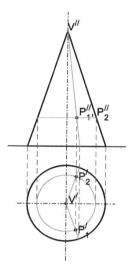

Figure 1.54 Point on cone surface — Finding the horizontal projection by a section parallel to the base

Solution 1 — Using Sections Parallel to the Base

Let us suppose that in Fig. 1.54 we are given the coincident frontal projections P_1'' and P_2'', and the knowledge that the former projection is that of a point on the visible side of the cone, while the latter is the projection of a point on the hidden part. We are asked to find the horizontal projections of the points P_1, P_2. We know that all cone sections parallel to the base, or, equivalently, perpendicular to the cone axis, are circles. In particular we are interested in the section on which the points P_1 and P_2 lie. Therefore, we draw the section in the horizontal plane using the simple construction shown in the figure. From the frontal projections, P_1'', P_2'', we draw the dashed line perpendicular to the ground line; it intersects the circle in the points P_1', P_2'. These are the projections we looked for. If the given projections are the horizontal ones, we begin by drawing the circle that passes through P_1', P_2' and is concentric with the base. Next we draw the frontal projection of this circle and find its intersection with the perpendicular to the ground line drawn from P_1', P_2'.

Solution 2 — Using a Generatrix

In Fig. 1.55, given the common point of the frontal projections we draw the line that passes through that point and the projection of the vertex, V''. This line represents the projections of two generatrices, one on the visible

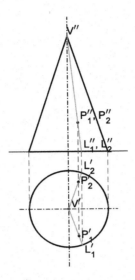

Figure 1.55 Point on cone surface — Finding the horizontal projection with the help of a generatrix

part of the cone, the other on the hidden part. From the intersection of this line with the ground line we draw a line perpendicular to the ground line; it intersects the base circle in two points, L_1', L_2'. We draw lines that connect the latter points with the centre of the base, which coincides with V', the horizontal projection of the vertex. The intersections of the latter lines with the vertical line drawn from P_1'', P_2'' are the projections we looked for. If we are given the horizontal projections we have to reverse the procedure and start from the horizontal projection.

1.16 CONIC SECTIONS

1.16.1 Introduction

The curves generated by cutting the surface of a right, circular cone with planes are called **conic sections**; they present particular interests in mathematics and have many important applications in science and technology. The study of conic sections started in ancient Greece with Manaechmus, around 360–350 BC, and was continued by other Greek mathematicians, among them Euclid (born c. 300 BC) and Archimedes (born c. 290–280, died c. 211 BC). The major work on conics is due to Apollonius of Perga (c. 240–190 BC) who wrote eight books (today we would rather call them

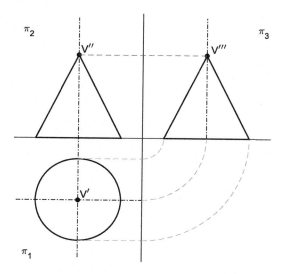

Figure 1.56 A right, circular cone

chapters) seven of which survived. Apollonius has given those curves the names that are in use to this day. The kind of curve produced by the cutting plane depends on the position of this plane relative to the cone surface, a fact summarized in Fig. 1.66. All conic sections can be described by second-degree algebraic equations. In particular cases the sections can *degenerate* into straight lines or a point. The reciprocal statement is true: all second-degree algebraic equations describe conic sections. Closer to our times, the Belgian-French mathematician Dandelin (1794–1847) gave newer, remarkably elegant proofs that relate the conic sections to their definitions as loci. In the following sections we are going to explain how cutting the cone shown in Fig. 1.56 by planes with different slopes generates the various conic sections.

1.16.2 The Circle

If the sectioning plane is perpendicular to the axis of the cone the section is a **circle**. This is case 'b' in Fig. 1.66. The plane cuts two opposed generatrices; therefore, the resulting curve is closed. The circle is defined as the locus of the points, in a plane, such that their distance from a given point, called *centre*, is constant. The constant distance is the *radius* of the circle. For the particular case in which the origin of coordinates coincides with the centre of the circle we obtain the canonical equation of this curve directly

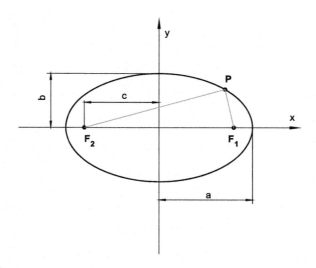

Figure 1.57 The definition of the ellipse

from the above definition

$$x^2 + y^2 = r^2 \tag{1.5}$$

1.16.3 The Ellipse

If the slope of the cutting plane is less than that of the generatrix, the section is an **ellipse**. This situation corresponds to case 'c' in Fig. 1.66. As the plane cuts two opposed generatrices, we understand that the resulting curve is closed. For the definition of the ellipse let us refer to Fig. 1.57. Given the two points F_1 and F_2, the *ellipse* is the locus of the points P such that the sum of their distances to the points F_1 and F_2 is constant. The points F_1 and F_2 are the *foci* of the ellipse (singular *focus*). Let the x-coordinates of the two foci be c and $-c$, while their y-coordinates are 0. Writing in algebraic terms that the sum of the distances of the point P to the two foci is equal to the constant $2a$ yields

$$\sqrt{(x-c)^2 + y^2} + \sqrt{(x+c)^2 + y^2} = 2a \tag{1.6}$$

Isolating the second term in the left-hand side of the equation and squaring yields

$$(x+c)^2 + y^2 = 4a^2 - 4a\sqrt{(x-c)^2 + y^2} + (x-c)^2 + y^2$$

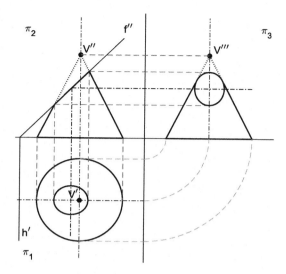

Figure 1.58 Truncated cone with elliptic upper face

After expanding the parentheses and simplifying we obtain

$$cx - a^2 = a\sqrt{(x-c)^2 + y^2}$$

After squaring again, simplifying, and collecting similar terms we remain with

$$x^2(a^2 - c^2) + a^2 y^2 = a^2(a^2 - c^2)$$

Let $a^2 - c^2 = b^2$. The equation reduces to

$$b^2 x^2 + a^2 y^2 = a^2 b^2$$

It remains to divide both parts of the equation by $a^2 b^2$ and we obtain the canonic form

$$\frac{x^2}{a^2} + \frac{y^2}{b^2} = 1 \qquad (1.7)$$

The constant $2a$ is the length of the horizontal axis of symmetry of the ellipse known as *major axis*. The constant $2b$ is the length of the vertical axis of symmetry called *minor axis*. When $a = b = r$ the ellipse becomes a circle with radius r.

Fig. 1.58 shows the cone in Fig. 1.56 truncated so that the edge of the upper face is an ellipse. It is assumed that the part of the cone above the

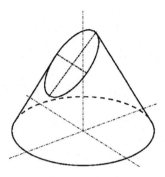

Figure 1.59 Truncated cone with elliptic upper face — Axonometric view

cutting plane was taken away. An axonometric view of the same truncated cone is shown in Fig. 1.59.

One of Kepler's laws states that the planets move along orbits that are ellipses with the sun in one of the foci. Elliptical mirrors have interesting reflection properties. Examples of extension of those properties to acoustics are the *whisper galleries*, such as the famous one in St. Paul's Cathedral of London.

1.16.4 The Parabola

If the slope of the sectioning plane is exactly equal to that of the generatrix, the resulting section is a **parabola**. This is case 'd' in Fig. 1.66. We see that the sectioning plane cuts only one nappe of the conic surface. Therefore, the curve has only one branch that extends to infinity. For the definition of the parabola we refer to Fig. 1.60. Given the point F, called *focus*, and a straight line, called *directrix*, the parabola is the locus of the points that are equally distant from the focus and from the directrix. Let **P** be a point of the parabola. We choose the midpoint between **F** and the directrix as origin of coordinates, and the parallel to the directrix, through this origin, as the y-axis. We draw $\overline{PL}$ perpendicular to the directrix. Let the x-coordinate of the focus be p. Writing that the distance from **P** to **F** equals the distance from **P** to **L** we obtain

$$x + p = \sqrt{(x - p)^2 + y^2}$$

Squaring and simplifying yields the canonical equation of the parabola

$$y^2 = 4px \qquad (1.8)$$

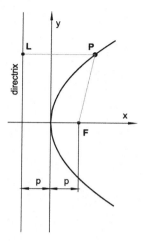

Figure 1.60 The definition of the parabola

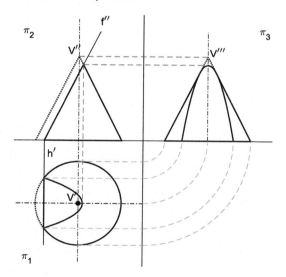

Figure 1.61 Truncated cone with parabolic section

Fig. 1.61 shows the cone in Fig. 1.56 truncated so that the edge of the plane face is a parabola. It is assumed that the part of the cone above the cutting plane was taken away. An axonometric view of the same truncated cone is shown in Fig. 1.62.

Rotating a parabola around its axis generates a surface called *paraboloid*. In his second book on *Floating bodies*, Archimedes studied the stability of

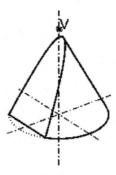

Figure 1.62 Truncated cone with parabolic section — Axonometric view

floating paraboloids (Nowacki, 2002). Another application in Naval Architecture is due to the English Thomas Simpson (1710–1761) who derived a method for numerical integration in which a given curve is approximated by a parabola (see, for example, Biran and López-Pulido, 2014; Biran and Breiner, 2002). Galileo Galilei (Italian, 1564–1642) has shown that the trajectory of projectiles in void is a parabola (see Example 4.6). Paraboloidal reflectors are used to produce light beams and to concentrate radio and TV signals received from communication satellites.

1.16.5 The Hyperbola

If the slope of the sectioning plane is larger than that of the generatrix, the section is a **hyperbola**. This situation corresponds to case 'e' in Fig. 1.66. As the sectioning plane cuts both nappes, the resulting curve has two branches that extend to infinity. For the definition of the hyperbola we use Fig. 1.63. Given the two points F_1 and F_2, the *hyperbola* is the locus of the points P such that the difference of their distances from the two points is constant. The points F_1 and F_2 are the *foci* of the hyperbola. Let the x-coordinates of the two foci be c and $-c$, while their y-coordinates are 0. Writing in algebraic terms that the difference of the distances of the point P from the two foci is equal to the constant $2a$ we obtain

$$\sqrt{(x-c)^2 + y^2} - \sqrt{(x+c)^2 + y^2} = 2a \qquad (1.9)$$

We leave to the reader to treat this equation in the same manner we did with the equation of the ellipse, but using now the substitution $c^2 - a^2 = b^2$. The result is the canonic equation

$$\frac{x^2}{a^2} - \frac{y^2}{b^2} = 1 \qquad (1.10)$$

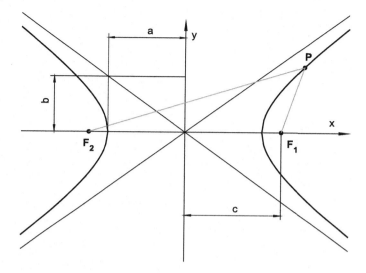

Figure 1.63 The definition of the hyperbola

As the branches of the hyperbola tend to infinity they approach two straight lines called the *asymptotes of the hyperbola*; their equations are

$$y = \frac{b}{a}x, \; y = -\frac{b}{a}x \tag{1.11}$$

Fig. 1.64 shows the cone in Fig. 1.56 truncated so that the edge of the plane face is a hyperbola. It is assumed that the part of the cone above the cutting plane was taken away. An axonometric view of the same truncated cone is shown in Fig. 1.65.

For a property based directly on its definition, the hyperbola was extensively used in a system of electronic navigation. Like the ellipse and the parabola, the hyperbola has important properties of ray reflection.

Before going on to a different subject, let us look again to Fig. 1.66. There are two examples of degenerate conic sections:

Case a — The section is a point; it can be regarded as a circle with radius equal to zero, or as an ellipse with both axes reduced to zero.

Case f — The section is a pair of intersecting straight line. This situation can be the limit of case (e) when the cutting plane passes through the vertex. Then, the hyperbola coincides with its asymptotes.

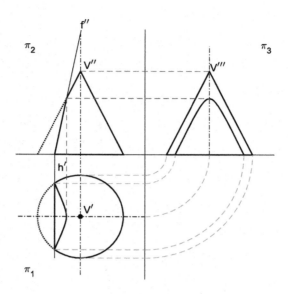

Figure 1.64 Truncated cone with hyperbolic section

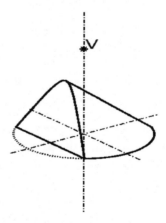

Figure 1.65 Truncated cone with hyperbolic section — Axonometric view

1.17 WHAT IS AXONOMETRY

The preceding sections covered the basics of object representation by or-thogonal projections on two planes that are perpendicular one to the other. For this method we use the term *orthographic projections*. The projection planes are also planes of coordinates. In three–dimensional space we have three axes of coordinates and each one is perpendicular to the other two.

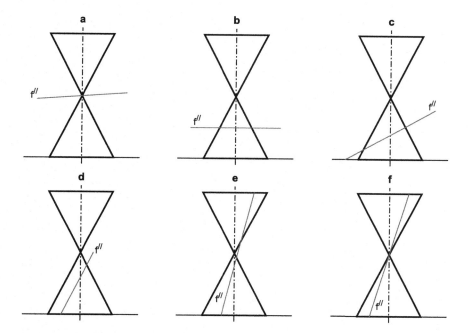

Figure 1.66 The family of conic sections

In the sketch, in the horizontal plane, π_1, we see the projections of the x and y axes as perpendicular one to the other, while in the frontal plane, π_2, the projections of the x and z axes are perpendicular one to the other. As the orthogonal projection is a special case of the parallel projection, it inherits the properties of the latter. In particular, distances and angles parallel to a projection plane are projected on that plane in true size. It is possible to measure directly such distances in any direction. To fully benefit of this property objects should be placed with the main faces (or features) parallel to projection planes. In general, drawing and measuring in orthographic projections present no special difficulties. On the other hand, to perceive the shape of a three-dimensional object one has to look at two projections and synthesize mentally the complete image. This operation requires certain *spatial abilities* and some training. Moreover, excepting certain simple objects, it is not possible to show all dimensions in one projection. In general, the dimensions must be distributed over at least two projections. For simple objects it is possible to avoid the above drawbacks by projecting on a single plane that makes an angle with two or all three coordinate planes. For reasons explained below we call this method **axonometry**. The axonometry can be also used for showing pictorially the main features of complex objects.

Oblique axonometry uses parallel projection and the projection rays are oblique to the projection plane. Two popular variants are the *cavalier* and the *cabinet projections*. Such projections are used in architecture, and in catalogues of machine details. In this book we consider only *orthogonal axonometric projections* in which the projection rays are perpendicular to the projection plane. In general, the projections of the coordinate axes are not perpendicular one to another, and each axis is projected at its particular scale. Therefore, it is possible to measure only in directions parallel with the projected axes, hence the name axonometry, a term composed of two Greek words that mean 'axes' and 'measuring'.

In Section 1.5 we have written that for reconstituting a three-dimensional object one needs two different images. This observation is valid also for axonometric projections. An axonometric projection of an object and of a system of coordinates is ambiguous as to the position of the object with respect to the axes of the coordinates. To solve the ambiguity one has to add a projection of the object on one of the coordinate planes. Obviously, providing a stereo pair of axonometric images would also solve the problem.

1.17.1 The Law of Scales

In an axonometric projection the axes of coordinates are projected at scales that may differ from one axis to another. We are going to show that these scales are not independent, but fulfill a simple relationship. To be more specific, the sum of squares of the scales equals 2. In Fig. 1.67 the projection plane is inclined with respect to all coordinate planes; it is identified by its traces h, f, and s. The traces h and f intercept the x-axis in the point X_0, the traces h and s intercept the y-axis in the point Y_0, and the traces f and s intercept the z-axis in the point Z_0. The orthogonal projection of the origin of coordinates on the oblique projection plane is O'.

We take apart the triangle OX_0O' and show it in Fig. 1.68. As $\overline{OO'}$ is perpendicular to the projection plane, it is perpendicular to all lines belonging to that plane and in particular to $\overline{X_0O'}$. Let the notation of angles be $\widehat{OX_0O'} = \alpha_1$, $\widehat{X_0OO'} = \alpha$. The segment $\overline{OX_0}$ is projected as $\overline{O'X_0}$. This means that distances measured parallel to the x-axis are projected at the scale

$$S_x = \frac{\overline{O'X_0}}{\overline{OX_0}} = \cos\alpha_1 \tag{1.12}$$

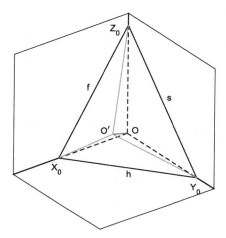

Figure 1.67 Axonometric projections of coordinate axes

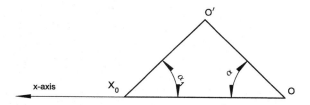

Figure 1.68 Axonometric projection of the x-axis

Similarly, in the triangle OY_0O' the segment $\overline{OO'}$ is perpendicular to $\overline{Y_0O'}$. Let the notation of angles be $\widehat{OY_0O'} = \beta_1$, $\widehat{Y_0OO'} = \beta$. The segment $\overline{OY_0}$ is projected as $\overline{O'Y_0}$. This means that distances measured parallel to the y-axis are projected at the scale

$$S_Y = \frac{\overline{O'Y_0}}{\overline{OY_0}} = \cos\beta_1 \tag{1.13}$$

Finally, in the triangle OZ_0O' the segment $\overline{OO'}$ is perpendicular to $\overline{Z_0O'}$. Let the notation of angles be $\widehat{OZ_0O'} = \gamma_1$, $\widehat{Z_0OO'} = \gamma$. The segment $\overline{OZ_0}$ is projected as $\overline{O'Z_0}$. This means that distances measured parallel to the z-axis are projected at the scale

$$S_z = \frac{\overline{O'Z_0}}{\overline{OZ_0}} = \cos\gamma_1 \tag{1.14}$$

Squaring both sides of Eqs (1.12) to (1.14), and considering the trigonometric relationships in the three triangles, we obtain

$$S_x^2 = \cos^2 \alpha_1 = 1 - \sin^2 \alpha_1 = 1 - \cos^2 \alpha \qquad (1.15)$$
$$S_y^2 = \cos^2 \beta_1 = 1 - \sin^2 \beta_1 = 1 - \cos^2 \beta$$
$$S_z^2 = \cos^2 \gamma_1 = 1 - \sin^2 \gamma_1 = 1 - \cos^2 \gamma$$

Adding these equations side by side yields

$$S_x^2 + S_y^2 + S_Z^2 = 3 - (\cos^2 \alpha + \cos^2 \beta + \cos^2 \gamma) \qquad (1.16)$$

The quantities between parentheses are the *direction cosines* of the normal $\overline{OO'}$ and their sum equals 1. Indeed, the projections of this normal on the three axes are $\overline{OO'} \cos\alpha$, $\overline{OO'} \cos\beta$, $\overline{OO'} \cos\gamma$. The length of $\overline{OO'}$ can be calculated from its projections by Pythagoras' theorem in three dimensions

$$\overline{OO'}^2 = (\overline{OO'} \cos\alpha)^2 + (\overline{OO'} \cos\beta)^2 + (\overline{OO'} \cos\gamma)^2$$

which gives

$$\cos^2 \alpha + \cos^2 \beta + \cos^2 \gamma = 1$$

Substituting this value into Eq. (1.16) we obtain the *fundamental law of axonometry*

$$S_x^2 + S_y^2 + S_Z^2 = 2 \qquad (1.17)$$

1.17.2 Isometry

When the isometric projection is displayed by means of specialized software it is easy to choose any position of the projection plane and let the computer draw the projection as required by Eq. (1.17). Doing the same in manual work could be very difficult. Therefore, there is an interest in simplified procedures. The first idea that may cross our mind is to let all three scales be equal, that is

$$S = S_x = S_y = S_z = \sqrt{\frac{2}{3}} \simeq 0.82 \qquad (1.18)$$

Intuition may tell us that Eq. (1.18) is fulfilled when, in Fig. 1.67,

$$\overline{OX_0} = \overline{OY_0} = \overline{OZ_0} \qquad (1.19)$$

We can prove Eq. (1.19) deriving from Fig. 1.68 the equality

$$\overline{OX_0} = \overline{OO'}/\sin\alpha_1$$

Squaring both sides and after some substitutions we get

$$\overline{OX_0}^2 = \frac{\overline{OO'}^2}{\sin^2\alpha_1} = \frac{\overline{OO'}^2}{1-\cos^2\alpha_1} = \frac{\overline{OO'}^2}{1-S_x^2}$$

In the same way we write

$$\overline{OY_0}^2 = \frac{\overline{OO'}^2}{1-S_y^2}$$

$$\overline{OZ_0}^2 = \frac{\overline{OO'}^2}{1-S_z^2}$$

Dividing the above equations side by side and extracting the square roots yields

$$\frac{\overline{OX_0}}{\overline{OY_0}} = \sqrt{\frac{1-S_y^2}{1-S_x^2}}$$

$$\frac{\overline{OY_0}}{\overline{OZ_0}} = \sqrt{\frac{1-S_z^2}{1-S_y^2}}$$

$$\frac{\overline{OZ_0}}{\overline{OX_0}} = \sqrt{\frac{1-S_x^2}{1-S_z^2}}$$

Taking into account the equality of scales, that is Eq. (1.18), we obtain Eq. (1.19). Then, the right-angled triangles $X_0 OY_0$, $Y_0 OZ_0$, and $Z_0 OX_0$ are congruent and $\overline{X_0 Y_0} = \overline{Y_0 Z_0} = \overline{Z_0 X_0}$. As shown in Fig. 1.70, the triangle of traces, $X_0 Y_0 Z_0$, is equilateral and its angles are all equal to 60°. The projected origin is the centre of the circumscribed circle and the projected axes bisect the angles of the triangle into angles of 30°. The angles between the projected axes are all equal to 120°, as drawn in Fig. 1.69. We draw these axes at 30° from a horizontal. As a first and classical example, Fig. 1.71 shows the *isometric cube*. The circles inscribed in the faces of the cube are projected as ellipses. This cube appears in the German standard DIN 5, Part 1 (1970).

In manual work it is impractical to multiply all dimensions by the scale 0.82. Therefore, various standards allow to draw at the scale 1:1. Comparing such a drawing with the corresponding orthographic projections causes

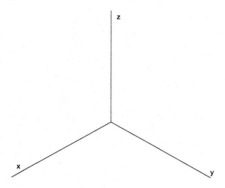

Figure 1.69 Coordinate axes in isometric projection

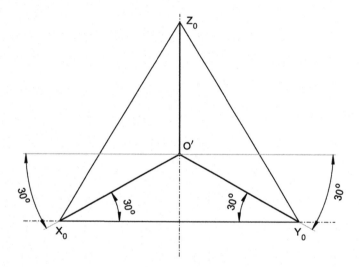

Figure 1.70 The triangle of traces in isometric projection

a sensation of enlargement. Some CAD programs that can display isometric projections let the user choose between the 1:1 and the $\sqrt{2/3}$ scales.

Example 1.1 (Isometric drawings of piping systems). An outstanding application of the isometric projection is in the drawing of piping systems. The advantages are obvious even for such a simple example as that shown in Fig. 1.72. This drawing was carried on with the help of a CAD program, using an *isometric grid*. Pipes are represented as single lines following the axes of the actual pipes. This single view is sufficient for understanding how the

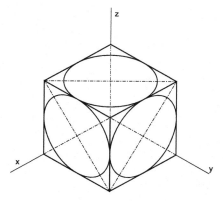

Figure 1.71 The isometric cube

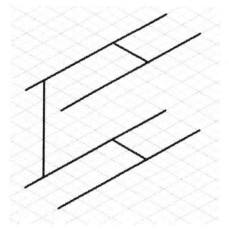

Figure 1.72 Piping in isometric projection

pipes are connected and for writing all dimensions. Fig. 1.73 shows the same system in orthographic projections. To completely define the system we used a front view, a top view, and two sections. To understand the connections between the four pipes that run in parallel with the x-axis, one has to see all four projections and mentally synthesize the knowledge.

Marine systems consist of dozens of pipes, valves, and other fittings. Therefore, one can appreciate that it would be practically impossible to design and build such systems without isometric drawings. A detailed example of a piping system, and valuable explanations are included in the German standard DIN 5, Part 1 (1970).

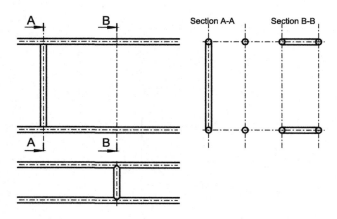

Figure 1.73 Piping in orthographic projection

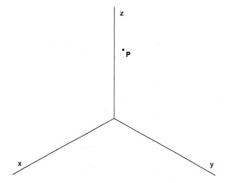

Figure 1.74 An ambiguous isometric projection

1.17.3 An Ambiguity of the Isometric Projection

Fig. 1.74 shows in isometric projection a system of coordinates and a point, **P**. This projection is not sufficient to define the position of the point **P** with respect to the system of coordinates. As seen in Fig. 1.75, to the given projection of the point **P** we can associate as horizontal projection any point lying on the vertical through **P**, below the intersection h_0 with the y-axis. Three examples are $\mathbf{P}'_0$, $\mathbf{P}'_1$ and $\mathbf{P}'_2$. The frontal projections, $\mathbf{P}''_0$, $\mathbf{P}''_1$ and $\mathbf{P}''_2$ can be obtained by an obvious construction and thus we completely define the points $\mathbf{P}_0$, $\mathbf{P}_1$ and $\mathbf{P}_2$ that all can correspond to the given isometric projection **P**.

To conclude the discussion we mention that for solids that extend in one direction much more than in the other two, the isometric projection

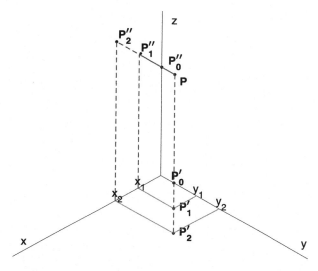

Figure 1.75 An ambiguity of the isometric projection

may look unnatural. The cause is that from our visual experience we expect parallel lines to converge, unless they are parallel to the image plane. In such cases it is recommended to use the *dimetric* projection in which only two scales are equal. Good definitions and illustrations can be seen, for example, in the German standard DIN 5, Part 2 (1970). When all three scales are different, we speak about a *trimetric* projection. These matters are detailed in some textbooks on engineering drawing (for example Chirone and Torn-incasa, 2005), or in specialized books, such as the superbly illustrated one of Pavanelli et al. (2003).

1.18 DEVELOPED SURFACES

1.18.1 What Is a Developed Surface

A cube, a cylinder, or a cone can be built by bending a flat sheet of cardboard or metal. To do this we must draw a certain figure on the flat sheet. We call this figure the *development* of the cube, cylinder, or cone. Because we can develop the envelopes of cylinders and cones we say that these surfaces are *developable*. Only a limited number of surface kinds are developable. The first necessary, geometrical condition for enabling development is that the surface be *ruled*. A ruled surface is one on which it possible to lay straight lines that lie completely on it. For a cylinder or a cone such lines are their generatrices. Rotating the hyperbola in Fig. 1.63 around the x-axis generates a hyperboloid with one nappe. This is a ruled surface, but one that does not fulfill a second condition that we do not describe here. Therefore, the hyperboloid with one nappe is not developable. A necessary and sufficient condition that developable surfaces meet is described in Section 6.9. An example of important surface that is not developable is the sphere. It is obvious that no straight line can be laid on a sphere. An important property of developed surfaces is that the distance between two points is conserved. More specifically, let P_1 and P_2 be two points on the surface to be developed, and P_{1D}, P_{2D} the corresponding points on the developed surface. Then, the distance between the points P_1 and P_2, on the given surface, is equal to the distance between the points P_{1D}, P_{2D} on the developed figure. We use this property in the developing process.

Many pieces of equipment are produced by bending metallic sheets. Therefore, the techniques for calculating and drawing the developed surfaces are described in some textbooks of engineering drawing, such as Chirone and Tornincasa (2005), or Chapter 15 in Simmons et al. (2012). Two examples of books entirely dedicated to the subject are Böge (1959) and Bundjulov et al. (1964).

In shipbuilding it would be ideal to develop the hull surface. For certain ship forms it is possible to develop only parts of a shell, for other forms it is not possible at all. Therefore, only approximate developments are possible. This subject is dealt with in Section 2.9.

1.18.2 The Development of a Cylindrical Surface

The developed envelope of a right, circular cylinder is a rectangle. Thus, in Fig. 1.76, given the base radius, r, and the cylinder height, h, we draw the developed surface as a rectangle with base $2\pi r$ and height h. This solution is valid only if the thickness of the material is negligible. Otherwise the theory teaches us that when bending the metal sheet into a cylindrical solid, the external surface is stretched, while the inner surface is compressed. This behaviour is concisely described in Section 5.5. There is a surface on which lengths are not changed. For a cylinder as considered in Fig. 1.77, this surface corresponds to the mean diameter of the cylinder. More specifically, given the inner diameter of the cylinder, d_i, and the cylinder thickness, t, the length that does not change is πd_m, where

$$d_m = d_i + t/2 + t/2 = d_i + t$$

According to Bundjulov et al. (1964), this rule is exact as long as $d_i > 10t$.

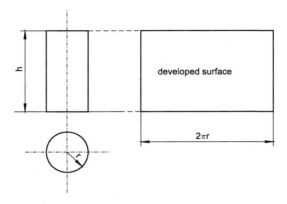

Figure 1.76 The development of a right, circular cylinder

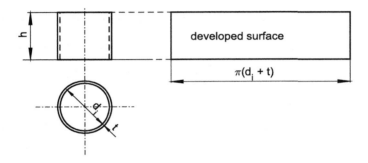

Figure 1.77 The development of a right, circular cylinder with finite thickness

1.18.3 The Development of a Conic Surface

The developed envelope of a right, circular cone is a circular sector as shown in Fig. 1.78. To find the angle γ of the sector we observe that the circumference of the cone base is equal to the length of the circular arc that delimits the developed surface and write

$$2\pi r = \gamma g$$

where γ is the angle of the circular sector, and g is the length of the generatrix

$$g = \sqrt{h^2 + r^2}$$

The angle γ must be measured in radians. In engineering drawings angles are measured with the aid of protractors that are graduated in degrees.

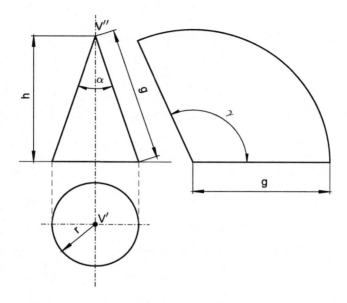

Figure 1.78 The developed surface of a right, circular cone

To convert radians to degrees we use the identity

$$\frac{\gamma_{radians}}{\pi} = \frac{\gamma_{degrees}}{180}$$

Finally, the value of the angle γ in degrees is given by

$$\gamma = 360\frac{r}{g} \tag{1.20}$$

1.19 SUMMARY

The human eye, like any photographic or TV camera, sees the world in central projection. All projection rays pass through one point, the *projection centre*. The only properties that are conserved (invariants) between an object and its image are:

1. the parallelism of lines parallel to the image plane;
2. the cross-ratio of four collinear points.

   Drawing and measuring in central projection require special geometrical constructions or calculations. To simplify the work we assume that the projection centre is sent to infinity. Then, the projection rays become parallel. Therefore, we call this method *parallel projection*. We obtain three new invariants:

1. sizes of objects (distances and angles) parallel to the image plane;
2. parallelism of lines;
3. the simple ratio of three collinear points.

The parallel projection is a particular case of the central projection and inherits the invariants of the central projection. The parallel projection has more useful properties than the central projection; it is easier to draw an object in parallel projection than in central projection and it is easier to measure lengths in parallel projection.

Let us further assume that the projection rays are perpendicular to the image plane. We are dealing now with the *orthogonal*, or *orthographic projection* and obtain a further invariant: the size of a right angle with a side parallel to the image plane. The orthographic projection is a particular case of the parallel projection and inherits the invariants of the parallel projection.

In ISO 5456-1:1996(E) (1996), Part 1, we find a summary of the properties of different projection methods. A technical vocabulary of terms relating to projection methods in English, French, German, Italian and Swedish can be found in ISO 10209-2:1993(E/F).

We said that when we draw in parallel projection we accept the seemingly unnatural assumption that we look from infinity. Nature, however, provides us with a real example of parallel projection: shadows cast by objects illuminated by the sun. The distance between us and the sun is so large that its rays are practically parallel. Shadows of objects illuminated by a small source of light are approximations of central projections.

Monge developed a system for drawing in orthogonal projections. This is the *descriptive geometry* that allows us to completely define 3D objects by 2D projections. A point is defined by projections on two planes perpendicular one to the other. One of the planes is rotated so that the right angle between the two planes becomes 180° and the two projections lie in one plane. The two projections contain all three coordinates of the point. Points are projected as points, straight lines are projected as straight lines or points. Planes can be easily defined and identified by their intersections with the projection planes that are usually the coordinate planes. We call these intersections *traces*. Particularly interesting are planes parallel to the coordinate planes; we use them to cut the ship-hull surface and obtain the *ship lines*.

In orthographic projections it is convenient to place the object so that as many representative faces as possible are parallel to the projection planes, let's say the coordinate planes. Instead of using a set of orthographic pro-

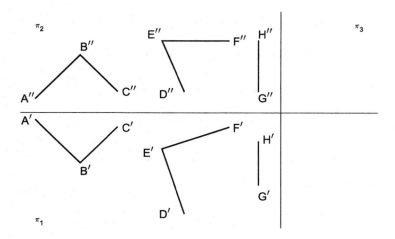

Figure 1.79 True length, right angle

jections, sometimes it is sufficient to draw a single projection on a plane
not parallel to any coordinate plane, more specifically to any representative
face of the object. We are dealing then with *axonometry*. For simple objects
a single axonometric projection completely defines the shape of the object
and contains all dimensions. On the other hand, in axonometric projections the axes of coordinates are projected at scales that may differ from one
axis to the other. Therefore, distances can be measured only in parallel with
the projected axes, hence the name axonometry. The scales at which the
three axes are projected are not independent. More specifically, the sum of
the squares of scales equals 2. To simplify the work we may assume that the
three scales are equal. In this case the projection is called *isometric* and the
angles between the projected axes are equal to 120°.

Some surfaces, for example cylinders and cones, can be produced by
bending a plane sheet of metal. We say that those surfaces are *developable*.
An important example of surface that is not developable is the sphere. The
extent to which the hull surface is developable and what to do when it is
not is of paramount importance in shipbuilding where we are talking about
shell expansion.

1.20 EXERCISES

Exercise 1.1 (True length, right angle). This exercise refers to Fig. 1.79.
1. Which line segment is projected in true length on π_1?

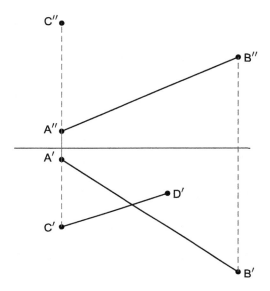

Figure 1.80 To Exercise 1.2

2. Which line segment is projected in true length on π_2?
3. Which line segment is projected in true length on π_3?
4. Which angle is right?

Exercise 1.2 (Intersecting lines). Fig. 1.80 shows the horizontal and frontal projections of the points A, B and C, and the horizontal projection, D', of the point D. Assuming that the lines AB and CD intersect, find the frontal projection, D'', of the point D.

Exercise 1.3 (Square in horizontal plane). Fig. 1.81 is an axonometric view of a square lying in a horizontal plane. Draw the sketch containing the orthographic projections of the square on π_1, π_2 and π_3.

Exercise 1.4 (Square in frontal plane). Fig. 1.82 is an axonometric view of a square lying in a frontal plane. Draw the sketch containing the orthographic projections of the square on π_1, π_2 and π_3.

Exercise 1.5 (Point on pyramid face). In Fig. 1.83 we show:
• the horizontal and frontal projections of a pyramid with base $ABCD$ and vertex V;
• the frontal projection of a point, P, that belongs to the face ABV of the pyramid.

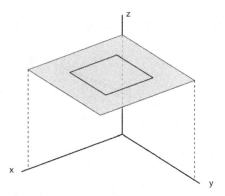

Figure 1.81 A square in a horizontal plane

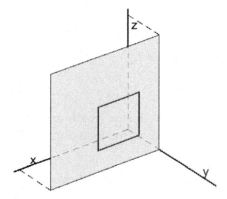

Figure 1.82 A square in a frontal plane

Your tasks are:
1. to draw the side projection of the pyramid;
2. to draw the horizontal and side projections of the point P.

Hint. As the points V and P belong to the face ABV, the straight line defined by these points belongs to the same face. Let the intersection of the line VP with the pyramid edge AB be L. The projections of the point P lie on the corresponding projections of the line VP.

Exercise 1.6 (An ice-cream cone). Fig. 1.84 shows an 'ice-cream cone' composed of
1. a right, circular cone;
2. a hemisphere.

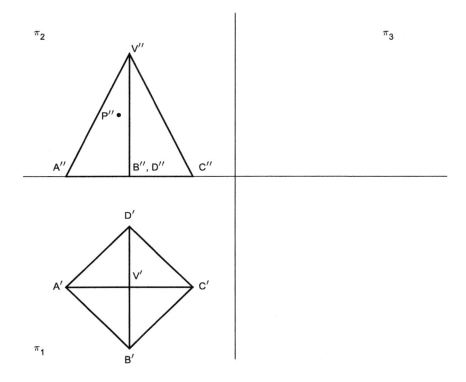

Figure 1.83 A point on a face of a pyramid

The points A to E belong to the visible part of the combined surface, while the points F to J are on the hidden part. The figure shows the frontal projections of these points. All given points lie in the plane defined by the traces h, f. You are asked to draw:
1. the side projection of the combined surface;
2. the horizontal projections of the given points;
3. the side projections of the same points;
4. the horizontal and side projections of the section produced by the plane defined by the traces h, f.

Exercise 1.7 (Cone with hole). Fig. 1.85 shows the frontal projection of a cone with a hole defined by the points A to H. Your task is to draw:
1. the horizontal projections of the points A to H;
2. the side projections of the points A to H.

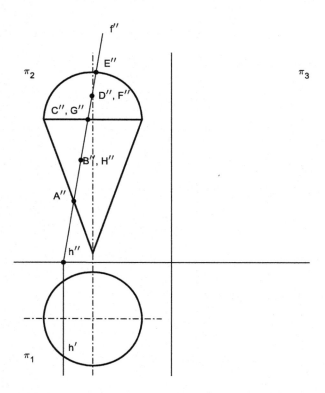

Figure 1.84 An ice-cream cone

Exercise 1.8 (Isometric to orthographic). Fig. 1.86 shows the isometric projection of a simple solid. Draw the corresponding orthographic projections.

Exercise 1.9 (Another solid). Fig. 1.87 shows the isometric projection of a solid composed of a half-cone and a prism. Draw the corresponding orthographic projections.

Exercise 1.10 (Developing a conic surface). Referring to Section 1.18.3 prove that the angle, γ, of the developed surface depends only on the angle α of the cone vertex.

APPENDIX 1.A THE CONNECTION TO LINEAR ALGEBRA AND MATLAB

Using linear algebra we may greatly simplify geometrical calculations. On the other hand, many concepts of linear algebra can be explained and vi-

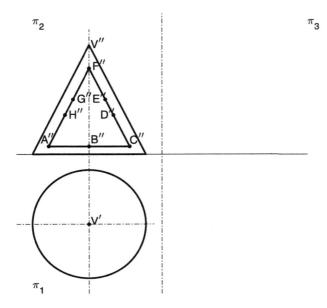

Figure 1.85 Cone with hole

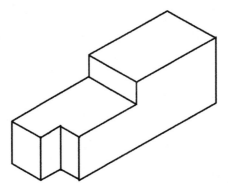

Figure 1.86 An isometric projection

sualized by geometrical examples. The interplay and parallelism between these two branches of mathematics is so natural and helpful that large parts of them could (and should) be taught together. Thus, for example, in Italian universities the courses of linear algebra bear names that include the term *geometria*, and the book of linear algebra of Marco Abate (1996) has the title *Geometria*. As another example, the title of the book of Farin and Hansford (2005) explicitly reminds the interplay between the two branches

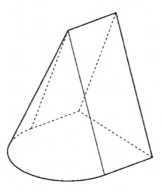

Figure 1.87 A solid composed of half-cone and a prism

of mathematics: *Practical linear algebra — A geometry toolbox.* The fact that for many geometrical objects and operations we can find a parallel in linear algebra allows us to use suitable computer software for geometric modelling. Therefore, in this book we make massive use of MATLAB, an environment initially developed for linear algebra, but today so versatile and powerful that it is used in practically all branches of mathematics and engineering.

A point can be represented in MATLAB as an array whose elements are the coordinates of that point. We can use either row or column arrays; both variants appear in the literature. In this book we adopt the column-array representation. Thus, a point **P** can be defined in the usual cartesian, orthogonal system of coordinates by

$$\begin{bmatrix} P_x \\ P_y \\ P_z \end{bmatrix}$$

An example in MATLAB is

```
P1 = [ 5
3
6 ]
P1 =

     5
     3
     6
```

A shorter way to input a column array is

```
P1 = [ 5; 3; 6 ];
```

Ending the command with a semicolon suppresses its echoing. The reason for choosing the column-array notation is that several points defined as above can be easily *concatenated* in MATLAB to define a line or a closed polygon. For example, let us define a second point

```
P2 = [ 6; 4; 7 ];
```

Then, the segment $\overline{P_1 P_2}$ is represented by

```
P1P2 = [ P1 P2 ];
```

By definition, the horizontal projection of any point lies in the plane π_1 whose z-coordinate is 0. Therefore, to obtain the horizontal component of any point it is sufficient to set its z-coordinate to 0. In linear algebra, accepting the above convention for points, this operation can be carried on by left-multiplying the array of coordinates by a corresponding *projection matrix*. For example, for the point $\mathbf{P_1}$ defined above we write

$$\mathbf{P_1'} = \begin{bmatrix} 1 & 0 & 0 \\ 0 & 1 & 0 \\ 0 & 0 & 0 \end{bmatrix} \begin{bmatrix} 3 \\ 5 \\ 6 \end{bmatrix} = \begin{bmatrix} 3 \\ 5 \\ 0 \end{bmatrix}$$

The matrix for horizontal projection is entered in MATLAB with the command

```
Hproj = [ 1 0 0; 0 1 0; 0 0 0 ];
```

It is easy to see that the same matrix projects the straight-line segment $P_1 P_2$ on the π_1 plane. Assuming that $P_1 P_2$ and *Hproj* have been entered we carry on the operation as

```
P1P2h = Hproj*P1P2
P1P2h =
       5     6
       3     4
       0     0
```

The generalization to any object defined by points is straightforward. In Exercise 1.11 we invite the reader to derive the matrices for frontal and side projections and to experiment with them.

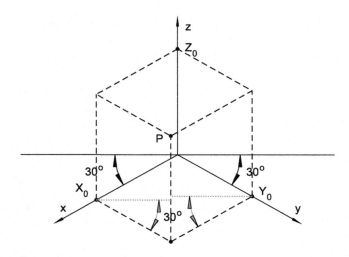

Figure 1.88 Transforming coordinates for the isometric projection

What about other projections? For example, let us suppose that given a point

$$P = \begin{bmatrix} X_0 \\ Y_0 \\ Z_0 \end{bmatrix}$$

in three-dimensional space, we want to find its coordinates in a two-dimensional isometric projection. To solve the problem we refer to Fig. 1.88. First, we see the projected axes of coordinates; they correspond to Fig. 1.69. To find the position of the projected point **P** we go to the left, along the x-axis, a distance equal to X_0. Next, we turn to the right and continue a distance Y_0 on a path parallel to the y-axis. Finally, we go up, in parallel to the z-axis, a distance equal to Z_0. Translating these motions into equations yields

$$x = (-X_0 \cos 30° + Y_0 \cos 30°)\sqrt{\frac{2}{3}}$$

$$y = (-X_0 \sin 30° - Y_0 \sin 30°)\sqrt{\frac{2}{3}} + Z_0\sqrt{\frac{2}{3}}$$

For use with MATLAB we can define the transformation matrix

$$\sqrt{\frac{2}{3}} \begin{bmatrix} -cosd(30) & cosd(30) & 0 \\ -sind(30) & -sind(30) & 1 \end{bmatrix} \tag{1.21}$$

This is a matrix of coordinate transformation, not a projection matrix. It is easy to check that the operation is not *idempotent*, a propriety explained in Exercise 1.12. Left multiplication of this matrix by a three-row array of three-dimensional coordinates yields a two-row array of coordinates in the isometric projection.

Exercise 1.11 (Matrices for frontal and side projections). Derive the matrices for orthogonal projection on the π_2 and π_3 planes and apply them to the points $\mathbf{P_1}$ and $\mathbf{P_2}$, and to the straight-line segment $\mathbf{P_1P_2}$ defined above. This is the place to mention that it is not difficult to develop also matrices for the central and parallel projections.

Exercise 1.12 (A property of projection matrices). In this appendix we obtained the horizontal projection of the point $\mathbf{P_1}$ as

$$\underline{P}'_1 = H_{proj}\mathbf{P_1}$$

Show that

$$H_{proj}\mathbf{P}'_1 = \mathbf{P}'_1$$

This means that

$$H_{proj}^2 = H_{proj}$$

Verify that the above properties hold also for the projection matrices developed in Exercise 1.11. In fact, the property holds for any projection matrix. We say that projection matrices are *idempotent*.

In this book we use MATLAB as the main tool for illustrating concepts and implementing procedures. The reader who is not familiar with this software can find help in the excellent documentation provided by the MathWorks and in some of the many books that treat the subject, for example Biran (2011).

APPENDIX 1.B FIRST STEPS IN MULTISURF

In this appendix we show how the operation of projection is implemented in a software developed mainly for Naval Architectural use. We have chosen MultiSurf, a product of AeroHydro, Inc. This is a program for **surface modelling** whose theoretical background is well documented (see Letcher et al., 1995; Mortenson, 1997; Letcher, 2009) and its functions can be readily related to concepts treated in this book. The reader who intends to

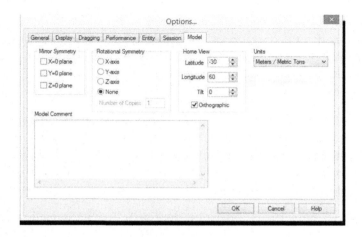

Figure 1.89 Options dialog screen

use this software should run the examples detailed in the tutorials provided by the distributor of the program and refer to the manual. Below we use typewriter characters to identify terms that appear on the MultiSurf screen and in the documentation of the software, for example Properties.

After starting MultiSurf select from the main menu File and from the dropdown menu choose New. A box of options appears as shown in Fig. 1.89. In the next chapter we show how to set the options for a hull surface. In this example we change only the units to those useful for ships, i.e. m and t. Click OK. An axonometric view of the coordinate axes appears in the drawing window. Go now to the main menu at the top and click on Insert. In the menu that opens click on Point. A submenu opens; select Point. This opens the Properties manager window shown in Fig. 1.90. It is convenient to dock it on the left side of the screen and adjust its dimensions to the necessary minimum. Change Name to P, and dx, dy and dz to 1. The line labeled Point shows an asterisk, '∗', which means that the values in the d boxes are *absolute* coordinates measured from the origin of coordinates. After changing one of the d-values, a small, green check mark appears at the right. Click this mark or move to another field to have the entry take effect. After entering all new values click the green check mark in the left, bottom corner of the Properties manager. The point P appears in the drawing window; to display its name check on the Toggle Point Names button of the toolbar (see Fig. 1.91).

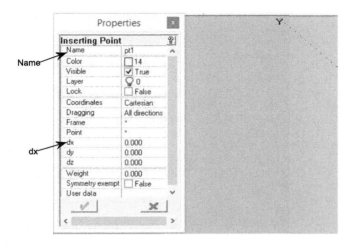

Figure 1.90 The Properties manager

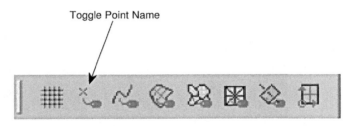

Figure 1.91 The Toggle-Names buttons

Now, let us draw the horizontal projection of the point P. Click again Insert and Point. In the menu that opens choose this time Projected Point (B in Fig. 1.92). In the Properties manager change the name to P' according to our conventions of notation. In the field labeled Point we must specify P, the name of the *parent point*. To do this click in the right-hand column of the line Point and next click on the axonometric view the point P to select it. In the field Mirror we must set *Z = 0, the equation of the plane on which we want to project. We find this plane in the Available Entities manager. Click to set the value and use the green radio button to confirm. Finally, click the green check mark in the bottom, left corner of the Properties manager. Continue by inserting the frontal projection, P" with *Y = 0, and the side projection, P"', with *X = 0. The point P and its projections on the coordinate planes are now all on the screen. You may rotate the figure by using the arrow keys, and zoom in or out with the

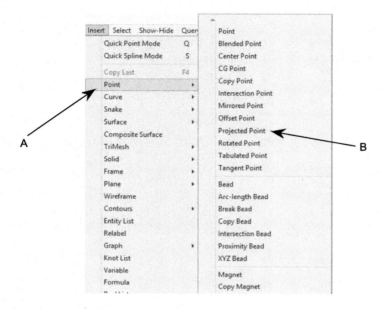

Figure 1.92 Inserting a projected point

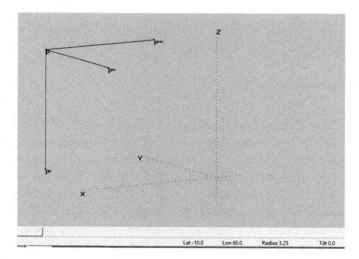

Figure 1.93 A view of the point P and its projections

scrolling wheel of the mouse. Another possibility to rotate the image is to press down the right mouse button, choose Rotate in the menu that opens, and then keep pressed the left mouse button while moving the mouse. The

Figure 1.94 The view icons

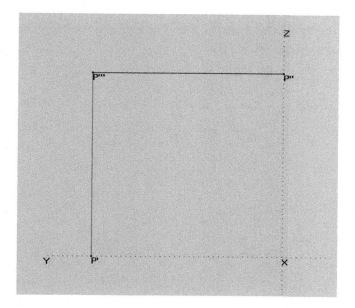

Figure 1.95 The projection on the X=0 plane

position in space of the four points is not very clear, and this reminds our comments on the ambiguity of the axonometric projections. To better understand the relative positions of the four points we may connect them by straight lines. For example, hold the Ctrl key and click the points P and P'. In the Properties manager the head line changes to Multiple. Click Insert and choose Curve. In the submenu click on Line. The Properties manager mentions the names of the two points that define the straight line. In continuation insert the lines PP" and PP"'.

Use the arrow keys or the mouse to bring the coordinate axes in the position displayed in Fig. 1.93. The Status bar at the bottom displays four values that define the view orientation and distance. Tilt should be 0, which means that the z-axis is vertical. Now, the orientation of the projection lines is clear, but the positions of the four points relative to the coordinate planes is still ambiguous. This ambiguity is solved in orthographic projections. Look for the view icons shown in Fig. 1.94. Clicking the second icon from left displays the projection on the system $*X=0$ plane as we see it in Fig. 1.95. In descriptive geometry we would call the view in Fig. 1.93 *side projection*, or *projection on π_3*. In engineering drawing we call this view *side elevation*. In the drawing of ship lines this projection generates the *body plan*. To see the projections on the other coordinate planes click the fourth and the fifth icons from the left. The rightmost icon returns us 'home'. This is the axonometric view.

CHAPTER 2

The Hull Surface — Graphic Definition

Contents

2.1 INTRODUCTION

Men built boats, and later ships, many thousands of years ago, but developed methods for defining their shapes only a few hundred years ago. Man-made artifacts have been dated as 50000-years old in Australia, others, discovered in Crete, in the years 2008–2009, were appreciated as being 120000 years old. As these regions were separated from mainland by that time, men had to use some floating devices to reach those places. Traces of such devices have not been discovered. The oldest archeological finds belong to the West-European Mesolithic period; they are pirogues made from one tree trunk (in French *monoxyle*) and were used in fluvial navigation. The oldest dating is around 8265 BC, in the Netherlands. Boats and paddles from the following centuries were found in Germany, Denmark, on the Seine in France, and in the UK. Wikipedia cites iconographic descriptions from the Nubian desert (cca. 5000 BC), and from the island of Syros in the Aegean Sea (2800 BC). The oldest ship relic is the burial vessel of Cheops (2650 BC). From the centuries that followed we have literature and archeological finds about navigation, sea and fluvial transportation, and marine battles from Egypt, Greece, the Middle East, and even from as far as China.

Geometry for Naval Architects
https://doi.org/10.1016/B978-0-08-100328-2.00011-0

Copyright © 2019 Elsevier Ltd.
All rights reserved.

The earliest ships were *sewn*. This term must not be taken in its literal sense, but refers to vessels whose parts were kept together by vegetal tendons. Later wooden vessels were built by the method known as **shell-first**. The construction began by laying down the keel, next the shell planking. Finally, the **frames** were fitted inside the resulting surface to provide the structural strength. The builder had to figure out the resulting shape, but, in fact, the shape resulted from the curvature assumed by the planks. McGrail (2001) comments, 'How the plank-first builder gets the shape of the hull is a mystery, in the medieval sense of that word. It is also an art, since the builder uses his personal abilities and experience. Sometimes it is described as "building by eye"...'

At some point an advanced method appeared; it is called **skeleton first** (also *frames first*, in French *membrure première*, in German *Skelettbauweise* or *Spantenbauweise*), and it is the way in which almost all ships are built today. The construction begins by laying down the keel, assembling on it the frames and finally covering the skeleton with planks. When and where this method appeared is a hot subject of research. Symposia were dedicated to it (see, for example, Nowacki and Valleriani, 2003). A wreck dating from the third century AD was found off Guernsey and it has details showing that the vessel was built skeleton-first at least partially (Barker, 2003). Similar indications appear on a wreck found at Port Berteau and dated around 600 AD (Letuppe, 2010). As Barker writes, 'These examples are however only representative of a much larger and steadily increasing group of "Romano-Celtic" vesels from north-west Europe.' Evidence of skeleton-first building was also identified on a wreck from around 500 discovered at Dor (Tantura), in Israel, and on a wreck found at Serçe Liman, Turkey, and dated 1024-5 (Valenti, 2009; Letuppe, 2010).

In the skeleton-first method the frames define the hull surface. Frames designed arbitrarily would lead to a hull surface that is not *fair*, a term we'll explain in Section 2.6, or may even be impossible to build. The external curve of a frame is the local, transverse section of the hull surface. For a *smooth* surface the set of sections fulfill certain geometrical conditions of continuity. Therefore, the frames should be properly **designed**. We do not know how this goal was achieved for the oldest skeleton-first ships. That was a period of transition between two methods. At some point a so-called *Mediterranean method* emerged and the first archeological evidence of this technique is the 'Culip VI' wreck built around 1300 and discovered off the coast of Catalonia (Rieth, 2003a). From the time of the Renaissance we have written descriptions of the techniques employed in Venice, then a leading maritime power. The method spread to the Iberian peninsula and to France.

In those times the hull surface was not designed by drawing at reduced scale, but by shaping structural members at full size, on the lofting ground. A short, clear description of the process is given by Castro (2007). The construction began by laying down the keel and assembling forward the stem, and aft the sternpost. One, two or three identical *master frames* were assembled in the middle region of the keel. The shape of those frames was obtained by combining arcs of circle with common tangents at the joining points, usually with a flat bottom in the middlebody. Additional frames were mounted towards the extremities. Their shape resembled those of the master frames, but the dimensions were modified. One change was *narrowing*, another one, *rising*. These changes were performed either by empirical geometric rules — Castro calls them *geometrical algorithms* — or with the aid of a set of specially built instruments. Other researchers mention an additional change, a progressive rotation of the upper lines towards the extremities. Finally, *ribbands* (*lisses* in French, *armadouras* in Portuguese) were stretched from the stern, over the frames, till the sternpost and they helped to define the shapes of the bow and stern frames. To make a connection with the ship-lines drawing treated in this chapter, we may say that the *narrowing* can be seen in the horizontal projection (waterlines plan), while the *rising* appears in the frontal projection (sheer plan). As to the ribbands, they define sections by oblique planes (see Fig. 1.41) that are the *diagonals* described in this chapter in Section 2.5. A description in Portuguese of an interesting variant of the process is due to Loureiro (2006). McGrail (2001) found that similar methods are used even today in Southern India and attributes them to the Portuguese influence, other researchers identified such procedures in Brazil and in the building of fishing boats in Greece.

Let us discuss a bit the idea of the frame shape composed of circular arcs. In Fig. 2.1 we show two examples inspired from a drawing that appears in *L'architecture navale du Sr Dassié*, 1677. These frames are typical for that period. We reconstructed the sections in MATLAB. The master frame 1 is composed of four pieces. The flat bottom marked S_1 is a straight line. The *bilge* S_2 is an arc of circle with centre in O_1. By construction, the tangent to the arc in the starting point continues the straight line S_1. The arc S_3 belongs to a circle with centre at O_2. The point where S_2 joins S_3 lies on the line that passes through O_1 and O_2. In consequence, the tangents to the two circular arcs in the joining point have the same direction. The last piece, S_4, is an arc of circle with centre at O_3. The two arcs join in the point

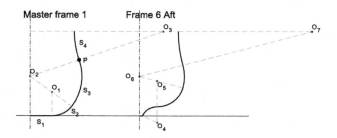

Figure 2.1 Frame shapes composed of circular arcs

P and their common normal lies on the line $O_2 O_3$. It follows that the two arcs have a common tangent in P. We leave to the reader to analyze in the same way the composed curve of frame 6 aft. These constructions reveal facts of paramount importance. First, hundreds of years ago shipbuilders understood that the shape of a ship section cannot be that of a single simple curve. Today we would say that the shape of a real ship frame cannot be represented by a single elementary equation and this idea leads to what is called now *piecewise interpolation*. Second, certain conditions of *continuity* are fulfilled at the points where two elementary curves are joined. We skip over what mathematicians call today C_0-*continuity*, and go to the condition of common tangents in the points where two arcs are joined. These two ideas are basic in the modern theory of *splines*.

Barker (2003) quotes English texts on shipbuilding that appeared from 1570 on. In 1711 William Sutherland used for the first time the term **whole-moulding**. Writes Barker, 'It is a moot point whether "whole-moulding" takes its name from forming the whole hull from a single mould (which is never actually achieved in full), or from the more restricted fact that the whole midship section is created from a single mould.' Rieth (2003b) analyzes the method described by Sutherland and details the three instruments on which it is based. The first one is the **master mould** (in French *maître gabarit*) that defines the shape of the master frames. The other two instruments are used to derive other frames by modifying the master frame. One instrument is the **rising square** (in French *tablette d'acculement*), the other a wooden scale called in French *trébuchet* and used to narrow the breadths (see also Rieth, 2001).

Barker (2006) writes that 'there are very few technical or working drawings extant from shipbuilding before 1570... The earliest ...are indeed little more than sketches to record a few key dimensions, and illustrate ships' fitted.' Barker discusses the ship drawings we have from that time until

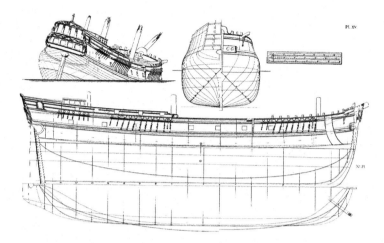

Figure 2.2 A drawing from Chapman's *Architectura Navalis Mercatoria* (1768). By courtesy of Sjöhistoriska museet, Stockholm

the middle of the 17th century and remarks that in that period 'building methods, in essence, did not *require* formal drawings at all'. Then, when were scale drawings used for the first time in practical shipbuilding? Danish sources claim that Ole Judichær did it as early as in 1692. We find more and more examples as we advance in the 18th century. In Fig. 2.2 we show an example from a beautiful collection of line drawings by the famous Swedish Naval Architect Fredrik Henrik af Chapman (1721–1808). On the drawing there are scales of length in Swedish, English and French units of that time.

The surface of a conventional-ship hull is too complex to allow definition by simple mathematical expressions, although attempts have been made to achieve this. As early as the 18th century, Fredrik Chapman thought about representing ship lines by arcs of parabolas. Around 1915 Admiral David Taylor (USA, 1864–1940) renewed the research on *mathematical forms* and developed a fully fledged method. In this method the waterlines were described by polynomials, separately for the forebody and the afterbody. Taylor himself noted that, 'practically all U.S. naval vessels designed during the last ten years have had mathematical lines' (quoted in Hamlin, 1988, p. 14). In particular, the method was used for designing the hulls of the Taylor's Standard Series of ship models. There was no practical continuation in the decades that followed, in spite of the work of several researchers. Weinblum (1953) describes succinctly those attempts. Until the advent of computers and adequate software the only possibility was to define the hull

Table 2.1 The terminology of the lines drawing

English	French	German	Italian
lines drawing	plan des formes	Linienriß	piano di costruzione
body plan	transversal, vertical	Spantriß	piano trasversale
waterlines plan	horizontal	Wasserlinienriß	piano orizzontale
sheer plan	longitudinal	Längsriß	piano longitudinale
station	couple de tracé	Kostruktionsspant	ordinata
waterline	ligne d'eau	Wasserlinie	linea d'acqua
buttock	longitudineau	Schnitt	longitudinale
diagonal	lisse plane, livet	Sente	diagonale
mould loft	salle à tracer	Schnürboden	sala a tracciare

surface graphically. As we have seen in Chapter 1, simple surfaces, such as those of a cube, a parallelepiped, a cylinder, or a cone, can be defined and visualized by multiple projections. More complex surfaces, however, cannot be fully defined by projections only; they require also sections. This way of defining the hull surface is the subject of this chapter. Although the computers provide today more tools for the design and the visualization of the hull surface, the traditional line drawings remain the simplest and most important means for presentation. Once done manually, today they are produced with the aid of computers and dedicated software.

2.2 THE LINES DRAWING

2.2.1 A Simple, Idealized Hull Surface

Let us begin with a simple hull composed of elementary surfaces. The middle body is a half of a cylinder, the forebody a half of a right, circular cone, and the poop a quarter of a sphere. The radius of the sphere is equal to that of the cylinder directrix and to the radius of the cone base. The three traditional projections of this surface are shown in Fig. 2.3. This is not the shape of a real ship. However, we think that it may be easier to visualize the concepts introduced here if we apply them first to simple surfaces, two of which were treated in the preceding chapter. Even in this grossly simplified case it takes some time to mentally synthesize the true three-dimensional shape out of the available two-dimensional projections. For real ship forms it would be impossible. The traditional method for solving the problem is to cut the hull surface with sets of parallel planes and project the resulting sections on the coordinate planes.

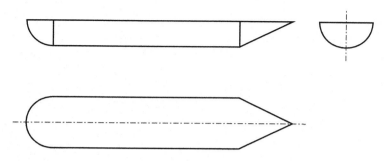

Figure 2.3 A simple hull

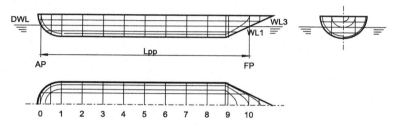

Figure 2.4 The lines drawing of the simple hull

To start we define a *design water line*. For merchant ships the standard procedure is to choose the *summer load line*, that is the highest water line up to which it is allowed to load the ship during the summer. In Fig. 2.4 the design water line is marked DWL. The design waterline intersects the hull outline in two points; we draw there perpendiculars to DWL. The aftermost perpendicular is called *aft perpendicular* and is marked AP, the other vertical line is the *forward perpendicular* marked FP. Other definitions are possible and we'll return to them later.

The distance between perpendiculars is called *length between perpendiculars* and its standard notation is L_{pp}. We divide the length between perpendiculars into equal intervals, in our example ten. Then, the interval length is obviously $dL = L_{pp}/10$. In the first interval at aft the hull shape changes rapidly. To catch this change we subdivide the first interval into two equal subintervals. As the hull continues forward of FP, we add at the right a subinterval of length $dL/2$. At each division we cut the hull surface with a transverse plane. The traces of the cutting planes are shown in the frontal and horizontal projections. In this example all transverse sections are semicircles. As the hull surface is symmetric with respect to a longitudinal plane (the *centreplane*) it is sufficient to draw only half-sections. If the drawing format allows it, a frequent practice is to show the sections in the upper, right-hand part of the lines drawing. At the right of the trace of the plane of symmetry (the dash-dot line) we show the sections of the forward half of the hull, while at the left we draw the sections of the aft half of the hull. The transverse sections are called **stations** and their set constitutes the **body plan**. Starting from aft we name the sections as

$$0, \frac{1}{2}, 1, 2, \ldots, 10, 10\frac{1}{2}$$

The next step is to cut the hull surface with horizontal planes. The traces of these planes appear in the frontal projection and in the body plan. In our example we divide the hull *depth* into four equal intervals and, as usual, we show the resulting sections in the left-hand, lower part of the lines drawing. The horizontal sections are called **waterlines** and the representation of their set is the **waterlines plan**. For reasons of symmetry we draw again only half sections. In this example the waterlines are quarters of circle at aft, straight lines in the middlebody, and arcs of hyperbolas at the bow. The waterlines plan includes the projection of the **deck at side**. The qualifier 'at side' characterizes the line that results from the intersection of the deck and side surface. The deck can have a transverse curvature, a case that we will treat in Section 2.3. Starting from the lowest one and going upward we note the waterlines as *WL*1, *WL*2, *WL*3, *WL*4.

The third set of sections is generated by cutting the hull surface with longitudinal planes. The traces of the cutting planes appear in the body plan and in the waterlines plan. The resulting sections are known as **buttocks**; they are shown in the frontal projection and their set is called **sheer plan**. Buttocks are generally noted by Latin numerals, for example 'Buttock I, Buttock II, ...'. In our example we show only one buttock; it begins as a quarter of circle at aft, continues as a straight line in the middlebody, and ends as an arc of hyperbola at the bow.

We can define the various hull sections more rigorously if we assume a system of coordinates. Generally, the x-axis runs along the ship, the y-axis is transversal, that is a horizontal perpendicular to the x-axis, and the z-axis is vertical. The position of the origin of coordinate, and the direction of the axes of coordinates differ according to practice; we'll return later to this subject. Now, we can say that the stations are constant-x sections, the waterlines, constant-z sections, and the buttocks, constant y-sections. Concisely, we can say that we represent the hull surface by curves of constant values for one of the coordinates. Each section is contained in a plane parallel to one of the projection planes. Therefore, each section appears as curved in one projection, and as straight lines in the other two. We can see here an analogy with *contour maps* in which a part of the earth surface is represented by lines of constant altitude. In MATLAB, after defining a surface in 3D space, it is possible to plot lines of constant-z with the aid of the functions `contour`, `contour3` and `contourf`. `Contour` is also the term used in MultiSurf for sections parallel to a given reference.

In Fig. 2.4 let the common radius of the aft hemisphere, middle cylinder and bow cone be $R = 5$ m. We assume that the DWL is located 2.5 m above the base line. The length between perpendiculars is $L_{pp} = 44.330$ m. With the origin of coordinates at AP, the x-coordinates of the displayed stations are

Station No.	Station x, m	Station No.	Station x, m
0	0.000	6	26.598
1/2	2.217	7	31.031
1	4.433	8	35.464
2	8.866	9	39.897
3	13.299	10	44.330
4	17.732	10 1/2	46.547
5	22.165		

Figs 2.5 to 2.7 show axonometric views of the three sets of sections that compose the lines drawing. Figs 2.4 to 2.7 are drawn in MATLAB. For the last three we used the function `plot3`. Translation of the main terms in this section appear in Table 2.1.

2.3 MAIN DIMENSIONS AND COEFFICIENTS OF FORM

The terms defined in this chapter belong to the basic terminology of Naval Architecture. Therefore, let us start with the beginning. In English, for a ship we use the pronoun *she*. Forward we have the **bow**, aft, the **stern**. Looking from the stern toward the bow, the right-hand side of the vessel is the **starboard side**, the left-hand side is the **port side**. The concepts of **main dimensions** and **coefficients of form** are related to the lines drawing. In this chapter we give the definitions of the most important terms. For more figures, comments, and translations into other languages we refer the reader to Chapter 1 in Biran and López-Pulido (2014). The lines are designed in the first iterations of the ship-design process when structural details are not yet known. In later iterations the hull surface defined

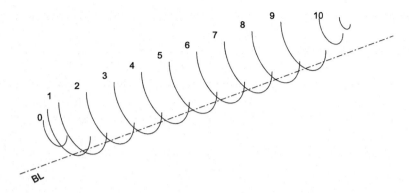

Figure 2.5 The stations of the simple hull

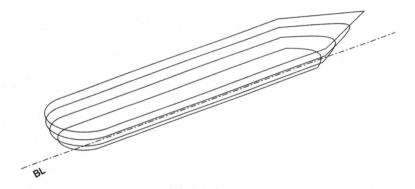

Figure 2.6 The waterlines of the simple hull

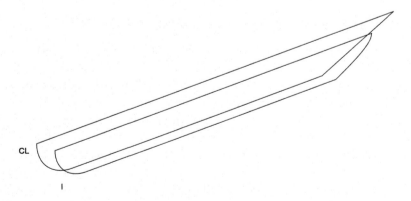

Figure 2.7 The centreplane section and a buttock of the simple hull

by these lines is the internal surface of the plating. This surface is smooth, a qualifier that should be taken here in the most general sense. For reasons of strength, usually the plates have different thicknesses. Then, the external surface of the shell is not smooth and defining the main dimensions on it would be awkward. The internal surface of the shell is the **moulded surface** of the ship and dimensions defined on it are known as **moulded dimensions**.

In Section 2.2.1 we defined the length between perpendiculars, L_{pp}. The midpoint of this length is called *midships*, and the station that contains it is the **midship section**. The symbol used in English literature for this term is shown in Fig. 2.9 in the appropriate place. In German literature we find a simplified symbol consisting of a circle and an inscribed X. French Naval Architects place a perpendicular in the midship section and call it *perpendiculaire milieu* (Rondeleux, 1911; Hervieu, 1985). Italian Naval Architects call this 'perpendicolare al mezzo' and note it by MP (Miranda, 2012). The following main dimensions are defined in the midship section.

Breadth, B — This is the breadth of the DWL at draught T. Alternative terms are *moulded breadth* and *moulded beam*. More than often B is different from the maximum breadth of the waterline that can occur in another section. Also, the deck breadth may be larger than the moulded breadth and this is the case of vessels with side *flare*.

Draught, T — This is the distance between the lowest point of the moulded hull and the DWL. One common definition is 'the distance between the upper surface of the keel and the design waterline'.

Depth, D — This is the distance between the lowest point of the moulded hull and the deck-at-side line.

Freeboard, f — This is the distance between the DWL and the deck-at-side line. In algebraic notation we can write $f = D - T$.

Part of the bottom of many ships is flat and horizontal. Other ships display a **rise of floor** measured in Fig. 2.8 as r_f. An alternative term is *deadrise*. The corresponding German word is 'Aufkimmung' (Rupp, 1981), and the Italian term is 'stellatura' (Miranda, 2012). The deck of many ships display a curvature in the transverse sections. This property is called **camber** and it is measured in Fig. 2.8 as c. The standard value for c is 1/50 of the local deck breadth. The transverse section of a deck with camber is a parabola. A complete view in the sheer plan would show two deck lines:

- the *deck at centre*, which is the intersection of the deck surface with the centreplane;
- the *deck at side*, which is the intersection of the deck and side surfaces.

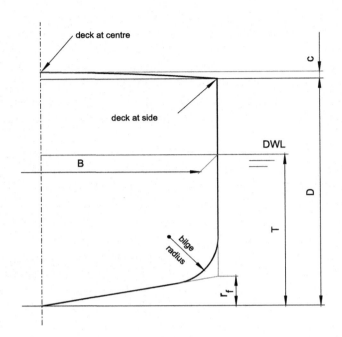

Figure 2.8 Dimensions defined in the midship section

The above lines meet at their ends. Usually, only the deck–at–side line is shown. Certain ships do not have camber because this feature would interfere with their function. Examples of such vessels are containerships and aircraft carriers.

In Section 2.2.1 we defined the aft perpendicular, AP, as the vertical line passing through the intersection of the centreplane contour of the moulded–hull surface and the DWL. This definition is mainly employed for warships. For merchant ships it is usual to assume the axis of the rudder as aft perpendicular, as shown in Fig. 2.9. As to the forward perpendicular, we can now be more specific and explain that it is usual to place it at the intersection of the external surface of the stem and the DWL. This definition is an exception as it uses a point that is not on the moulded surface. Other lengths defined in Fig. 2.9 are

- the **length overall**, L_{OA}, measured between the ship extremities above water;
- the **waterline length**, L_{WL}, a property important in hydrodynamical calculations.

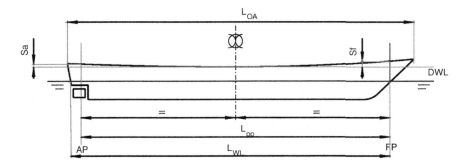

Figure 2.9 Definitions of lengths and sheer

For ships with submerged appendages, such as a bulbous bow or a sonar, it may be necessary to define a *length overall submerged* and a draught under the appendage. Many times the projection of the deck line in the sheer plan is curved. This curvature, called **sheer**, helps in preventing the water ingress on deck. The curved line is composed of two arcs of parabola that have a common, horizontal tangent in the midship section. Referring to Fig. 2.9 the standard values of the sheer at *AP* and *FP* are

$$S_a = 25 \left(\frac{L_{pp}}{3} + 10 \right), \quad S_f = 50 \left(\frac{L_{pp}}{3} + 10 \right) \tag{2.1}$$

where S_a and S_f are measured in mm, while L_{pp} is measured in m. Again, there is no sheer in ships where this feature would interfere with their function.

The **moulded volume of displacement** at a given draught, T_0, is the volume enclosed by the moulded hull surface up to the waterplane corresponding to T_0 and that waterplane. We use for it the notation ∇. To classify ship lines and to relate them to certain hydrostatic and hydro-dynamic properties Naval Architects use a set of non-dimensional numbers called **coefficients of form**. The most important coefficient is the **block coefficient** defined as

$$C_B = \frac{\nabla}{LBT} \tag{2.2}$$

The **waterplane coefficient** is the number

$$C_{WL} = \frac{A_W}{LB} \tag{2.3}$$

where A_W is the *waterplane area*. The definition of the **midship coefficient** is

$$C_M = \frac{A_M}{BT} \qquad (2.4)$$

where A_M is the area of the midship section. The **prismatic coefficient** is given by

$$C_P = \frac{\nabla}{A_M L} \qquad (2.5)$$

and the **vertical prismatic coefficient by**

$$C_{VP} = \frac{\nabla}{A_W T} \qquad (2.6)$$

Researchers define sometimes separate coefficients for the fore- and afterbody. An advantage of using non–dimensional numbers is that their values do not depend on the system of coordinates used in the design, as long as it is consistent. More advantages appear in the research and presentation of relationships between various ship parameters.

The volume of displacement is related to the displacement mass by Archimedes' principle

A body immersed in a fluid is subjected to an upwards force equal to the weight of the fluid displaced

The force predicted by this principle is the *buoyancy*. If the body floats freely, for equilibrium of forces the weight of the body must equal the force exercised by the fluid. Then, noting by Δ the mass of the body, and by ρ the density of the fluid, we can write

$$\Delta = \rho C_B LBT \qquad (2.7)$$

Eq. (2.7) sub–estimates the mass of real ships because it is based on moulded dimensions. The thickness of the shell and various *appendages*, such as the rudder, propeller and propeller struts, contribute additional volumes, i.e. an additional buoyancy force. This problem is discussed in Biran and López-Pulido (2014), Section 4.3.

2.4 SYSTEMS OF COORDINATES

In European practice and international standards the x-axis is positive forward. Then, if the reference frame is right-handed, the y-axis points to port

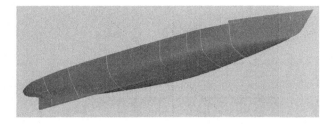

Figure 2.10 The hull M959 — A shaded view

and the z-axis is vertical upwards. We are aware of at least an old program for hydrostatic calculations that used a left-handed system of coordinates and this is today the case for some graphical software. The origin of coordinates recommended by some standards is at AP. This is useful when the distribution of masses is referred to the same origin. For hydrostatic calculations it is convenient to place the origin in the midship section. Then, one can immediately appreciate the position of points like the *centre of buoyancy* and the *centre of floatation*, terms that will be defined in Chapter 3. In MultiSurf the x-axis is defined as positive towards aft, the y-axis positive towards starboard, and the z-axis positive upwards. This reference frame is right-handed. The position of the origin displayed in the documentation of the software is at FP. We like, however, to place here too the origin in the midship section.

2.5 THE HULL SURFACE OF A REAL SHIP

After having introduced the lines drawing by means of a simple example, let us see a realistic one. The hull shape we use is derived from that of a passenger-ship model tested in the Rome basin (Vasca di Roma) and listed as *Carena C959* (see INSEAN, 1965). We modified in part the model; therefore, we call it M959. Fig. 2.10 shows a shaded view of the hull, the lines drawing appears in Fig. 2.11. We developed the model in MultiSurf. The lines generated by this software are not annotated. Therefore, we exported the model to Turbocad and added only a few annotations as the figure is rather small and we wanted to keep it readable. We marked the numbers of sections in the waterline plan, and the Roman numbers of buttocks in the body and sheer plans. As to waterlines, we annotated only WL0 in the sheer and waterline plans. In practice the lines are drawn on

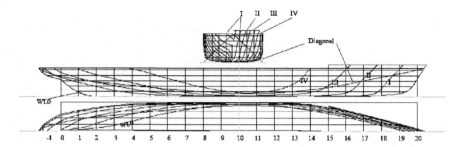

Figure 2.11 The lines of the hull M959

large formats, for example 2A0. All stations, waterlines and buttocks are annotated in all projections.

In many cases the designers add a fourth set of sections, the **diagonals**. These are obtained by cutting the hull surface with planes perpendicular to the yOz coordinate plane, but making an angle with the other two coordinate planes. In the chapter on descriptive geometry we describe these special planes in Figs 1.38–1.40. In Fig. 2.11 we show only one diagonal. Diagonals are drawn as much as possible as normals to sections in the region of their maximum curvature. In the body plan the diagonals appear as pairs of lines symmetrical with respect to the centreplane. Each pair represents in fact one line, the trace of the sectioning plane. The other view of the diagonal is the one in its plane and it is shown under the waterlines plan. MultiSurf does not show this *true-length* projection, but displays the projections on the coordinate planes, as shown in Fig. 2.11.

The various sections were generated as `contours` by using the Multi-Surf facilities described in Section 2.10. We can view the sets of sections by selecting them on the screen and hiding the unselected ones. Proceeding so we show in Fig. 2.12 an axonometric view of twenty-two stations, in Fig. 2.13 the axonometric view of five waterlines, and in Fig. 2.14 a view of the diagonal in its plane. The diagonal was generated by a cutting plane inclined at 45° with respect to the horizontal and frontal coordinate planes. Therefore, to see the diagonal in true-length projection we rotated it by 45°.

2.6 CONSISTENCY AND FAIRNESS OF SHIP LINES

The lines drawing must fulfill certain conditions of **consistency** and **fairness**. By consistency we mean the compliance with geometrical relation-

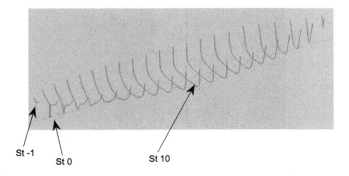

Figure 2.12 The stations of the hull M959

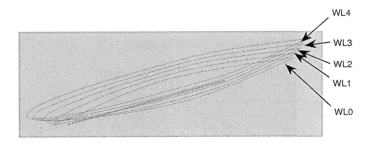

Figure 2.13 The waterlines of the hull M959

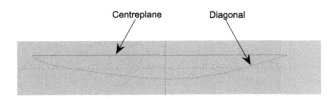

Figure 2.14 A diagonal of the hull M959

ships deduced in descriptive geometry. We summarize them below showing also the numbers of the corresponding sections in Chapter 1.

Section 1.8   Two projections completely define a point in 3D space. Any additional projection is derived from the first two.

Sections 1.8, 1.9   Any projection on a coordinate plane defines two coordinates of a given point. A given point coordinate can be read in two projections on coordinate planes.

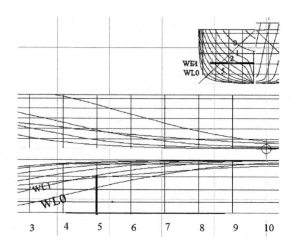

Figure 2.15 The half-breadth in the body and waterline plans

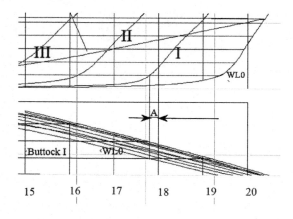

Figure 2.16 The intersection between a buttock and a waterline

Section 1.10.2 If two straight lines intersect, the projections of the point of intersection on the frontal and horizontal coordinate planes lie on a perpendicular to the ground line. The extension to the intersection of two curves is straightforward and it is based on the same reasoning as for straight lines.

To exemplify the above conditions let us begin with Fig. 2.15 where we look for the *half-breadth* of Station 5 in waterline WL1. We measure this length in the body and the waterlines plans; it is marked by a wider line in both projections. Next, we consider in Fig. 2.16 the intersection

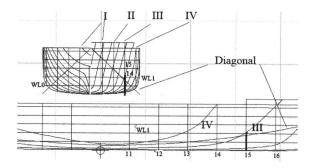

Figure 2.17 The intersection between a station and a buttock

of waterline WL0 and buttock I. We see the projections of the intersection point in the sheer and waterlines plans. The two projections lie on a vertical line. This means that the distance of these projections from the plane of station 18 is the same in the two projections. In the figure it is noted by *A*. Finally, in Fig. 2.17 we look for the height of buttock III in the plane of station 15. This length can be measured in the body and the sheer plans where it is marked by a wider line. When the ship lines were drawn manually, the designer had to care for their consistency. Today this property is automatically achieved by surface-modelling software.

If the notion of consistency can be defined easily and even in precise geometrical terms, it is not so for the notion of fairness. As an outstanding Naval Architect said many years ago, it is difficult to define what a fair line is, but anyone can easily identify a line that isn't fair. For an example let us see the lines in Fig. 2.18. In the sheer plan the keel and deck lines look fine. So do the stations in the body plan. From these stations we derive the waterlines. There is something unpleasant in them. A closer look reveals unexpected changes of curvature. We say that such lines are not *faired*. A person familiar with ship lines would look back to the body plan and identify there the cause: an irregularity in the distances between stations. This case justifies what we claimed in Section 2.1, 'the frames should be properly **designed**'. Assembling the shell over the shown stations will yield a hull surface that looks like that of a ship that was involved in a collision. To manually correct the unfairness the designer should start from the waterlines plan. Any correction done there will modify the lines in the other views and introduce there unfairness. Therefore, it could be wise to correct only part, let's say half of the unfairness in the waterlines plan, then half of the damage induced by this action in one of the other

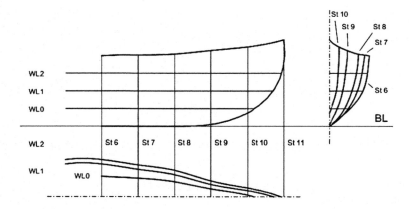

Figure 2.18 Ship lines that are not fair

plans, for example in the sheer plan. This last correction could affect the fairness in the waterlines plan. The designer must return and correct half of the new distortion. Working thus between the various plans, fairness can be obtained after several iterations. However, this means fair at the scale of the drawing; we would say today at the *resolution* of the drawing. For production the lines must be redrawn at full size. This operation generally reveals deviations from fairness that cannot be detected at reduced scale. One cause is the limited 'resolution' of the human eye. Another cause is that drawn lines, however thin, have a certain width themselves. Let us suppose, for example, that the scale of the lines drawing is 1:200, as is the case for large ships. At full size the width of the drawn line may mean a large deviation from fairness. This explains why the ship lines had to be faired again when drawn at full size on the moulding loft. Branco and Gordo (2008) explain in their lecture notes that the manual drawing of the ship lines in the mould loft could have kept busy two good lofters during four weeks. This stage represented a true bottleneck of the construction process. An alternative method was developed in Germany; in the 1950s it spread from what was then popularly called East Germany, correctly DDR, to Sweden and also to very few shipyards in the UK. The ship lines were drawn at the scale 1:10 under magnifying glasses. The lines were photographed on glass plates that were brought into the workshop and projected directly on the plates. Moulding lofts have been replaced today by computer screens and the fairing process is carried on with the aid of dedicated software.

To close this section let us remind related terms and comments in other languages. We like the Italian terms *bilanciamento* for the process that results in consistency, and *avviamento* for fairing. Weinblum (1953) writes:

> 'Beim Entwurf des Linienrisses spielen die Begriffe des harmonischen Verlaufs der Linien und ihres guten Straakens eine wichtige Rolle. Sie sind aus baulichen, ästethischen und hydrodynamischen Erwägungen entstanden; letzere waren jedoch recht vage und z.T. sogar anfechtbar.
>
> Beide Begriffe — Straaken und Harmonie — sind nicht scharf formuliert; trotzdem verkörpern sie einen wertvollen Erfahrungsschatz.'

In our translation this sounds, 'When designing the lines drawing the two concepts of harmonic run and good fairing play an important role. They developed from constructive, aesthetical, and hydrodynamical considerations; however, the latter were vague and, in part, even contestable. Both notions, fairing and harmony, are not formulated rigorously, nevertheless they embody a valuable treasure of experience'.

Interesting, 'straken', the German term for 'to fair', derives from the English word 'strake', which means a row of plates. The Portuguese term for 'fair' is 'desempolado'. The word 'empola' itself means 'blister', 'bubble'. Then, as Branco and Gordo (2008) remark, the Portuguese term means literally 'to eliminate blisters'. There is one particular comment of Branco and Gordo that points to the subjective aspect of the notion of *fair*. In our somewhat liberal translation it is, '... the same curve can be considered fair by one lofter and not fair by another. In practice, we never get the same fair hull surface if the work is given separately to two lofters, although they start from the same initial data.'

Our discussion of fair lines implied the notion of continuity. The requirement for continuity of the functions that describe the lines — in mathematical terms C_0 *continuity* — is obvious and general. The first and the second derivatives of the above functions, however, must be only piecewise continuous. The best known example is that of planing hulls that frequently look as in Fig. 2.19. In one or more points the stations have two tangents. In other words, the first derivatives of the two curves that meet in such points are different. The points of discontinuity lie on curves called **chine lines**. The chine lines themselves must be fair.

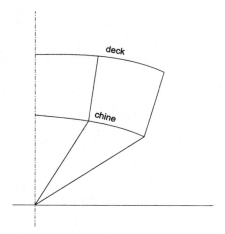

Figure 2.19 A chine line

2.7 DRAWING INSTRUMENTS

English	spline	duck
French	latte	
German	Straklatte	Strakgewichte (Molche)
Italian	listello	peso

To draw ship lines one needs more instruments than the straightedge and the compass known since antiquity. The first instrument specific to Naval Architecture is the **spline**, an elastic strip used for long lines with reduced curvature, such as waterlines are. According to the *Online Etymology Dictionary* the term is first recorded in this sense in 1756, in an East Anglian dialect. *Webster's Ninth New Collegiate Dictionary* indicates the same year. Smaller splines were used on drawing boards for ship lines at reduced scale, much larger splines were employed on the moulding loft for ship lines at the 1:1 scale. During a long period wood was the preferred material, but the literature mentions also metallic splines and when plastic materials appeared they replaced the wood. To force a spline to pass through given points it was held in place by weights. Because of their form these weights are called in English *ducks*. In Fig. 2.20 we show a spline fitted to a waterline and two of the ducks that force the spline. In 1946 Schoenberg (Romanian-American, 1903–1990) adopted the term spline for mathematical functions used in interpolating points and drawing curves. We treat the subject in Part 3 of this book. A comprehensive review of the subject can be found in

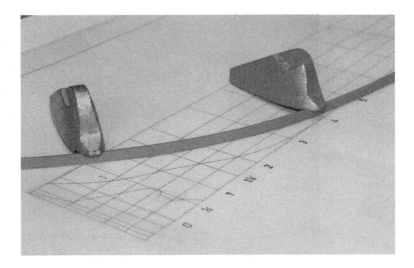

Figure 2.20 A spline and two ducks

Nowacki (2002). In Nowacki's paper we also can see an example of spline manipulated with mechanical screws instead of ducks.

Splines cannot be used for large curvatures. According to Lessenich (2009), instruments for drafting curves 'came into use about the last ten or fifteen years of the 18th century'. Lessenich describes several sets of curves, some of them still in use. One kind is the *Copenhagen set* with the oldest, still existing exemplars dated 1817. Examples belonging to the 20th century are the *English set*, the *Hamburg set*, the *German set*, *French curves*, and the *Burmester curves*. In Fig. 2.21 we show four pieces of a Hamburg set.

2.8 TABLE OF OFFSETS

In addition to the line drawings it is usual to deliver a numerical description of the hull. The main component is the set of *half-breadths* of waterlines measured at each station. The half breadths are in fact the y-coordinates of the waterlines measured at each station. The data are presented in a table with two entries. For example, the half-breadth of WL3 at St8 is entered at the intersection of the column containing the half-breadths of St8 with the line that contains the half-breadths of WL3. The set of half-breadths is called **offsets** and the table that shows them is known as **table of offsets**. An example of table of offsets is shown in Biran and López-Pulido (2014), p. 15. It is helpful to add more data, such as the heights above base line of

Figure 2.21 Drawing curves

Station	Half-breadths							Heights			
	WL0	WL0	WL1	WL2	WL3	WI4	Deck	Keel	Deck	Buttock I	Buttock II
1											
2											
3											

Figure 2.22 The heading of a table of offsets

the keel, buttocks, and the deck at side at each station. Then, the information can be arranged as in the table shown in Fig. 2.22 where each line contains the data of one station.

Preliminary offsets were delivered together with the drawing of the ship lines, *final offsets* after fairing the lines in the mould loft. Today, software such as MultiSurf automatically generates the table of offsets once the hull-surface design is completed.

2.9 SHELL EXPANSION AND WETTED SURFACE

The ship lines faired at full size in the mould loft were used to produce templates for cutting floors, bending frames, and cutting and forming the plates of the shell. The shapes of frames are transversal sections of the hull surface. Generally, these sections are not the stations that appear in the lines drawing. These lines are drawn in the first stages of the ship design, before the structural design. Then, the stations are positioned at equal intervals along the ship. This procedure enables an immediate visualization and there are people who, after looking for a few seconds, begin to feel the 3D shape of the lines. Another reason is that equally spacing the stations simplifies the rules for numerical integration (see, for example, Chapter 3 in Biran and López-Pulido, 2014), a consideration that may lose importance in advanced Naval-Architectural software. The position of structural frames, however, is based on strength and arrangement considerations, and on the rules of classification societies.

The shell of a steel ship is composed of strakes of steel plates. The plates are supplied as planar material; they must be cut and formed to suit the hull surface. Parts of this surface may be planar, for example at the bottom, *transom* sterns, or parts in the middle of the ship. For many ships, mainly large ships, the shape of transverse sections is constant within a length of the middle body. We call such parts *parallel middlebody* and they are cylindrical surfaces in the general sense of the term (see Section 1.14). The sides of parallel middlebodies are plane. The detail drawing of plates in the planar parts of the hull surface is trivial. A second category is that of regions of the hull surface that are developable. A representative example is the bilge region of a parallel middlebody. As written above, such surfaces are cylindrical and we explained how to develop them in Section 1.18.2. The plates that belong to this category are formed by simple bending, without stretching or pressing.

The third category is that of parts of the hull surface that are not developable. Such surfaces are also said to have *double curvature*. Plates that belong to this category must be stretched or pressed. The shape to be marked and cut on the flat plate was approximated by manual procedures such as described by Selkirk and Nieremair (1977), or Hamlin (1988). To explain the method let us begin with Fig. 2.23. We show there a part of the lines of a real ship. Let us assume that we want to develop the plate limited by the waterlines 4 and 6 and the stations 7 and 9. In the electronic version of the book this plate appears in yellow. We begin by dealing with the curvature of the stations. We measure the lengths of the arcs of stations 7, 8, and 9 between the waterlines 4 and 6. These lengths are called **girths**. When measured from the centreplane to a certain waterline *WLi*, these lengths are called *half girths to waterline i*. We plot the measured distances in Fig. 2.24, at the actual locations of the respective stations and obtain what Selkirk and Nieremair (1977) call *transverse expansion of half girths*. The

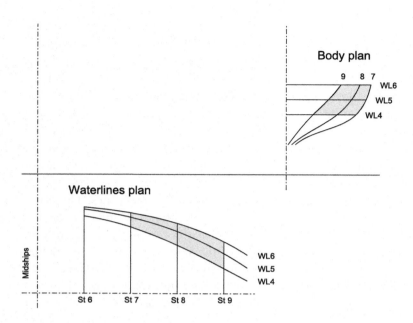

Figure 2.23 Part of a ship-lines drawing

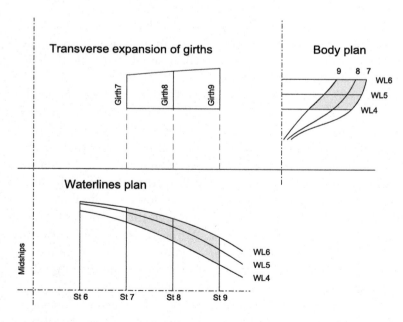

Figure 2.24 The transverse expansion of girths

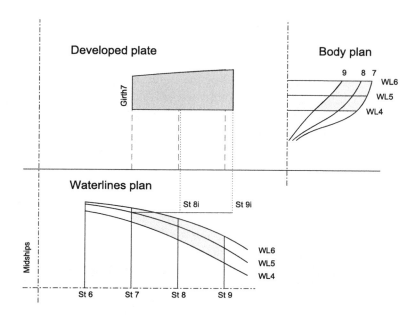

Figure 2.25 The plate after rectification of waterline 5

length could be measured by setting a certain length on a *divider* and checking how many times it enters along the given arc. Another possibility was to
carry on the measurement with the aid of a *curbimeter*, an instrument used
in topography. The same authors claim that when the transverse expansion
is carried on for the whole shell, we obtain an area that can differ by only
2% from the exact value. To improve the approximation we deal also with
curvature in the horizontal plane. For this we *rectify* the waterlines. This
term comes through the French 'rectifier' from the Latin 'rectus', which
means right. We measure the length of a waterline and mark it on a straight
line. The procedure was carried on by setting a spline along the waterline
and marking the first point of the arc to be measured and the places of
stations. Keeping the first point fixed, the spline was set free. In Fig. 2.25
we do this to WL5, the 'mean' waterline of the plate to be developed. As
we see, Station 8 moves to the position marked *St 8i*, and Station 9 to the
position marked *St 9i*. The improved approximation of the plate development appears in the electronic edition in transparent red. In Section 1.18.1
we explain that the main property of the developed surface is that it preserves the distances between two given points. In the procedure we have
just described the distances between points along the contours of stations

are preserved, and so are the distances along waterline 5. What happens to other distances? How correct is the procedure? Aren't there other procedures? The answer to such questions belongs to differential geometry and we are going to deal with this in Part 2 of the book. However, even differential geometry is not sufficient to completely solve the development of double-curvature plates because these parts of the hull must be stretched. Then, the distances between two points of the flat plate usually change on the formed plate. In other words, the plate undergoes a plastic deformation and this belongs to elasticity.

After developing all plates, half of the hull surface is shown in a 2D drawing called **shell expansion**. This drawing helps in planning the continuation of the construction process. If an average plate thickness is estimated in the beginning, multiplying this value by the area of the expansion yields a first indication of the mass of plates and enables a preliminary order of material. The submerged area can be identified and, by separating the areas that are alternately submerged or in air, and the area that stays mainly in the air, it is possible to estimate the necessary quantities of various paints and the requirements of cathodic protection against corrosion. Finally, the estimation of the submerged area yields the value of the *wetted surface* needed in the calculation of frictional resistance. An exact calculation of this latter value requires mathematical means that will be explained in Chapter 3. Several manual procedures for developing the hull surface are described by Dormidontow (1954).

2.10 AN EXAMPLE IN MULTISURF

In this example we show how to design in MultiSurf the simple hull used in Section 2.2.1, and how to produce the corresponding lines drawing. For the assumed composite surface we need only straight lines and elementary curves such as those described in Chapter 1. The lines drawing of real ships require other curves that will be introduced in Part 2 of the book. Therefore, only after treating those curves we will be able to show how to define the hull of a real ship with the aid of a computer program.

Let us start again MultiSurf and choose New from the File menu. In the Options dialog window that opens we choose now Mirror Symmetry → Y = 0 plane, and again Units → Meters/Metric Tons. The first choice enables us to define only one half of the hull surface; the program knows how to complete the other half as symmetric with respect to the plane $Y = 0$ (the centreplane). We accept in this example the standard coordinate axes of MultiSurf, that is x positive towards aft. Looking in the sense of the x-axis y is positive towards left. Finally, z is positive upwards. Thus, our system of coordinates is right-handed. This convention is usual in some American software and then the preferred origin of coordinates is on the forward perpendicular, *FP*. Our choice is to place the origin in the horizontal plane tangent to the hull surface. Thus, the x-axis lies along the base line, *BL*. With these assumptions, the x-coordinates of the stations displayed in Fig. 2.4 are

Station No.	Station x, m	Station No.	Station x, m
0	44.330	6	17.732
1/2	42.114	7	13.299
1	39.897	8	8.866
2	35.464	9	4.433
3	31.031	10	0.000
4	26.598	10 1/2	−2.217
5	22.165		

For this example it is convenient to define first the middlebody; it is half of a cylinder and its directrix is a semicircle. To draw the directrix we are going to use the Arc curve type that allows us to define an arc of circle starting from a point P_1, with the centre in a point P_2, and ending at the point P_3. Placing the directrix at the forward end of the middlebody, the coordinates of these points are

$$P_1 = \begin{bmatrix} 5 \\ 0 \\ 0 \end{bmatrix}, \; P_2 = \begin{bmatrix} 5 \\ 0 \\ 5 \end{bmatrix}, \; P_3 = \begin{bmatrix} 5 \\ 5 \\ 5 \end{bmatrix} \qquad (2.8)$$

Insert the above points. To use *absolute* coordinates take care that, in the Properties manager, the entry line Point contains an asterisk, '*'. Keep Ctrl pressed and choose the three points P_1, P_2, P_2 in this order. The Properties manager will show the headline Multiple. Continue with Insert, Curve, Arc. In the Properties manager go to the entry line Type. In the pull-down menu choose option 2, From point1, with Point2 as centre, to Point3 (see Fig. 2.26). Click the green check mark.

The surface we want to define is ruled; to obtain it in MultiSurf we need two supporting curves. We just defined the first. One possibility to define the second curve is to insert an arc based on the points with absolute

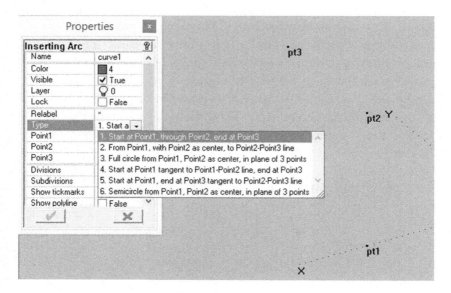

Figure 2.26 Inserting a circular arc

coordinates

$$P_4 = \begin{bmatrix} 45 \\ 0 \\ 0 \end{bmatrix}, \ P_5 = \begin{bmatrix} 45 \\ 0 \\ 5 \end{bmatrix}, \ P_6 = \begin{bmatrix} 45 \\ 5 \\ 5 \end{bmatrix} \qquad (2.9)$$

Another possibility is to define the above points with reference to P_1, P_2, P_3. Thus, for P_4 choose insert → Point. In the Properties manager click the line Point, and in the Available Entities manager click pt1, as shown in Fig. 2.27. Next, in the Properties manager set dx to 40.000, while dy and dz remain 0.0000. These are the coordinates of pt4 relative to pt1. Please mind that changing the absolute coordinates of pt1 will change also the absolute coordinates of pt4. In the terminology of computer sciences pt4 is a **child** of pt1, while pt1 is the **parent** of pt4.

After inserting the points pt5 and pt6 we can insert an arc of circle proceeding in the same way as for curve1 (see Fig. 2.26). Select curve1 and curve2 and then Insert → surface → Ruled Surface. In the Properties manager change only what is necessary to make it look as in Fig. 2.28. The same figure shows also the resulting middlebody surface. When the surface is selected, an arrow that appears near the centroid of the surface indicates the direction of the normal to the surface.

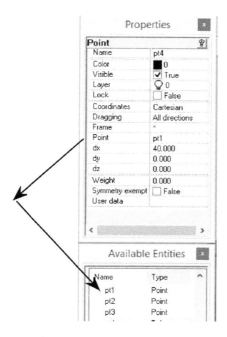

Figure 2.27 Inserting Point4 as child of Point1

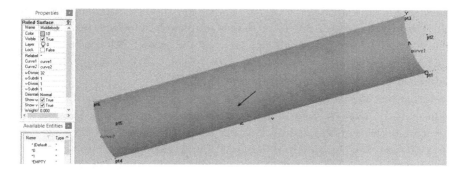

Figure 2.28 Inserting the middlebody surface

In the next step we are going to build the bow surface. It is a part of a conic surface, again a ruled surface. As for the middlebody, we need two curves as support. The first is curve1, the second will be a straight line beginning at pt3 and ending at the foremost point of the hull. Thus, we define this conic surface with the aid of an arc of its directrix, and a generator. Fig. 2.29 shows the Properties manager of the foremost point

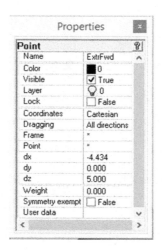

Figure 2.29 Inserting the foremost point

that we name `ExtrFwd`. We insert now the cone directrix by choosing `pt3` and `ExtrFwd` and then `Insert → Curve → Line`. Choose `curve1` and the directrix and then `Insert → Surface → Ruled Surface`.

In the third step we generate the poop surface by rotating `curve2` by 90°. The necessary actions are `Insert → Surface → Revolution Surface`. Use the values displayed in Fig. 2.30. To view the complete hull surface press `F5`. Use the ← and → keys to rotate the view of the model. Alternatively, select from the main menu `View → Symmetry Images`. For orthographic projections click the various icons shown in Fig. 1.94.

Now, that the hull surface is defined, we can generate the sections that will appear in the lines drawing. First, we produce the body plan. Let us begin, for example, with the stations $10\frac{1}{2}$ and 10. The actions are `Insert →` `Contours → Contours`. As shown in Fig. 2.31, in the entry line `Cut type` of the `Properties` manager choose `Offsets from Mirror/Surface`. In the next entry line click in the red field and in the `Available Entities` manager choose `*X=0`. This setting specifies contours parallel to the plane $X = 0$. Setting `First index` to 0, and `Last index` to 1 means that we are going to define two contours. For q0 specify the x-coordinate of station $10\frac{1}{2}$. Setting `q-Int` to 2.216 means that the next station is situated at $x = 0$. When the cursor is on the entry line `Surface` choose `Surface 1` by clicking either in the drawing, or in the `Available Entities` manager. This action specifies the surface to be cut. Continue in a similar manner for the contours of the middlebody and poop surfaces. To generate the waterlines set `Mirror/`

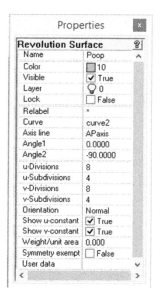

Properties	☒
Revolution Surface	?
Name	Poop
Color	☐ 10
Visible	☑ True
Layer	◯ 0
Lock	☐ False
Relabel	×
Curve	curve2
Axis line	APaxis
Angle1	0.0000
Angle2	-90.0000
u-Divisions	8
u-Subdivisions	4
v-Divisions	8
v-Subdivisions	4
Orientation	Normal
Show u-constant	☑ True
Show v-constant	☑ True
Weight/unit area	0.000
Symmetry exempt	☐ False
User data	˅

Figure 2.30 Inserting the poop as a revolution surface

Properties	☒
Contours	?
Name	BowSections
Color	☐ 12
Visible	☑ True
Layer	◯ 0
Lock	☐ False
Cut type	Offset from Mirror/Surfa
Mirror/surface	*X=0
First index	0
Last index	1
q0	-2.217
q-Int	2.216
Surfaces/TriMeshes	(1)
Weight/unit length	0.000
Symmetry exempt	☐ False
User data	˅

Figure 2.31 Inserting forward stations

surface to *Z=0 (see Fig. 2.32), and for buttocks choose as Mirror/surface *Y=0 (Fig. 2.33). To see the complete hull press F5; the result is shown in Fig. 2.34.

We have completed the lines drawing with the aid of a computer program. An experienced person would need approximately the same time to

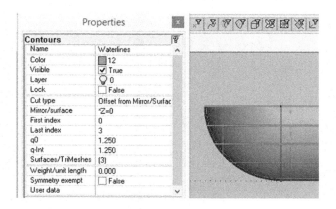

Figure 2.32 Inserting waterlines

Contours		
Name	Buttocks	
Color	⬛12	
Visible	✔ True	
Layer	🔘 0	
Lock	☐ False	
Cut type	Offset from Mirror/Surface	
Mirror/surface	*Y=0	
First index	0	
Last index	1	
q0	0.000	
q-Int	1.250	
Surfaces/TriMeshes	(3)	
Weight/unit length	0.000	
Symmetry exempt	☐ False	
User data		

Figure 2.33 Inserting buttocks

use the methods described in Chapter 1 and draw manually the lines. Then, is there any advantage in using the computer? The answer is not just 'Yes', but 'Yes, there are multiple advantages'. First of all, software like MultiSurf allows the user to easily change the lines. For an example, using the model of our simple hull, turn the view to the orthogonal projection on the xOz plan by clicking on the fourth view icon from left (see Fig. 1.94). Pick up the right-most point (named ExtrFwd) and move it horizontally to the left and to the right. Al entities defined in relation to this point will change too. Another kind of changes can be performed with Edit → Transform.

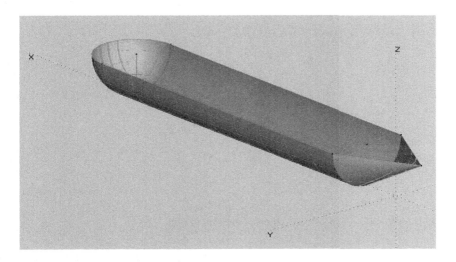

Figure 2.34 View of the simple hull

In the pull-down menu choose Scale. A dialogue box opens; using it one can produce a hull affine to the given one. The subject of *affine hulls* is discussed briefly in Biran and López-Pulido (2014), Section 4.6. In particular, choosing the same scale for the three coordinate axes produces a hull that is *geometrically similar* to the initial hull, shortly *Geosim*.

Another advantage of designing the hull with the aid of the computer is that once we have the hull model, it is easy to interface it with software modules that perform various calculations needed in ship design and production. To see this, return again to our simple model. In the View menu choose Display → Offsets. A dialogue box opens, check in it Add Deck. Do so because the model developed until now is 'open' above and this can cause trouble in hydrostatic calculations. The screen will display the view shown in Fig. 2.35. Ignore the error message because we only deal with an example. We see the stations that serve for hydrostatic calculations. While this figure is open, choose Hydrostatics in the menu Tools. A dialogue box opens. You may either leave the default value 1.026 for Specif. Weight, or change it to a *displacement factor* as recommended in Biran and López-Pulido (2014), pp. 107–8. The field Zcg refers to the z-coordinate of the centre of gravity, more often noted by Naval Architects as $\overline{KG}$ or VCG. Set the value to 3.5 m. The field Sink refers to the draught, usually noted by T. Enter 2.5 m. Click OK and browse through the resulting, extensive output. Among others, for example, we read the *volume of displacement* $\nabla = 660.3 \text{ m}^3$.

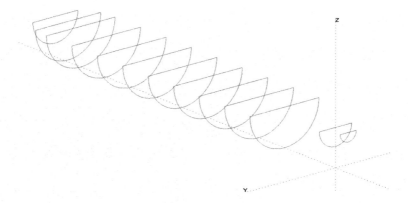

Figure 2.35 MultiSurf — The Offsets screen of the simple hull

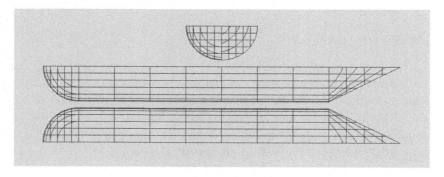

Figure 2.36 MultiSurf — The lines drawing of the simple hull

To obtain the lines drawing shown in Fig. 2.36 carry on the sequence

$$View \rightarrow Display \rightarrow ShipLines$$

Two dialogue windows will open. You may accept default values or modify them.

2.11 SUMMARY

Men built boats thousands of years ago, but developed methods for defining their shapes only in the last centuries. The need to define the hull surface arose together with the appearance of the construction method called *skeleton first*. In this procedure the keel is laid down first, the frames are assembled on it, and, last, the shell is mounted over the frames. The external shape of a frame is that of the local, transverse section of the hull surface. In the first centuries of the skeleton-first method no scale drawings were used. The frames were shaped at full scale using templates and certain geometrical rules. The first evidence of ship lines drawn at scale and used for practical shipbuilding are from the end of the 17th century. At the beginning of the 19th century the descriptive geometry became a tool in the drawing of ship lines.

In the traditional method the hull surface is described by sets of planar sections parallel to the coordinate planes. Generally, the x-axis runs along the ship length, the y-axis is transversal, and the z-axis, vertical. A first set of sections are *contours* drawn at constant x-values. These transverse sections are called *stations* and their set is known as the *body plan*. A second set consists of contours drawn at constant z-values; they are called *waterlines* and their set is the *waterlines plan*. The third set is composed of contours drawn at constant y-values, they are the *buttocks*, and their set is called *sheer plan*. Often a fourth kind of sections is added. These are the *diagonals* generated by planes that are perpendicular to the yOz plane and are inclined with respect to the horizontal projection plane. To illustrate the above concepts we give as a first example an idealized hull composed of a quarter of a sphere, a cylinder half, and a cone half. A second example is that of a real-ship hull.

The ship lines must fulfill certain conditions of *consistency* and be *fair*. The former conditions originate from descriptive geometry. Given a point in 3D space, each projection on a coordinate plane shows two of its coordinates. Conversely, each point coordinate can be measured in two projections. Therefore, the coordinate values of a given point should be identical in two projections. If consistency can be defined simply and rigorously, it is not so for the notion of fairness. This notion is related to the geometrical concepts of continuity and curvature, and has also an aesthetical aspect.

The ship lines are designed at reduced scale, in the design office. The construction process required lines drawn at full scale. Lines that looked well at reduced scale may appear as not faired at full scale. Therefore, the common practice was to redraw and fair again the lines at full scale, in the mould loft, a time-consuming operation. In the 1950s some shipyards used lines drawn at the scale 1:10 under magnifying glasses, and photographed on glass plates. In the workshop the photographed lines were projected on the steel plates. Today the whole process is carried on by specialized software. The drawing of ship lines required specialized drawing instruments. The most specific and important one was the *spline*, an elastic wooden, metallic or plastic strip. Special curves were introduced at the end of the 18th century; the best known examples are the *Copenhagen* and the *Hamburg* sets.

The shell is made of plates that are supplied as flat material. Often, parts of the hull surface are plane; their production is simple. Other parts, and always the hull extremities, are not plane. Then, the plates must be formed to suit the hull surface. If the curved regions are developable, the plates must be bent only. Such is the case in the bilge region of parallel middlebodies. In fact, such regions are cylindrical. Hull regions with *double curvature* are not developable. A method for an approximate development of such plates is described shortly in this chapter. It consists in *expanding* the *girths* of transverse sections and *rectifying* an average waterline.

The chapter ends by showing how to draw the simple, idealized hull with the aid of MultiSurf.

2.12 EXERCISES

Exercise 2.1 (A diagonal of the simple hull). Consider the lines drawing of the simple hull shown in Fig. 2.4 and a diagonal generated by a plane that passes through the common axis of the spherical, cylindrical, and conic parts, and is inclined at 45° with respect to the horizontal projection plane. Draw this diagonal.

Exercise 2.2 (Deck lines). Explain why the deck-at-centre and the deck-at-side lines intersect at the ship extremities.

Exercise 2.3 (Coefficients of form). Prove that the coefficients of form defined by Eqs (2.2) to (2.6) are non-dimensional.

Exercise 2.4 (Prismatic coefficients). Prove that

$$C_p = \frac{C_B}{C_M}, \quad C_{VP} = \frac{C_B}{C_{WL}} \tag{2.10}$$

Exercise 2.5 (Completing lines and developing a plate). Fig. 2.37 shows more lines of the ship Lido 9 (Biran and López-Pulido, 2014) than Fig. 2.23. You are required to carry on the tasks listed below.
1. Draw station $9\frac{1}{2}$ in the waterlines plan.
2. Draw *Wl3* in the waterlines plan, between Stations 6 and $9\frac{1}{2}$.
3. Draw buttock 1 in the sheer plan, between Stations 6 and $9\frac{1}{2}$.
4. Draw the diagonal (marked 'Diagonal' in the body plan), between Stations 6 and 10.

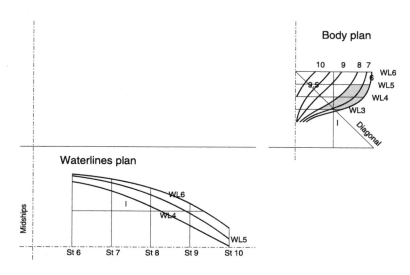

Figure 2.37 Part of a ship-lines drawing

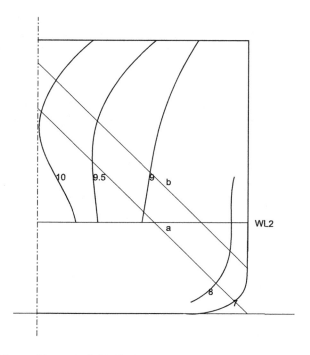

Figure 2.38 Cargo ship, part of ship lines

5. This question refers to the part of the hull surface limited by *WL3, WL5, St 6* and *St 8*. This part appears in the body plan as a yellow patch in the electronic version of the book, as a grey patch in the printed version. Draw:

   5.1 — the expansion of girths;

   5.2 — the expansion after rectification of the suitable waterline.

Exercise 2.6 (Completing lines 2). Fig. 2.38 is based on the table of offsets of a cargo ship found in Sessa (2008). The main data are

Length between perpendiculars, L_{pp}	120 m
Moulded beam, B	18 m
Depth	12 m

The figure shows parts of the Stations 7, 8, 9, $9\frac{1}{2}$ and 10, and the trace of WL2 in the body plan. The ship has a *bulbous bow*. Your tasks are listed below.

1. Find the scale of the body plan. For obvious reason do not expect it to be standard.
2. Draw at the same scale the waterline 2 between Stations 7 and 10.
3. Draw at the same scale the diagonals *a* and *b* between Stations 7 and 10.

CHAPTER 3

Geometric Properties of Areas and Volumes

Contents

Geometry for Naval Architects
https://doi.org/10.1016/B978-0-08-100328-2.00012-2

Copyright © 2019 Elsevier Ltd.
All rights reserved.

3.1 INTRODUCTION

The evaluation of the hydrostatic and stability properties of floating bodies requires the calculation of areas, centroids, and second moments of plane sections, of volumes and their centroids, and of centres of gravity and mass moments of inertia. For some of these terms and their translations into French, German and Italian see Table 3.1. Such calculations involve many integrations. This chapter deals with the definitions of the above concepts and the mathematics related to them. We also indicate how the general definitions and methods are applied in Naval Architecture. As explained in Chapter 2, in the traditional method the hull surface is defined graphically by plane sections, and numerically in tabular form. Therefore, traditionally the required integrations are carried on numerically. The treatment of this subject is not included in this book and we only give a few examples in Section 3.9. For a detailed treatment we refer the reader to Chapter 3 in Biran and López-Pulido (2014). In the general treatment of the properties of areas of plane sections we use the coordinates x and y. In specific applica-

Table 3.1 Properties of plane areas and volumes

English	French	German	Italian
area	aire	Flächeninhalt	area
first moment	moment statique	Moment erster Ordnung, statisches Moment	momento statico
centroid	centre de gravité	Schwerpunkt	baricentro
second moment	moment quadratique moment d'inertie	Flächenträgheitsmoment, Moment 2. Grades	momento d'inerzia
product of inertia	produit d'inertie, moment d'inertie centrifuge	Zentrifugalmoment, Deviationsmoment	momento centrifugale
radius of gyration	rayon de gyration	Trägheitsradius	raggio d'inerzia
principal axes	axes principaux	Hauptachsen	assi principali
volume	volume	Volumen	volume

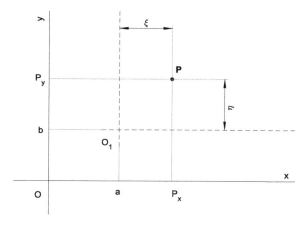

Figure 3.1 Translated system of coordinates

tions in Naval Architecture these coordinates are valid for waterplanes. For transverse sections, such as stations, we use the coordinates y and z. Our notation for second moments of areas follows the German practice. In our opinion this notation is more 'homogeneous'. For example, we write I_{xx} rather than I_x and this is consistent with the usual notation I_{xy} for the product of inertia. In Section 3.5.4 we introduce the notion of tensor of inertia, and in Section 3.5.7 we treat concisely the principal axes and moments of inertia as a problem of eigenvalues. Although we do not discuss the theory behind them, these two sections bring the discussion to a higher level and allow the reader to use better computational tools, such as those provided by MATLAB. Skipping these sections would not impair the understanding of this chapter and its application in the practice of Naval Architecture.

3.2 CHANGE OF COORDINATE AXES

The mathematical definitions and methods treated in this chapter involve sometimes the operations of **translation** and **rotation** of coordinate axes. This section is a basic introduction to the subject. Translation and rotation are examples of **geometrical transformations**, more specifically **affine transformations** that will be treated more systematically in Chapter 8.

3.2.1 Translation of Coordinate Axes

Fig. 3.1 shows a system of coordinates, x and y, with the origin in O, and a point **P** with coordinates P_x, P_y. Let us suppose that the given system of

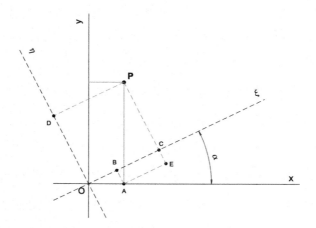

Figure 3.2 Rotated system of coordinates

coordinates is transformed by a *translation* of a units to the right and b units upward. The axes of the new system are shown as dashed lines. Let the coordinates of the point **P** in the translated system be ξ, η; they are related to those in the initial system by the general equations

$$\xi = x - a$$
$$\eta = y - b \tag{3.1}$$

In matrix form these relationships are

$$\begin{vmatrix} \xi \\ \eta \end{vmatrix} = \begin{vmatrix} x \\ y \end{vmatrix} - \begin{vmatrix} a \\ b \end{vmatrix} \tag{3.2}$$

3.2.2 Rotation of Coordinate Axes

We consider the point **P** shown in Fig. 3.2. Let its coordinates relative to the x- and y-axes shown as solid lines be P_x, P_y. We assume now that the system of coordinates is rotated in the positive trigonometric sense by an angle α. The rotated axes are shown as dashed lines. Let the coordinates of the same point in the rotated system be P_ξ, P_η. We draw $\overline{PE}$ and $\overline{AB}$ perpendicular to the rotated axis $O\xi$ and write

$$P_\xi = \overline{OC} = \overline{OB} + \overline{BC}$$

where $\overline{OB} = P_x \cos \alpha$, and, by construction, $\overline{BC} = \overline{AE} = P_y \sin \alpha$. It follows that

$$P_\xi = P_x \cos \alpha + P_y \sin \alpha$$

The coordinate P_η is given by

$$P_\eta = \overline{EP} - \overline{EC} = P_y \cos \alpha - P_x \sin \alpha$$

Summing up, the general relationships between the coordinates in the rotated system and those in the initial system are

$$\xi = x \cos \alpha + y \sin \alpha$$
$$\eta = -x \sin \alpha + y \cos \alpha \tag{3.3}$$

or, in matrix form,

$$\begin{vmatrix} \xi \\ \eta \end{vmatrix} = \begin{vmatrix} \cos \alpha & \sin \alpha \\ -\sin \alpha & \cos \alpha \end{vmatrix} \begin{vmatrix} x \\ y \end{vmatrix} \tag{3.4}$$

3.3 AREAS

3.3.1 Definitions

In a given plane let A be a region bounded by a closed curve C. The **area** of A is a measure of the extent of A. By definition the area of a square with unit side has the value 1 and its unit is the squared length unit. Examples of *unit-square* areas are:

- the area of a square with side 1 cm is 1 cm^2;
- the area of a square with side 1 m is 1 m^2 and this is the SI unit used in Naval Architecture;
- the area of a square with side 1 ft is 1 ft^2.

The area of any other figure is equal to the number of unit squares it contains. An immediate consequence of this definition is that the area of a rectangle with base b and height h has the value bh.

The calculation of areas is based on axioms that usually are not formulated explicitly. To state the axioms we must first introduce the notion of *congruence*. We say that two figures are **congruent** if one of them can be superposed over the other without changing the distances between its points. The geometrical transformations that enable us to bring a figure over another under the above condition are translations, rotations, and reflections.

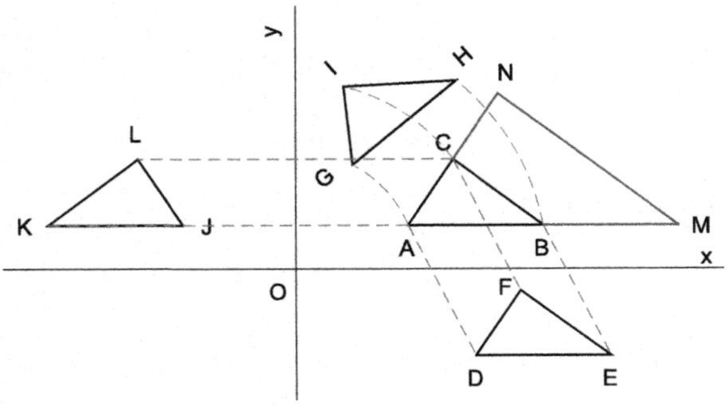

Figure 3.3 Congruent and similar triangles

For example, in Fig. 3.3 the triangle ABC can be superposed over the triangle DEF by a translation, over the triangle GHI by a rotation, and over the triangle JKL by a reflection in the Oy axis. We assume the following axioms:

1. if two figures are congruent their areas are equal;
2. if a figure can be decomposed into disjoint, 'elementary' figures, its area is equal to the sum of the areas of the elementary figures.

Let us consider also the case of geometrically similar figures. For example, in Fig. 3.3 the triangle AMN is geometrically similar to the triangle ABC. In this particular case we can write

$$\frac{\overline{AM}}{\overline{AB}} = \frac{\overline{MN}}{\overline{BC}} = \frac{\overline{NA}}{\overline{CA}} = 2$$

The area of the triangle AMN equals 2^2 times the area of the triangle ABC. The generalization of this result is obvious and it has a particular application in the calculation of the area of a figure whose dimensions are read from a scaled drawing: *if a figure has the area A_f when drawn at the scale 1:n, its actual area is $n^2 A_f$.*

In Fig. 3.4 we consider an area A, enclosed by a curve, and an element of area $dA = dxdy$. The area A is calculated as

$$A = \int\int_A dA = \int\int_A dxdy \qquad (3.5)$$

In the literature we find also the notation

$$A = \int_A dA$$

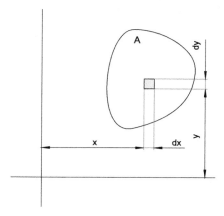

Figure 3.4 For the definitions of area properties

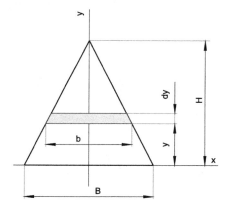

Figure 3.5 Isosceles triangle 1

The notation with two $\int$ symbols may be more expressive because, in general, A can be calculated as a double integral. By a convenient choice of dA, other than $dxdy$, the area can be calculated as a single integral, a procedure we are going to use in most of the examples that follow.

3.3.2 Examples

Example 3.1 (Area of isosceles triangle). Figs 3.5 and 3.6 show an isosceles triangle with base B and height H. We want to find an expression for its area.

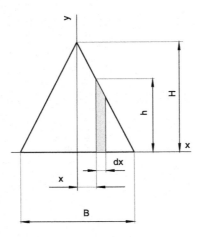

Figure 3.6 Isosceles triangle 2

Solution 1. The axis Oy divides the triangle into two right-angled trian-gles. Any of these triangles can be superposed over the other by reflection in the Oy axis. The right-angled triangles are congruent and their areas are equal. If we reflect one of these triangles about the Ox axis and translate it so that its hypotenuse coincides with that of the other triangle, we obtain a rectangle with base $B/2$ and height H. The area of the rectangle is $BH/2$ and so is the area of the given isosceles triangle.

Solution 2. As shown in Fig. 3.5 we choose as elemental area dA (called also *element of area*) the region with base b and height dy (painted in yellow in the electronic version of the book). The area of the triangle is given by

$$A_t = \int_0^H b\,dy$$

By Thales' theorem

$$b = \frac{B}{H}(H - y)$$

Substituting this into the integral yields

$$A_t = \int_0^H b\,dy = \frac{B}{H}\int_0^H (H - y)\,dy = \frac{BH}{2}$$

Solution 3. We refer now to Fig. 3.6 and divide the given triangle into two right-angled triangles. As they are congruent it is sufficient to calculate the area of one of them and multiply the result by two. This time we choose as elemental area hdx and write

$$A_t = 2\int_0^{B/2} h\,dx$$

By Thales' theorem

$$\frac{h}{H} = \frac{B/2 - x}{B/2}$$

which yields

$$h = H\left(1 - \frac{2x}{B}\right)$$

Substituting this value and carrying on the integral we obtain

$$A_t = 2H\left|x - \frac{x^2}{B}\right|_0^{B/2} = \frac{BH}{2}$$

Solution 4. In this solution we calculate A_t as the area under a given curve. The equations of the triangle sides are

$$-\frac{x}{B/2} + \frac{y}{H} = 1; \quad \frac{x}{B/2} + \frac{y}{H} = 1$$

We calculate the area of the triangle as

$$A_t = \int_{-B/2}^{B/2} y\,dx = \int_{-B/2}^{0} H\left(1 + \frac{2x}{B}\right) dx + \int_0^{B/2} H\left(1 - \frac{2x}{B}\right) dx = \frac{BH}{2}$$

Example 3.2 (Area under parabola). Fig. 3.7 shows a parabola that passes through the points

$$\left|\begin{matrix} -C \\ 0 \end{matrix}\right|, \left|\begin{matrix} 0 \\ H \end{matrix}\right|, \left|\begin{matrix} C \\ 0 \end{matrix}\right|$$

It is easy to check that the equation of this parabola is

$$y = -\frac{H}{C^2}x^2 + H$$

The area under the parabola is given by

$$A_P = \int_{-C}^{C} \left(-\frac{H}{C^2}x^2 + H\right) dx = H\left|-\frac{x^3}{3C^2} + x\right|_{-C}^{C} = \frac{2}{3}(2CH)$$

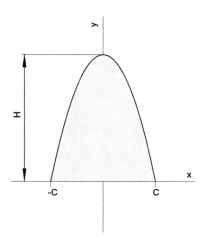

Figure 3.7 Area under parabola

The area of the rectangle that circumscribes the given arc of parabola is $2CH$. We conclude that *the area under an arc of parabola equals two thirds of the area of the circumscribed rectangle*. This result can be used in the derivation of Simpson's rule of integration (see, for example, Biran and López-Pulido, 2014, pp. 83–4).

3.3.3 Examples in Naval Architecture

The calculation of the hydrostatic properties of a vessel include the calculation of waterplane and station areas. For a vessel in *upright condition* (i.e. heel angle equal to zero) the waterplane area is the area of a horizontal section, that is the area enclosed by a waterline; its general expression is

$$A_w = 2 \int_{x_1}^{x_n} y dx$$

Here x_1 is the x-coordinate of the aftermost point of the given waterline, x_n the x-coordinate of the foremost point, and y the half-breadth as function of x. In hydrostatic curves A_w is represented as a function of the draught, T.

The area of a station up to a given draught, T_0, is calculated as

$$A = 2 \int_{0}^{T_0} y dz$$

In particular, the midship-section area is noted by A_M. A **Bonjean curve** is a curve of a station area as function of the draught T. For more details see Section 3.4.3. The **sectional-area** curve represents the areas of stations, up to a given waterline, versus the ship length, that is the x-coordinate. The sectional-area curve is important in the first stages of the design and it is used in the transformation of ship lines, in particular by the Lackenby method (see Section 3.9).

The evaluation of the dynamical stability of a vessel requires the calculation of areas under the righting and heeling arms. For example, the area under the curve of the righting arm, between $0°$ and $30°$, is

$$\int_0^{\pi/6} \overline{GZ} d\phi$$

where $\overline{GZ}$ is the righting arm in m, and ϕ the heel angle in radians. This area is proportional to the restoring energy accumulated while heeling up to $30°$.

Waterplane and station areas are calculated also for heeled hulls. Then, the sections are no more symmetric and the integrations must be carried on taking into account the shape of the whole section.

3.4 FIRST MOMENTS AND CENTROIDS OF AREAS

3.4.1 Definitions

We refer again to Fig. 3.4. The **moment** of the area A with respect to the x-axis is

$$M_x = \int \int_A y dA \tag{3.6}$$

and the moment of the same area with respect to the y-axis is

$$M_y = \int \int_A x dA \tag{3.7}$$

To distinguish these moments from those introduced in Section 3.5 they are called in English literature *first moments* and in other languages *static moments*. If the coordinate axes are translated as in Fig. 3.1, the moments with respect to the new axes are

$$M_\xi = \int_A \int (y - b)\,dA = M_x - bA$$

$$M_\eta = \int_A \int (x - a)\,dA = M_y - aA \tag{3.8}$$

The moments equal zero when

$$a = \frac{M_y}{A}, \quad b = \frac{M_x}{A}$$

The point defined by the coordinates a and b, in the system shown in Fig. 3.4, is the **centroid** of the area A. Another term for this point is **barycentre**. In some languages the centroid is called *centre of gravity of the area*, by analogy with the centre of gravity of a plate made of homogeneous material.

If an area A can be decomposed into n disjoint areas A_i such that

$$A = \sum_{i=1}^{n} A_i$$

the moment of A with respect to a given axis is equal to the sum of the moments of the areas A_i with respect to the same axis. Noting, for example, by x_c the x-coordinate of the centroid of A, and by x_i the x-coordinate of the centroid of A_i, we can write

$$x_c = \frac{\sum_{i=1}^{n} x_i A_i}{A} \tag{3.9}$$

Note that we add moments and not coordinates of centroids.

3.4.2 Examples

Example 3.3 (Centroid of isosceles triangle). We continue here Example 3.1 and want to find the centroid of the triangle shown in Figs 3.5 and 3.6. As the triangle is symmetric about the Oy axis, the x-coordinate of the centroid is zero. To find the y-coordinate we calculate the moment about the x-axis:

$$M_x = \int_0^H yb\,dy = B\int_0^H \left(y - \frac{y^2}{H}\right)dy = B\left|\frac{y^2}{2} - \frac{y^3}{3}\right|_0^H = \frac{BH^2}{6}$$

The y-coordinate of the centroid is

$$y_C = \frac{BH^2}{6} / \frac{BH}{2} = \frac{H}{3}$$

3.4.3 Examples in Naval Architecture

The centroid of a vessel waterplane is called *centre of floatation* and we note it by F. For small angles of inclination the axis of inclination passes through F (see, for example, Biran and López–Pulido, 2014, pp. 42–45). In Section 3.3.3 we mentioned the Bonjean curves that represent the areas of stations as functions of draught. There is a second set of Bonjean curves that represent the first moment of sectional areas, with respect to the base line, as functions of draught. For real–life ships both sets of Bonjean curves are calculated numerically. To explain the curves in an easier manner we choose here an example that can be treated analytically, more specifically a station whose shape is the parabola

$$z = ay^2 \tag{3.10}$$

The Bonjean curve of area is defined by

$$A = \int_0^T y\,dz = \int_0^T (z/a)^{1/2}\,dz = \frac{2}{3\sqrt{a}} T^{3/2} \tag{3.11}$$

and the Bonjean curve of moment by

$$M = \int_0^T zy\,dz = \int_0^T z\left(\frac{z}{a}\right)^{1/2} dz = \frac{2}{5\sqrt{a}} T^{5/2} \tag{3.12}$$

The assumed station and the corresponding Bonjean curves for $a = 0.5$ are shown in Fig. 3.8. An interpretation of the Bonjean curve of area can be seen in Fig. 3.9. Fig. 3.10 shows that the area under the first Bonjean curve is the moment with respect to the baseline. Before the advent of digital computers one way of calculating the Bonjean curves consisted in the evaluation of areas with the help of a planimeter (see Section 3.10.1). Today we carry on the calculations on digital computers and a convenient algorithm is, for example, that described in Biran and López–Pulido (2014), Section 3.4. A MATLAB function that performs this *integration with variable upper limit* can be found on the sites of the above-mentioned book.

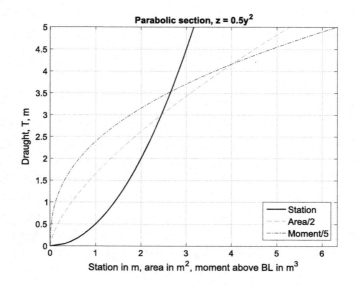

Figure 3.8 A parabolic station and its Bonjean curves

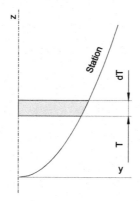

Figure 3.9 To the definition of the Bonjean curve of area

3.5 SECOND MOMENTS OF AREAS

3.5.1 Definitions

The **second moment** of the area A with respect to the x-axis is defined as

$$I_{xx} = \int\int_A y^2 \, dA \tag{3.13}$$

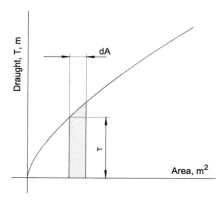

Figure 3.10 For the calculation of the Bonjean curve of moment

An alternative term employed especially in other languages is *moment of inertia of the area A with respect to the Ox axis*. This expression is based on the fact that the mathematical formulation is similar to that of a mass moment of inertia.

The second moment of the area A with respect to the y-axis is defined by

$$I_{yy} = \int \int_A x^2 dA \qquad (3.14)$$

and the **product of inertia** of the area A about the x- and y-axes is

$$I_{xy} = \int \int_A xy dA \qquad (3.15)$$

In other languages I_{xy} is called *centrifugal moment of inertia*. Finally, the **polar moment** is

$$I_p = \int \int_A r^2 dA = \int_A \int (x^2 + y^2) dA = I_{xx} + I_{yy} \qquad (3.16)$$

where r is the distance of dA from the origin O.

The moments about an axis, I_{xx} or I_{yy}, are sometimes called *axial moments*. Being a sum of squared distances times a positive area, these moments cannot be zero or negative. By the same reasoning, the moment about a point, I_p, also cannot be zero or negative. The moment with respect to two axes, I_{xy}, can be zero or negative. If one of the coordinate axes is an axis of symmetry of the figure, the product of inertia is zero. To prove this property

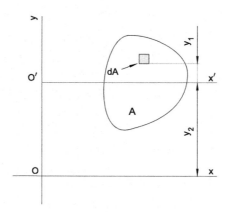

Figure 3.11 Parallel translation of axes

it is sufficient to note that, by virtue of the symmetry, for any element dA situated on one side of the axis there is a reflected element on the other side of the axis. The sum of the products $xy\,dA$ of these two elements is zero. The properties described in this paragraph can be used for an immediate, visual check of the plausibility of results of long calculations.

3.5.2 Parallel Translation of Axes

In Fig. 3.11 the axis $O'x'$ is parallel to the axis Ox. Considering the distance from the axis Ox to dA equal to $y_2 + y_1$, and taking into account that y_2 is a constant, we write

$$I_{xx} = \int_A \int (y_1 + y_2)^2 dA = \int_A \int y_1^2 dA + 2\int_A \int y_1 y_2 dA + \int_A \int y_2^2 dA$$
$$= I_{x'x'} + 2y_2 M_{x'} + y_2^2 A_F \tag{3.17}$$

Above, $M_{x'}$ is the first moment of the area A about $O'x'$. If and only if $O'x'$ passes through the centroid C of the area A, $M_{x'} = 0$ and Eq. (3.17) reduces to

$$I_{xx} = I_{x'x'} + y_2^2 A_F \tag{3.18}$$

This result is known as the **parallel axis theorem**. In German literature the result is called *Steiner's theorem* (Jakob Steiner, Swiss, 1796–1863), while in French and Italian literature it is attributed to Christiaan Huygens (Dutch, 1629–1695). An axis passing through the centroid of an area A is a **centroidal axis** of that area and the second moment with respect to it is known as a **centroidal moment**. A centroidal moment with respect to an axis $O'x'$ is smaller than any moment of inertia with respect to an axis parallel to $O'x'$. This property is obvious if we rewrite Eq. (3.18) as

$$I_{x'x'} = I_{xx} - y_c^2 A_F$$

where y_c is the y-coordinate of the centroid of A in the system of coordinates with origin in O. The extension of Eq. (3.18) to the second moment with respect to the y-axis is obviously

$$I_{yy} = I_{y'y'} + x_2^2 A_F \tag{3.19}$$

Here x_2 is the distance between the axis Oy and the centroidal axis $O'y'$. In Exercise 3.5 we ask the reader to prove that the corresponding equation for the product of inertia is

$$I_{xy} = I_{x'y'} + x_2 y_2 A_F \tag{3.20}$$

3.5.3 Rotation of Axes

In this section we rotate the axes of coordinates around the origin O by an angle α, and derive the expressions of the second moments about the rotated axes $O\xi$, $O\eta$. Applying Eqs (3.3) to Eq. (3.13) we get

$$I_{\xi\xi} = \int\int_A \eta^2 \, dA = \int\int_A (x^2 \sin^2 \alpha - 2xy \sin\alpha \cos\alpha + y^2 \cos^2 \alpha) \, dA$$
$$= I_{yy} \sin^2 \alpha - I_{xy} \sin 2\alpha + I_{xx} \cos^2 \alpha \tag{3.21}$$

From

$$\cos 2\alpha = \cos^2 \alpha - \sin^2 \alpha$$

we obtain

$$\cos^2 \alpha = \frac{1 + \cos 2\alpha}{2}, \quad \sin^2 \alpha = \frac{1 - \cos 2\alpha}{2}$$

Substituting these expressions into Eq. (3.21) yields

$$I_{\xi\xi} = \frac{I_{yy}}{2}(1 - \cos 2\alpha) - I_{xy} \sin 2\alpha + \frac{I_{xx}}{2}(1 + \cos 2\alpha)$$
$$= \frac{I_{xx} + I_{yy}}{2} + \frac{I_{xx} - I_{yy}}{2} \cos 2\alpha - I_{xy} \sin 2\alpha \tag{3.22}$$

Similarly, the second moment with respect to the η-axis is

$$
\begin{aligned}
I_{\eta\eta} &= \int_A \int \xi^2 dA = \int_A \int (x^2 \cos^2 \alpha + 2xy \sin \alpha \cos \alpha + y^2 \sin^2 \alpha) dA \\
&= I_{yy} \cos^2 \alpha + I_{xy} \sin 2\alpha + I_{xx} \sin^2 \alpha \\
&= \frac{I_{xx} + I_{yy}}{2} - \frac{I_{xx} - I_{yy}}{2} \cos 2\alpha + I_{xy} \sin 2\alpha
\end{aligned}
\tag{3.23}
$$

Finally, the product of inertia with respect to the rotated axes is given by

$$
\begin{aligned}
I_{\xi\eta} &= \int_A \int \xi \eta dA = \int_A \int (x \cos \alpha + y \sin \alpha)(-x \sin \alpha + y \cos \alpha) dA \\
&= \int_A \int (-x^2 \sin \alpha \cos \alpha + xy \cos^2 \alpha - xy \sin^2 \alpha + y^2 \sin \alpha \cos \alpha) dA \\
&= \frac{I_{xx} - I_{yy}}{2} \sin 2\alpha + I_{xy} \cos 2\alpha
\end{aligned}
\tag{3.24}
$$

To find the values of the angle α for which $I_{\xi\xi}$ reaches an extremum we calculate the derivative of $I_{\xi\xi}$ relative to α and equate it to zero

$$
\frac{dI_{\xi\xi}}{d\alpha} = -2\frac{I_{xx} - I_{yy}}{2} \sin 2\alpha - 2I_{xy} \cos 2\alpha = 0
$$

This gives

$$
\tan 2\alpha = -\frac{2I_{xy}}{I_{xx} - I_{yy}}
\tag{3.25}
$$

Similarly, for $I_{\eta\eta}$ we obtain

$$
\frac{dI_{\eta\eta}}{d\alpha} = \frac{I_{xx} - I_{yy}}{2} \sin 2\alpha + 2I_{xy} \cos 2\alpha = 0
$$

which yields again the result shown by Eq. (3.25).

Analyzing Eqs (3.22) to (3.25) we reach the following conclusions.

1. $I_{xx} + I_{yy} = I_{\xi\xi} + I_{\eta\eta} = I_p$.
2. Eq. (3.25) has as solutions two angles separated by 90°.
3. At each angle defined by Eq. (3.25) one of the moments of inertia reaches a maximum and the other a minimum.
4. At the angles defined by Eq. (3.25) the product of inertia equals zero.

To prove the first conclusion it is sufficient to add side by side Eqs (3.22) and (3.23). This result should have been expected from Eq. (3.16) as r, the distance between the origin of coordinates and dA, remains constant during rotation around the origin. We say that $I_{xx} + I_{yy}$ is an **invariant** of rotation and will return to this subject at a higher level in Section 3.5.4.

The second conclusion is based on the identity

$$\tan(2\alpha + \pi) = \tan 2\alpha$$

Thus, there are two axes for which the second moments with respect to them reach extremal values, and these axes are perpendicular one to the other; they are called **principal axes**.

The third conclusion is a consequence of the first. If the sum of the moments of inertia is constant, then, when one of them reaches a maximum, the other has a minimum.

The fourth conclusion can be proven by substituting Eq. (3.25) into Eq. (3.24). We leave this task to the reader (Exercise 3.6). We can state that *the product of inertia with respect to principal axes is null*. This property shows that the principal axes are in a certain sense a kind of axes of symmetry. Another consequence is that *when a figure has an axis of symmetry this is also a principal axis*. The moments of inertia with respect to the principal axes are called **principal moments of inertia**. To calculate the principal moments directly from the given moments we observe that

$$\sin 2\alpha = \frac{\tan 2\alpha}{\sqrt{1 + \tan^2 2\alpha}} = \frac{-\frac{2I_{xx}}{I_{xx} - I_{yy}}}{\sqrt{1 + \frac{4I_{xy}^2}{(I_{xx} - I_{yy})^2}}}$$

$$= \pm \frac{2I_{xy}}{\sqrt{(I_{xx} - I_{yy})^2 + 4I_{xy}^2}}$$

and

$$\cos 2\alpha = \frac{1}{\sqrt{1 + \tan^2 2\alpha}} = \frac{1}{\sqrt{1 + \frac{4I_{xy}^2}{(I_{xx} - I_{yy})^2}}}$$

$$= \pm \frac{I_{xx} - I_{yy}}{\sqrt{(I_{xx} - I_{yy})^2 + 4y_{xy}^2}}$$

Substituting these expressions into Eqs (3.22) and (3.23) yields

$$I_{x'x'} = \frac{I_{xx} + I_{yy}}{2} \pm \frac{I_{xx} - I_{yy}}{2} \cdot \frac{I_{xx} - I_{yy}}{\sqrt{(I_{xx} - I_{yy})^2 + 4I_{xy}^2}}$$

$$\mp I_{xy} \frac{2I_{xy}}{\sqrt{(I_{xx} - I_{yy})^2 + 4I_{xy}^2}}$$

$$= \frac{I_{xx} + I_{yy}}{2} \pm \frac{1}{2} \frac{(I_{xx} - I_{yy})^2 + 4I_{xy}^2}{\sqrt{(I_{xx} - I_{yy})^2 + 4I_{xy}^2}}$$

$$= \frac{I_{xx} + I_{yy}}{2} \pm \frac{1}{2}\sqrt{(I_{xx} - I_{yy})^2 + 4I_{xy}^2} \tag{3.26}$$

and

$$I_{y'y'} = \frac{I_{xx} + I_{yy}}{2} \mp \frac{I_{xx} - I_{yy}}{2} \cdot \frac{I_{xx} - I_{yy}}{\sqrt{(I_{xx} - I_{yy})^2 + 4I_{xy}^2}}$$

$$\pm I_{xy} \frac{2I_{xy}}{\sqrt{(I_{xx} - I_{yy})^2 + 4I_{xy}^2}}$$

$$= \frac{I_{xx} + I_{yy}}{2} \mp \frac{1}{2}\sqrt{(I_{xx} - I_{yy})^2 + 4I_{xy}^2} \tag{3.27}$$

The usual way of presenting these results is to state that the principal moments of inertia are given by

$$I_{max,\, min} = \frac{I_{xx} + I_{yy}}{2} \pm \sqrt{\left(\frac{I_{xx} - I_{yy}}{2}\right)^2 + I_{xy}^2} \tag{3.28}$$

3.5.4 The Tensor of Inertia

Given the moments of inertia, I_{xx}, I_{yy}, and the product of inertia, I_{xy}, of an area A, the corresponding **tensor of inertia** is defined by

$$J = \begin{vmatrix} I_{xx} & -I_{xy} \\ -I_{xy} & I_{yy} \end{vmatrix} \tag{3.29}$$

For a rotation of the axes Ox, Oy by an angle α, to the axes $O\xi$, $O\eta$, the tensor J transforms as

$$J(\xi, \eta) = RJR^T \tag{3.30}$$

where R is the matrix of rotation and R^T its transposed. To expand Eq. (3.30) we use Eq. (3.4) and write

$$\begin{vmatrix} I_{\xi\xi} & -I_{\xi\eta} \\ -I_{\xi\eta} & I_{\eta\eta} \end{vmatrix} = \begin{vmatrix} \cos\alpha & \sin\alpha \\ -\sin\alpha & \cos\alpha \end{vmatrix} \begin{vmatrix} I_{xx} & -I_{xy} \\ -I_{xy} & I_{yy} \end{vmatrix} \begin{vmatrix} \cos\alpha & -\sin\alpha \\ \sin\alpha & \cos\alpha \end{vmatrix}$$

(3.31)

After multiplying the first two matrices we can write

$$J(\xi,\eta) = \begin{vmatrix} I_{xx}\cos\alpha - I_{xy}\sin\alpha & -I_{xy}\cos\alpha + I_{yy}\sin\alpha \\ -I_{xx}\sin\alpha - I_{xy}\cos\alpha & I_{xy}\sin\alpha + I_{yy}\cos\alpha \end{vmatrix} \begin{vmatrix} \cos\alpha & -\sin\alpha \\ \sin\alpha & \cos\alpha \end{vmatrix}$$

The result of this multiplication is a *2-by-2* matrix. The element in the position 11 is

$$I_{xx}\cos^2\alpha - I_{xy}\sin 2\alpha + I_{yy}\sin^2\alpha$$

that is $I_{\xi\xi}$ as given by Eq. (3.21). The element in position 12 is

$$I_{xx}\sin\alpha\cos\alpha + I_{xy}\sin^2\alpha - I_{xy}\cos^2\alpha + I_{yy}\sin\alpha\cos\alpha$$

that is $I_{\xi\eta}$ as given by Eq. (3.24). We leave to the reader to prove that the element in position 21 is also equal to $I_{\xi\eta}$ as given by Eq. (3.24), and the element in position 22 is $I_{\eta\eta}$ as given by Eq. (3.23). This proves Eq. (3.30).

In Section 3.5.3 we have seen that $I_p = I_{xx} + I_{yy}$ is an invariant of rotation. In Eq. (3.29) this sum is the *trace* of the matrix. In other words, the *trace is a tensorial invariant*. Another tensorial invariant is the determinant of the matrix. To prove this statement we start from Eq. (3.30) and write

$$det(J(\xi,\eta)) = det(R) \times det(J) \times det(R^T)$$

As the determinants of the rotation matrix and of its transpose equal one, we conclude that

$$det(J(\xi,\eta)) = det(J)$$

(3.32)

3.5.5 Radius of Gyration

The **radius of gyration** with respect to the axis Ox is defined as

$$i_{xx} = \sqrt{\frac{I_{xx}}{A}}$$

(3.33)

Similarly, the radius of gyration with respect to the axis Oy is

$$i_{yy} = \sqrt{\frac{I_{yy}}{A}} \tag{3.34}$$

and the polar radius of gyration

$$i_p = \sqrt{\frac{I_p}{A}} \tag{3.35}$$

This radius is related to the others by the equation

$$i_p^2 = i_{xx}^2 + i_{yy}^2 \tag{3.36}$$

The concept of radius of gyration is used, for example, in buckling calculations. The standards of profiles usually indicate the radii of gyration of the sectional areas.

3.5.6 The Ellipse of Inertia

Given the moments of inertia with respect to the axes Ox and Oy, Eq. (3.21) yields the moment $I_{\xi\xi}$ with respect to an axis that makes an angle α with the axis O_x. In this section we consider the centroidal moment $I_{\xi'\xi'}$ and its polar representation. To do this we draw from the centroid C a vector, $\overline{CP}$, having the direction of the axis $C\xi'$ and the magnitude

$$\overline{CP} = \frac{C}{\sqrt{I_{\xi'\xi'}}}$$

Here C is a constant that we will define immediately. In a system of coordinates X, Y with the origin in C and axes parallel to the given axes Ox, Oy, the point P is represented by

$$X = \overline{CP}\cos\alpha = \frac{C}{\sqrt{I_{\xi'\xi'}}}\cos\alpha, \quad Y = \overline{CP}\cos\alpha = \frac{C}{\sqrt{I_{\xi'\xi'}}}\sin\alpha$$

Extracting $\cos\alpha$ and $\sin\alpha$ from these equations and substituting into Eq. (3.21) yields

$$I_{\xi'\xi'} = I_{xx}\frac{I_{\xi'\xi'}}{C^2}X^2 - 2I_{xy}\frac{I_{\xi'\xi'}}{C^2}XY + I_{xx}\frac{I_{\xi'\xi'}}{C^2}Y^2$$

which simplifies to

$$I_{xx}X^2 - 2I_{xy}XY + I_{yy}Y^2 = C^2 \tag{3.37}$$

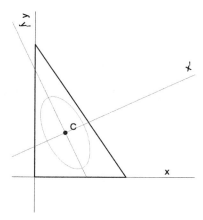

Figure 3.12 The ellipse of inertia of a right-angled triangle

At this point we set

$$C = i_{xx} i_{yy} \sqrt{A} \tag{3.38}$$

The constant C has the dimension of length elevated to the third power (shortly L^3) so that C^2 has the same dimensions as the three terms in the left-hand side of Eq. (3.37). Our choice will also lead to a nice result that we are going to show in the next section. We obtained the equation of an ellipse and we rewrite it as

$$\left| \begin{array}{cc} X & Y \end{array} \right| \left| \begin{array}{cc} I_{xx} & -I_{xy} \\ -I_{xy} & I_{yy} \end{array} \right| \left| \begin{array}{c} X \\ Y \end{array} \right| = C^2 \tag{3.39}$$

As an example, we plotted in Fig. 3.12 a right-angled triangle and its principal axes and ellipse of inertia. We see that the axes of the ellipse coincide with the principal axes of the triangle. This means that it is possible to simplify the equation of the ellipse if we rotate the axis CX and bring it over the axis Cx'. We are going to do this in the next section.

3.5.7 A Problem of Eigenvalues

With the notation defined by Eq. (3.29), Eq. (3.39) becomes

$$\left| \begin{array}{cc} X & Y \end{array} \right| \left| J \right| \left| \begin{array}{c} X \\ Y \end{array} \right| = C^2 \tag{3.40}$$

To get rid of the term in xy we must rotate the axes so that the terms $-I_{xy}$ become zero. In linear algebra this is achieved by *diagonalizing* the matrix. One way is to find a scalar λ and a vector $\mathbf{V}$ such that

$$J\mathbf{V} = \lambda\mathbf{V} \tag{3.41}$$

The scalar λ is called an **eigenvalue**, and $\mathbf{V}$ an **eigenvector** of J. Noting by I the unit *2-by-2* matrix, we rewrite Eq. (3.41) as

$$|J - \lambda I|\mathbf{V} = 0 \tag{3.42}$$

This form is equivalent to a system of two homogeneous linear equations that have non-trivial solutions only if the determinant of the matrix $|J - \lambda I|$ equals zero, that is

$$\lambda^2 - (I_{xx} + I_{yy})\lambda + (I_{xx}I_{yy} - I_{xx}^2) = 0 \tag{3.43}$$

The coefficient of the second term is the trace, and the third term the determinant of the matrix J. These quantities are invariants of rotation so that the equation holds for any angle of rotation. The solutions of Eq. (3.43) are

$$\lambda = \frac{I_{xx} + I_{yy} \pm \sqrt{(I_{xx} + I_{yy})^2 - 4(I_{xx}I_{yy} - I_{xy}^2)}}{2}$$
$$= \frac{I_{xx} + I_{yy}}{2} \pm \sqrt{\left(\frac{I_{xx} + I_{yy}}{2}\right)^2 + I_{xy}^2} \tag{3.44}$$

that is the principal moments as in Eq. (3.28).

The eigenvectors $\mathbf{V}_1$, $\mathbf{V}_2$ corresponding to the two values of λ indicate the directions of the principal axes. Let

$$\mathbf{V} = \begin{vmatrix} v_1 \\ v_2 \end{vmatrix}$$

Then, the angle of the rotated axis is given by $\tan\alpha = v_2/v_1$. To find this angle we rewrite Eq. (3.42) as two equations

$$I_{xx}v_1 - I_{xy}v_2 = \lambda v_1$$
$$-I_{xy}v_1 + I_{yy}v_2 = \lambda v_2 \tag{3.45}$$

Each equation gives a value for the ratio v_2/v_1 which defines the angle of the vector $\mathbf{V}$. Let us impose the condition that the two values of the angle are equal

$$\frac{v_2}{v_1} = \frac{I_{xx} - \lambda}{I_{xy}} = \frac{I_{xx}}{I_{yy} - \lambda} \tag{3.46}$$

Expanding Eq. (3.46) we obtain Eq. (3.43) that yields the eigenvalues λ. The conclusion is that for a given value of λ we have one value of the angle of the eigenvector. Knowing this, let us find now the angle of the principal axis $C\xi'$. For the sake of simplicity let $v_1 = 1$. From the first Eq. (3.45) we obtain

$$v_2 = \frac{I_{xx} - \lambda}{I_{xy}} = \frac{I_{xx} - (I_{xx} + I_{yy} \pm \sqrt{(I_{xx} - I_{yy})^2 + 4I_{xy}^2})/2}{I_{xy}}$$

$$= \frac{I_{xx} - I_{yy} \pm \sqrt{(I_{xx} - I_{yy})^2 + 4I_{xy}^2}}{2I_{xy}} = \tan\alpha \tag{3.47}$$

To compare this result with Eq. (3.25) we use the trigonometric formula

$$\tan 2\alpha = \frac{2\tan\alpha}{1 - \tan^2\alpha}$$

Let

$$Q = \sqrt{(I_{xx} - I_{yy})^2 + 4I_{xy}^2}$$

With this substitution we write

$$\tan 2\alpha = \frac{I_{xx} - I_{yy} \pm Q}{I_{xy}} \bigg/ \left(1 - \frac{(I_{xx} - I_{yy}) \pm Q)^2}{4I_{xy}^2}\right)$$

$$= 4I_{xy} \cdot \frac{I_{xx} - I_{yy} + Q}{4I_{xy}^2 - (I_{xx} - I_{yy})^2 - 2(I_{xx} - I_{yy})Q - Q^2} \tag{3.48}$$

After expanding Q and simplifying we obtain

$$\tan 2\alpha = -\frac{2I_{xy}}{I_{xx} - I_{yy}}$$

and this is exactly Eq. (3.25).

When reported to the principal axes, Cx', Cy', the equation of the ellipse of inertia reduces to

$$I_{x'x'}X^2 + I_{y'y'}Y^2 = C^2$$

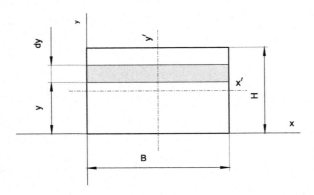

Figure 3.13 Rectangle I_{xx}

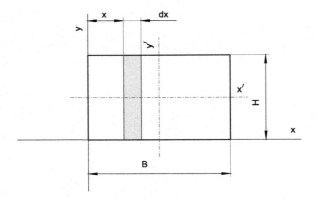

Figure 3.14 Rectangle I_{yy}

Substituting the expression of C and simplifying we obtain the elegant form

$$\frac{X^2}{i_{y'y'}} + \frac{Y^2}{i_{x'x'}} = 1 \tag{3.49}$$

3.5.8 Examples

Example 3.4 (The second moments of a rectangle). Figs 3.13 and 3.14 show a rectangle with base B and height H. To calculate the second moment of the rectangle area about the x-axis we refer to the first figure and write

$$I_{xx} = \int_0^H y^2 B dy = B \left. \frac{y^3}{3} \right|_0^H = \frac{BH^3}{3} \tag{3.50}$$

By parallel translation to the x'-axis passing through the centroid we get

$$I_{x'x'} = \frac{BH^3}{3} - \left(\frac{H}{2}\right)^2 BH = \frac{BH^3}{12} \tag{3.51}$$

To calculate the second moment of the rectangle area about the y-axis we refer to the second figure and write

$$I_{yy} = \int_0^H x^2 H dy = H \left.\frac{x^3}{3}\right|_0^B = \frac{B^3 H}{3} \tag{3.52}$$

By parallel translation to the y'-axis passing through the centroid we get

$$I_{y'y'} = \frac{B^3 H}{3} - \left(\frac{B}{2}\right)^2 BH = \frac{B^3 H}{12} \tag{3.53}$$

By virtue of symmetry, the centroidal product of inertia, $I_{x'y'}$, equals zero.

Example 3.5 (The second moments of an isosceles triangle). We continue here Example 3.1 and calculate the second moments of the triangle shown in Fig. 3.5. The moment of inertia about the x-axis is

$$I_{xx} = \int_0^H y^2 b dy = B \int_0^H y^2 \left(1 - \frac{y}{H}\right) dy = B \left.\frac{y^3}{3} - \frac{y^4}{4H}\right|_0^H = \frac{BH^3}{12} \tag{3.54}$$

By parallel translation to the centroid of the triangle we have

$$I_{x'x'} = \frac{BH^3}{12} - \left(\frac{H}{3}\right)^2 \frac{BH}{2} = \frac{BH^3}{36} \tag{3.55}$$

To calculate the second moment about the y-axis we refer to Fig. 3.6 and write

$$I_{xx} = 2 \int_0^{B/2} x^2 h dx = 2H \int_0^{B/2} x^2 \left(1 - \frac{2x}{B}\right) dx = 2H \left.\frac{y^3}{3} - \frac{y^4}{4H}\right|_0^{B/2}$$

$$= \frac{B^3 H}{48} \tag{3.56}$$

As the triangle is symmetric about the y-axis, the above moment of inertia is centroidal and, still more, it is also a principal moment of inertia.

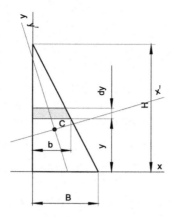

Figure 3.15 Right-angled triangle, 1st order of integration

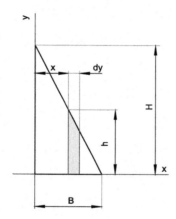

Figure 3.16 Right-angled triangle, 2nd order of integration

Example 3.6 (A right-angled triangle). In this example we consider the right-angled triangle shown in Figs 3.15 and 3.16 and calculate its product of inertia as a double integral. The elemental moment $xydxdy$ must be integrated with respect to x and y. Under certain conditions, usually fulfilled in Naval Architectural applications, the order of integration can be interchanged. This possibility is used in hydrostatic calculations even if the fact is not mentioned explicitly; we will give examples in Example 3.8. For the moment we are going to calculate the product of inertia using the two possible orders of integration. Our calculations use the second moments, I_{xx}, I_{yy}, calculated as in Eq. (3.55). We leave to the reader to check that this equation is valid also for the right angled triangle. We start from the definition

$$I_{xy} = \int \int_A xy \, dA$$

1 — Order dx, dy. We integrate first with respect to x. With the notations employed in Fig. 3.15 the second moment about the axis x is

$$I_{xx} = \int_0^H \left(\int_0^b xy \, dx \right) dy \tag{3.57}$$

Considering y constant, and reusing calculations carried on in Example 3.5, the integral between parentheses yields

$$\int_0^{B(1-y/H)} xy \, dx = y \left[\frac{x^2}{2} \right]_0^{B(1-y/H)} = \frac{B^2}{2} \left(y - \frac{2y^2}{H} + \frac{y^3}{H^2} \right)$$

Substituting this result into Eq. (3.57) we obtain

$$\int_0^H \frac{B^2}{2} \left(y - \frac{2y^2}{H} + \frac{y^3}{H^2} \right) dy = \frac{B^2}{2} \left| \frac{y^2}{2} - \frac{2y^3}{(3H)} + \frac{y^4}{4H^2} \right|_0^H$$
$$= \frac{B^2 H^2}{24} \tag{3.58}$$

2 — Order dy, dx. Now we integrate first with respect to y:

$$I_{xy} = \int_T \int xy \, dA = \int_0^B \left(\int_0^h xy \, dy \right) dx \tag{3.59}$$

Considering x constant, with the notations in Fig. 3.16 and reusing some expressions from Example 3.5, the integral between parentheses yields

$$x \left| \frac{y^2}{2} \right|_0^{H(1-x/B)} = x \frac{H^2}{2} \left(1 - \frac{x}{B} \right)^2$$

Substituting this result into Eq. (3.59) and carrying on the integration with respect to x we obtain

$$I_{xy} = \int_0^B x \frac{H^2}{2} \left(1 - \frac{x}{B} \right)^2 dx = \frac{H^2}{2} \left| \frac{x^2}{2} - \frac{2x^3}{3B} + \frac{x^4}{4B^2} \right|_0^B = \frac{B^2 H^2}{24} \tag{3.60}$$

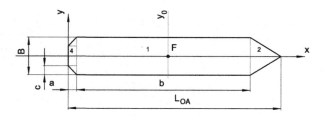

Figure 3.17 A composed area

The centroidal product of inertia is given by

$$I_{x'y'} = \frac{B^2 H^2}{24} - \frac{B}{3} \cdot \frac{H}{3} \cdot \frac{BH}{2} = -\frac{B^2 H^2}{72} \qquad (3.61)$$

We find the angle of the principal axes from

$$\tan 2\alpha = \frac{-\frac{B^2 H^2}{72}}{\frac{BH^3}{36} - \frac{B^3 H}{36}} = \frac{1}{2} \cdot \frac{BH}{H^2 - B^2} \qquad (3.62)$$

The centroid and the principal axes of the triangle are shown in Fig. 3.15.

Example 3.7 (A composed area). The polygon that appears in Fig. 3.17 can be regarded as an approximation of the waterplane of a monohull vessel. Let the input values, in m, be

$$L_{OA} = 135.0, \ B = 24.5, \ a = 5.5, \ b = 110.0, c = 5.8$$

To calculate the geometrical properties we divide the given area into five elementary areas for which we have ready formulae:
1. — a large rectangle;
2. — a triangle representing the bow;
3. — a small triangle at the port-side of the stern;
4. — a small rectangle at the stern;
5. — a small triangle at the starboard of the stern.
The calculations are carried on in the Excel spreadsheet reproduced in Fig. 3.18. We begin the file by entering the input data L_{OA}, B, a, b and c. The values of these constants are used several times, but they are entered only once. For example, the value of the constant L_{OA} is entered in the cell C2 and it will be called in the cells B11, G11 and H11 by the name C2. The advantages of this way of working are:

Avoid input errors — If the value of the same constant is entered several times, there is a risk to enter different values.

Simplifies changes — If the need arises to change the value of a constant, the change is done in one place only.

To simplify the use of formulae two *derived data* are calculated as

$$d = L_{OA} - a - b$$
$$e = B - 2c$$

We continue by building the spreadsheet on the same file. The heading has five lines as detailed below:

Lines 1 and 2 — The names of the quantities calculated in the cells of the corresponding columns.

Line 3 — The units of the quantities in the respective columns.

Line 4 — The ordinal numbers of the columns.

Line 5 — A symbolic explanation of the calculations performed in the corresponding columns. For example, in column 5 we read 2×3 and the meaning is that the cells of column 5 contain the products of the values stored in column 2 by the values stored in column 3.

Under the heading there are five lines corresponding to the five elementary areas into which the given figure is decomposed. Follows a line titled **Total**; it contains the sums of the columns, except for columns 3 and 4. The Total cell of the former contains the quotient of the sum of column 5 by that of column 2, the Total cell of the latter stores the sum of column 6 divided by the sum of column 2. These are the coordinates of the centroid. The centroid of the composed area, called *centre of floatation* for waterplanes of floating bodies, is shown as point F in Fig. 3.17; through it passes the centroidal axis Fy_0. The centroidal moments, that are also the principal moments, are calculated on the same file. Note that the parallel translation to the centroidal axes is carried on for sums only, and not for each elementary area in part. The last lines in Fig. 3.18 deal with the area and the principal moments of the rectangle circumscribed to the waterplane approximation. These values should be greater than those of the waterplane and so they are. This is an easy check of the plausibility of the results. If the test fails, an error occurred somewhere. If the check succeeds, it does not necessarily mean that the calculations are exact.

Let us assume now that half of the bow compartment, or *forepeak*, is damaged. Then, this compartment does not contribute anymore to the buoyancy and stability of the vessel and the remaining waterplane is that

	A	B	C	D	E	F	G	H	I	J	K	L
1	INPUT	m								Derived	m	
2		LOA	135.0	b		110.0				d	19.5	
3		B	24.5	c		5.8				e	12.9	
4		a	5.5									
5												
6			Centroid		First moments		Own second moments		Steiner term		Total second moments	
7	Item	Area	x-coordinate	y-coordinate	About y-axis	About x-axis	About y-axis	About x-axis	about y-axis	about x-axis	about y-axis	about x-axis
8		m²	m	m	m³	m³	m⁴	m⁴	m⁴	m⁴	m⁴	m⁴
9	1	2	3	4	5	6	7	8	10	11	13	14
10					2x3	2x4			5x3	6x4	7+10	8+11
11	1	2695.00	60.50	0.00	163047.50	0.00	2717458.33	134806.15	9864373.75	0.00	12581832.08	134806.15
12	2	238.88	122.00	0.00	29142.75	0.00	5046.23	5974.36	3555415.50	0.00	3560461.73	5974.36
13	3	15.95	3.67	8.38	58.48	133.71	26.80	29.81	214.44	1120.97	241.24	1150.78
14	4	70.95	2.75	0.00	195.11	0.00	178.85	983.90	536.56	0.00	715.41	983.90
15	5	15.95	3.67	-8.38	58.48	-133.71	26.80	29.81	214.44	1120.97	241.24	1150.78
16	Total	3036.73	63.39	0.00	192502.33	0.00	2722737.03	141824.03	13420754.69	2241.94	16143491.72	144065.97
17												
18		Reduce to centroidal axes										
19			Iyy=	3940494.50 m^4		Ixx	144065.97 m^4					
20		Circumscribed rectangle										
21			AreaR =	3307.50 m^2		IxR =	165443.91 m^4		IyR =	5023265.63		
22			Cwl =	0.92								

Figure 3.18 The spreadsheet of the composed area

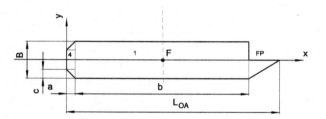

Figure 3.19 The damaged composed area

shown in Fig. 3.19. To recalculate the geometrical properties of the damaged waterplane we continue, on the same file, and add the spreadsheet shown in Fig. 3.20. The line, marked Intact repeats the total data of the intact waterplane. The second line, marked Forepeak, is used for the calculation of the properties of the flooded area. This time the line Total receives the differences of the two lines above it, except for columns 3 and 4 that receive the coordinates of the centroid in damage condition. Follows the parallel translation of axes to the centroid and after it the rotation to the principal axes. The final results are:

Angle of rotation of principal axes	−0.52 degrees
Maximum principal moment	3511187.90 m⁴
Minimum principal moment	140720.39 m⁴

Damaged waterplane

Item	Area	Centroid		First moments		Own second moments			Steiner term			Total second moments		
		x coordinate	y coordinate	About y-axis	About x-axis	About y-axis	About x-axis	Product of inertia	about y-axis	about x-axis	Product of inertia	about y-axis	about x-axis	Product of inertia
	m^2	m	m	m^3	m^3	m^4	m^4	m^4	m^4	m^4	m^4	m^4	m^4	m^4
1	2	3	4	5	6	7	8	9	10	11	12	13	14	15
				2x3	2x4				5x3	6x4		7+10	8+11	3*4*2
Intact	3036.73	63.39	0.00	192502.33	0.00	2722737.03	141824.03	0.00	13420754.69	2241.94	0.00	16143491.72	144065.97	0.00
Forepeak	119.44	122.00	4.08	14571.38	487.70	2523.12	995.73	792.52	1777707.75	1991.45	59499.78	1780230.87	2987.18	60292.30
Total	2917.29	60.99	-0.17	177930.95	-487.70	2720213.91	140828.30	-792.52	11643046.94	250.49	-59499.78	14363260.85	141078.79	-60292.30

Centroidal moments of inertia

Ixxc	140997.3	m^4	Iyyc	3510911.0	m^4	Ixyc	-30546.4	m^4

Principal axes and moments of inertia

tan2alpha	-0.018129		alpha =	-0.009	rad	-0.52	deg
(Ix + Iy)/2	1825954.1	m^4					
(Ix-Iy)/2	-1684956.9	m^4					
Term2	1685233.8	m^4					
Imax	3511187.9	m^4				Imin	140720.4	m^4

Figure 3.20 The Excel sheet for the damaged composed area

The centroid and the centroidal axis Fy' are shown in Fig. 3.19. As the angle of rotation and the y-coordinate of the centroid are small, and so is the drawing, the principal axes do not differ visibly from the centroidal ones.

At this point we have enough data to make some experiments. For example, let us check Eq. (3.30) and the tensorial invariants. We need a function that builds the rotation matrix. The following one will do the job.

```
function R = rotd(alpha)
%ROTD matrix for 2D rotation
%    Input: angle of counterclockwise rotation in degrees
%    Written by Adrian Biran, October 2015
R = [ cosd(alpha)    sind(alpha)
     -sind(alpha)    cosd(alpha) ];
end
```

We write now the following script on a file `DamageArea1.m`. The file imports data from the Excel spreadsheet `ComposedArea2` by using the function `xlsread`. Proceeding so and not manually we preserve the internal precision of the data and avoid input errors. We call `xlsread` with two arguments: the name of the spreadsheet file, including extension, and the cell range.

```
%DAMAGEAREA1 Given the tensor of the centroidal moments of
%    inertia of the damaged area, in the example of the composed
%    area, calculates the tensor of the principal moments.
%    Written by Adrian Biran, October 2015

% Extract data from Excel spreadsheet
```

```
alpha = xlsread('ComposedArea2.xlsx', 'H36:H36'); % angle
% of principal axes
Ixx  = xlsread('ComposedArea2.xlsx', 'B34:B34');
Iyy  = xlsread('ComposedArea2.xlsx', 'E34:E34');
Ixy  = xlsread('ComposedArea2.xlsx', 'H34:H34');
% Build the tensor of inertia
J    = [ Ixx -Ixy
         -Ixy  Iyy ];
R    = rotd(alpha);         % call rotation matrix
disp('Tensor of principal moments of inertia')
Jrot = R*J*R'
% Check tensorial invariants
disp('Compare traces')
sum(diag(J)) - sum(diag(Jrot))
disp('Compare determinants')
det(J) - det(Jrot)
%%%%%%%% EIGENVALUES %%%%%%%%%%%%
[ V, D ] = eig(J)
disp('Eigenvalues'), D
disp('Eigenvectors'), V
disp('angle = ')
alpha = atand(V(1, 2))
```

After having set the environment with the commands format long and format compact, we call the script and obtain the following output.

```
Tensor of principal moments of inertia
Jrot =
   1.0e+06 *
   0.140720389770684  -0.000000000000000
  -0.000000000000000   3.511187901580954
Compare traces
ans =
     0
Compare determinants
ans =
    1.220703125000000e-04
Eigenvalues
D =
   1.0e+06 *
   0.140720389770684                  0
                   0   3.511187901580954
```

```
Eigenvectors
V =
  -0.999958927325978    0.009063314023008
   0.009063314023008    0.999958927325978
alpha =
   0.519275423836731
```

We got the expected results with only a small difference in the deter-minants; it is due to errors in numerical calculations.

3.5.9 Examples in Naval Architecture

The second moments of the waterplane are used in the calculation of metacentric heights (see Biran and López-Pulido, 2014, Chapter 2). The *transverse metacentric radius* is given by

$$\overline{BM} = \frac{I_{xx}}{\nabla} \tag{3.63}$$

where ∇ is the volume of the submerged hull. The *longitudinal metacentric radius* is

$$\overline{BM_L} = \frac{I_L}{\nabla} \tag{3.64}$$

where I_L is the moment of inertia of the waterplane with respect to the principal axis that passes through the centre of floatation, F, and is parallel to the axis y. For an axis of inclination Fx_α that makes an angle α with the axis x the initial metacentric radius is

$$I_{x_\alpha} = \frac{I_{xx} \cos^2 \alpha - I_{xy} \sin 2\alpha + I_L \sin^2 \alpha}{\nabla}$$

As I_{xx} and I_L are principal moments, I_{xy} is zero and we remain with

$$\overline{BM_\alpha} = \overline{BM}_T \cos^2 \alpha + \overline{BM}_L \sin^2 \alpha \tag{3.65}$$

Another application that requires the calculation of moments of iner-tia of areas is in the calculation of the *sectional modulus*; it is discussed in Example 3.12.

3.6 VOLUME PROPERTIES

3.6.1 Definitions

In the three-dimensional space let V be a region bounded by a closed surface S. The **volume** of V is a measure of the extent of V. By definition the volume of a cube with unit side has the value 1 and its unit is the length unit raised to the 3d power. Examples of unit-cube volumes are:

- the volume of a cube with side 1 cm is 1 cm^3;
- the volume of a cube with side 1 m is 1 m^3. This is the SI unit used in Naval Architecture;
- the volume of a cube with side 1 ft is 1 ft^3.

The volume of any other solid is equal to the number of unit cubes it contains. An immediate consequence of this definition is that the volume of a parallelepiped with length L, breadth B, and height D has the value LBD.

Like the calculation of areas, the calculation of volumes is based on axioms that usually are not formulated explicitly. To state the axioms we must first extend the notion of *congruence*. We say that two solids are *congruent* if one of them can be superposed over the other without changing the distances between its points. As for areas, the geometrical transformations that enable us to bring a solid over another under the above condition are translations, rotations, and reflections. We assume the following axioms:

1. if two solids are congruent their volumes are equal;
2. if a solid can be decomposed into disjoint, 'elementary' solids, its volume is equal to the sum of the volumes of the elementary solids.

Let us consider also the case of geometrically similar solids. *If a solid has the volume V_s when drawn at the scale 1:n, its actual volume is $n^3 V_s$.*

3.6.2 Examples

Example 3.8 (The volume of a tetrahedron). Fig. 3.21 shows a tetrahedron whose base is a right-angled rectangle with sides A and B. In this example we are going to calculate the volume, V_T, of the tetrahedron in three ways, each time using another order of integration. Here we exemplify these methods on a simple solid and we can show rigorously that the three orders of integration yield the same result. We conclude this example by explaining how similar procedures are used in Naval Architecture to calculate the volume of displacement ∇ (volume of submerged hull).

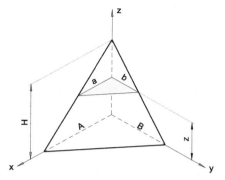

Figure 3.21 The volume of a tetrahedron — Horizontal sections

1 — We integrate first with respect to x and y.

$$V_T = \int_0^H \left(\int \int dx dy \right) dz$$

The integral between parentheses is the area of a horizontal section of the solid, that is a right-angled triangle with sides a and b (shown in yellow in the electronic edition) so that we rewrite the above equation as

$$V_T = \int_0^H \frac{ab}{2} dz \qquad (3.66)$$

Geometrical similarity yields

$$\frac{a}{A} = \frac{H-z}{H}, \quad \frac{b}{B} = \frac{H-z}{H}$$

Substituting these relationships into Eq. (3.66) we obtain

$$V_T = \frac{AB}{2} \int_0^H \left(1 - \frac{z}{H} \right)^2 dz = \frac{AB}{2} \left[z - \frac{z^2}{H} + \frac{z^3}{3H^2} \right]_0^H$$

$$= \frac{AB}{2} \cdot \frac{H}{3} \qquad (3.67)$$

The first factor is the base area

$$G = \frac{AB}{2}$$

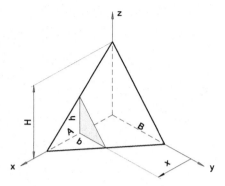

Figure 3.22 The volume of a tetrahedron — Transverse sections

We can state that the volume of the tetrahedron equals one third of the volume of the prism with base G and height H. The generalization of this result for pyramids and cones with base G and height H is straightforward.

2 — We integrate first with respect to y and z

$$V_T = \int_{x_0}^{x_n} \left(\int \int dy\,dz \right) dx$$
$$V_T = \int_0^H \frac{bh}{2} dx \tag{3.68}$$

The integral between parentheses is the area of a section parallel to the yOz plane. With the help of Fig. 3.22 and Thales' theorem we write

$$V_T = \int_0^A \frac{bh}{2} dx = BH \int_0^A \left(1 - \frac{x}{A}\right)^2 dx = \frac{AB}{2} \cdot \frac{H}{3} \tag{3.69}$$

3 — We integrate first with respect to x and z.

$$V_T = \int_0^B \left(\int \int dx\,dz \right) dy \tag{3.70}$$

The integral between parentheses is the area of a section parallel to the xOz plane and having the sides a and h (see Fig. 3.23). The calculation is

$$V_T = \int_0^H \frac{ah}{2} dy = \frac{AH}{2} \int_0^B \left(1 - \frac{y}{B}\right)^2 dy = \frac{AH}{2} H \left. \left| y - \frac{y^2}{B} + \frac{y^3}{3B^2} \right|_0^B \right.$$
$$= \frac{AB}{2} \cdot \frac{H}{3} \tag{3.71}$$

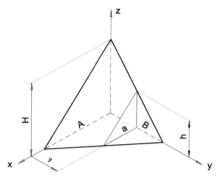

Figure 3.23 The volume of a tetrahedron — Longitudinal sections

For ships, the hydrostatic calculations include the calculation of the volume of the submerged hull, that is the volume of displacement, ∇. Like in the example detailed above, the volume ∇ can be obtained in three ways, as described below.

1. By integrating waterplane areas with respect to z.
2. By integrating station areas with respect to x. For ships in upright condition this method uses the Bonjean curves of areas.
3. By integrating areas of longitudinal sections with respect to y.

3.6.3 Moments and Centroids of Volumes

The moments of a volume V with respect to the coordinate planes are defined by

$$M_{yOz} = \int \int \int_V x\,dx\,dy\,dz$$

$$M_{xOz} = \int \int \int_V y\,dx\,dy\,dz$$

$$M_{xOy} = \int \int \int_V z\,dx\,dy\,dz \qquad (3.72)$$

An equivalent notation frequently encountered in the technical literature is

$$M_{yOz} = \int_V dV, \quad M_{xOz} = \int_V y\,dV, \quad M_{xOy} = \int_V z\,dV$$

The coordinates of the centroid of the volume V are

$$x_c = \frac{M_{yOz}}{V}, \quad y_c = \frac{M_{xOz}}{V}, \quad z_c = \frac{M_{xOy}}{V} \qquad (3.73)$$

If a volume V can be decomposed into n disjoint volumes V_i such that

$$V = \sum_{i=1}^{n} V_i$$

the moment of V with respect to a given plane is equal to the sum of the moments of the volumes V_i with respect to the same plane. Noting, for example, by x_c the x-coordinate of the centroid of V, and by x_i the x-coordinate of the centroid of V_i, we can write

$$x_c = \frac{\sum_{i=1}^{n} x_i V_i}{V} \tag{3.74}$$

and similar equations for the y- and z-coordinates. Note that we add moments and not coordinates of centroids.

Example 3.9 (The volume centroid of a tetrahedron). To find the z-coordinate of the centroid of the tetrahedron we must calculate first the moment with respect to the horizontal coordinate plane. It is convenient to consider that we integrate first with respect to x and y and then we work with the area $ab/2$ of a horizontal section such as that shown in Fig. 3.21. Using again geometrical similarity we write

$$\begin{aligned}
M_{xOy} &= \int_0^H z\frac{ab}{2}\,dz = \frac{1}{2}\int_0^H zA\left(1 - \frac{z}{H}\right)B\left(1 - \frac{z}{H}\right)dz \\
&= \frac{AB}{2}\left.\left|\frac{z^2}{2} - \frac{2}{3}\cdot\frac{z^3}{H} + \frac{z^4}{4H^2}\right.\right|_0^H = \frac{ABH^2}{12}
\end{aligned} \tag{3.75}$$

Finally, we divide the moment by the volume and obtain

$$z_c = \frac{ABH^2}{24}\bigg/\frac{ABH}{6} = \frac{H}{4} \tag{3.76}$$

We leave as an exercise to the reader to find the other coordinates of the centroid. Note that for each coordinate it is convenient to use a particular order of integration, as we did in Example 3.8.

Hydrostatic calculations include the centroid of the submerged hull volume. In English literature this point is called by abuse *centre of buoyancy* and its notation is B (see Biran and López-Pulido, 2014, Chapter 2). The x-coordinate, known as *longitudinal centre of buoyancy*, shortly *LCB*, is obtained by multiplying the areas of transverse sections by their distances from the origin, integrating the products along the length of the waterline and dividing the result by the volume of displacement. This amounts to

$$LCB = \frac{\int_{x_0}^{x_n} xA_x dx}{\nabla} = \frac{\int_{x_0}^{x_n} x \int \int dy dz}{\nabla}$$

The vertical coordinate of the centre of buoyancy is noted $\overline{KB}$, but is known also as *vertical centre of buoyancy* with the notation VCB. One way of calculating this coordinate is to read the Bonjean curves of moments, integrate the readings along the waterline, and divide the result by the volume of displacement

$$\overline{KB} = \frac{\int_{x_0}^{x_n} M dx}{\nabla} = \frac{\int_{x_0}^{x_n} (\int \int z dy dz) dx}{\nabla}$$

Another possibility is to multiply the waterplane areas by their height above the base line, integrate the products from BL to the given draught and divide by the volume of displacement

$$\overline{KB} = \frac{\int_0^T zA_w dz}{\nabla} = \frac{\int_0^T (\int \int z dy dx) dx}{\nabla}$$

As the integration must be carried on numerically, the latter procedure may be affected by errors at lower draughts.

3.7 MASS PROPERTIES

The term *mass properties* refers to *masses, centres of mass* and *moments of inertia*. Equivalently, one can talk about *weights* and *centres of gravity*. These properties are treated in detail in textbooks on mechanics, for example Meriam and Kraige (2006), Hahn (1992). Therefore, we mention here only a few aspects related to the geometrical properties discussed in this chapter.

If an object of volume V is made of homogeneous material of density ρ, its mass is ρV, and its centre of gravity coincides with the centroid of the volume. Let us consider a system of masses $m_1, m_2, \ldots, m_n$, and let the coordinates of the centre of gravity of the mass m_i be x_i, y_i, z_i. Then the centre of gravity of the system has the coordinates

$$x_G = \frac{\sum_1^n x_i m_i}{\sum_1^n m_i}, \quad y_G = \frac{\sum_1^n y_i m_i}{\sum_1^n m_i}, \quad z_G = \frac{\sum_1^n z_i m_i}{\sum_1^n m_i} \tag{3.77}$$

Note that we add moments and not coordinates. For a ship, using the convention of coordinates defined in Chapter 2, the terms are

$\sum_1^n m_i$ — *displacement mass* noted Δ. The SI unit is t;

x_G — *longitudinal centre of gravity* noted LCG;

y_G — *transverse centre of gravity* noted TCG. For a ship in intact condition and correctly loaded it should have the value 0;

z_G — *vertical centre of gravity* noted VCG, but in stability calculations $\overline{KG}$.

If one mass, m_j, is moved a distance δx_j in parallel to the x-axis, it is not necessary to repeat the whole calculation. As proved in Biran and López-Pulido (2014), Section 2.7, it is easier to calculate the change of coordinate as equal to the change of moment divided by the whole mass, for example

$$\delta X_G = \frac{\delta x_j m_j}{\sum_1^n m_i}$$

The moments of inertia are defined by equations similar to those of the second moments of areas using the element of mass, dm, instead of an element of area dA. We detail here only two of the moments of inertia that appear in the equations of ship motions. Thus, the moment of inertia with respect to the x-axis enters the equation of roll and is calculated as

$$I_{xx} = \sum_1^n (y_i^2 + z_i^2) m_i$$

It is convenient to use the corresponding radius of gyration. In Biran and López-Pulido (2014), Section 6.8, it is noted i_m and its definition is

$$i_m = \sqrt{\frac{I_{xx}}{\Delta}}$$

The moment of inertia with respect to the y-axis enters the equation of pitch and is calculated as

$$I_{yy} = \sum_1^n (x_i^2 + z_i^2) m_i$$

There is a theorem of parallel translation of axes that looks exactly like that for second moments of areas. The equations for rotations of axes are more complicated as they are three-dimensional. The notions of principal axes and principal moments of inertia are similar to those for second moments of areas. The tensor of inertia has nine components and it can be represented by an ellipsoid of inertia.

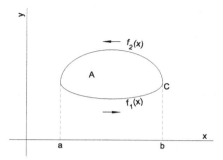

Figure 3.24 To Green's theorem, 1

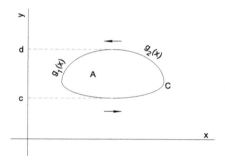

Figure 3.25 To Green's theorem, 2

3.8 GREEN'S THEOREM

The geometrical properties of areas, that is their areas and first and second moments, are defined as area integrals. Using a theorem due to George Green (English, 1793–1841) it is possible to calculate the properties as line integrals. This may simplify sometimes the work and is the base of important, practical applications. To explain the theorem we consider in Figs 3.24 and 3.25 an area A enclosed by a curve C. Let M and N be two functions defined over A and having continuous partial derivatives. Assuming that $a \leq x \leq b$, in Fig. 3.24 we describe the curve C by two functions, $f_1(x)$, $f_2(x)$, such that $f_1(x) \leq y \leq f_2(x)$. We calculate

$$\int\int_A \frac{\partial N}{\partial y} dxdy = \int_a^b \left(\int_{f_1}^{f_2} \frac{\partial N}{\partial y} dy \right) dx$$

$$= \int_a^b [N(x, f_2(x)) - N(x, f_1(x))]dx$$

$$= -\int_a^b N(x, f_1(x)) dx - \int_b^a N(x, f_2(x)) dx$$

$$= -\oint_C N(x, y) dx \tag{3.78}$$

We obtained a line integral. Mind that the integration is performed moving along the curve C as indicated by the arrow, that is keeping all the time the area A at the left. We continue now in Fig. 3.25 and, assuming $c \leq y \leq d$ we describe the curve C by the two functions, $g_1(x)$ and $g_2(x)$, such that $g_1 \leq x \leq g_2$. We write

$$\int\int_A \frac{\partial M}{\partial x} dxdy = \int_c^d \left(\int_{g_1}^{g_2} \frac{\partial M}{\partial x} dx \right) dy$$

$$= \int_c^d [M(x, g_2(x)) - M(x, g_1(x))] dy$$

$$= \int_c^d M(x, g_2(x)) dy + \int_d^c M(x, g_1(x)) dx$$

$$= \oint_C M(x, y) dy \tag{3.79}$$

Adding side by side Eqs (3.78) and (3.79) we obtain

$$\boxed{\int\int_A \left(\frac{\partial M}{\partial x} - \frac{\partial N}{\partial y} \right) dxdy = \oint_c [M(x, y) dy + N(x, y) dx]} \tag{3.80}$$

This is *Green's theorem*; two outstanding applications of it are the *planimeter* that we will discuss in Section 3.10.1, and digitizer-integrators an example of which we'll present in Section 3.10.2. However, first we are going to show some examples in which we use M and N functions proposed by Weisstein on the site of *MathWorld*.

Example 3.10 (Area by Green's theorem). Let $M = -y/2$, $N = x/2$. Then, the left-hand side of Eq. (3.80) becomes

$$\int\int \left(\frac{\partial N}{\partial x} - \frac{\partial M}{\partial y} \right) dxdy = \int\int_A \left(\frac{1}{2} + \frac{1}{2} \right) = \int\int_A dxdy$$

The right-hand side of Green's formula yields

$$\oint_C [Mdy + Ndx] = \frac{1}{2} \oint_C (xdy - ydx) \tag{3.81}$$

Table 3.2 Segments of isosceles triangle

No.	From $x =$	to $x =$	$y =$	$dy =$
1	$-B/2$	$B/2$	0	0
2	$B/2$	0	$H(1 - 2x/B)$	$-2(H/B)dx$
3	0	$-B/2$	$H(1 + 2x/B)$	$2(H/B)dx$

As an example consider the circle with parametric equations

$$x = r\cos t, \quad y = r\sin t$$

where t runs from 0 to 2π. Using Eq. (3.81) we calculate the area as

$$A = \frac{1}{2}\int_0^{2\pi}(r^2\cos^2 t + r^2\sin^2 t)\,dt = \frac{r^2}{2}[t]_0^{2\pi} = \pi r^2$$

Example 3.11 (Triangle properties by Green's theorem). In this example we return to the isosceles triangle of Examples 3.1, 3.3, and 3.5 and calculate its geometrical properties by Green's theorem. Starting from the left corner, and going in the counterclockwise sense, the 'curve' C is composed of three straight-line segments with the characteristics detailed in Table 3.2.

Area. We use the same functions and resulting equation as in Example 3.10. The first segment does not contribute. For the second segment the line integral is

$$\int_{B/2}^0 (x\,dy - y\,dx) = \int_{B/2}^0 -\frac{2H}{B}x\,dx - \int_{B/2}^0 H\left(1 - \frac{2x}{B}\right)dx$$

$$= -\left[\frac{H}{B}x^2\right]_{B/2}^0 - \left[H\left(x - \frac{x^2}{B}\right)\right]_{B/2}^0 = \frac{BH}{2} \qquad (3.82)$$

For the third segment we write

$$\int_0^{-B/2}(x\,dy - y\,dx) = \int_0^{-B/2}x\frac{2H}{B}\,dx - \int_0^{-B/2}H\left(1 + \frac{2x}{B}\right)dx$$

$$= -H[x]_0^{-B/2} = \frac{BH}{2} \qquad (3.83)$$

With the results of Eqs (3.82) and (3.83) we get, as expected,

$$A = \frac{1}{2}\left[\frac{BH}{2} + \frac{BH}{2}\right] = \frac{BH}{2}$$

Moment with respect to the x-axis. We use the functions

$$M = 0, \quad N = -\frac{y^2}{2}$$

For the left-hand side of Green's theorem we get

$$\int\int_A \left(\frac{\partial M}{\partial x} - \frac{\partial N}{\partial y} \right) dxdy = \int_A \int ydxdy$$

that is, indeed, M_x. The first segment does not contribute to the right-hand side of Green theorem. For the second segment we have

$$-\frac{1}{2} \int_{B/2}^0 y^2 dx = -\frac{1}{2} \int_{B/2}^0 H^2 \left(1 - \frac{2x}{B} \right)^2 dx$$

$$= \frac{H^2}{2} \left[x - \frac{2x^2}{B} + \frac{4x^3}{3B^2} \right]_0^{B/2} = \frac{BH^2}{12} \qquad (3.84)$$

The contribution of the third segment is

$$-\frac{1}{2} \int_0^{-B/2} y^2 dx = -\frac{1}{2} \int_0^{-B/2} -H^2 \left(1 + \frac{2x}{B} \right)^2 dx$$

$$= -\frac{H^2}{2} \left[x + \frac{2x^2}{B} + \frac{4x^3}{3B^2} \right]_0^{-B/2} = \frac{BH^2}{12} \qquad (3.85)$$

Summing up the moment is

$$M_x = \frac{BH^2}{12} + \frac{BH^2}{12} = \frac{BH^2}{6}$$

and the y-coordinate of the centroid is, as expected,

$$y_C = \frac{BH^2}{6} / \frac{BH}{2} = \frac{H}{3}$$

Moment with respect to the y-axis. The functions are

$$M = \frac{x^2}{2}, \quad N = 0$$

For the left-hand side of Green's theorem we get

$$\int\int_A \left(\frac{\partial M}{\partial x} - \frac{\partial N}{\partial y} \right) dxdy = \int_A \int xdxdy$$

that is, indeed, M_y. For the right-hand side of Green's theorem we have

$$\oint_C M dy + N dx = \frac{1}{2} \oint_C x^2 dy$$

The first segment gives no contribution. For the second segment we write

$$\frac{1}{2} \int_{B/2}^{0} x^2 (-2H/B) dx = -\frac{H}{B} \left. \frac{x^3}{3} \right|_{B/2}^{0} = \frac{B^3 H}{24}$$

The calculations for the third segment are

$$\frac{1}{2} \int_{0}^{-B/2} x^2 (2H/B) dx = \frac{H}{B} \left. \frac{x^3}{3} \right|_{0}^{-B/2} = -\frac{B^3 H}{24}$$

It follows that $M_y = 0$ and $x_c = 0$, as expected for a symmetrical figure.

Second moment with respect to the x-axis. The functions to use are $M = 0$ and $N = -y^3/3$. For the left-hand side of Green's theorem we have

$$\int \int_A \left(\frac{\partial M}{\partial x} - \frac{\partial N}{\partial y} \right) dx dy = \oint_C y^2 dx dy$$

that is, indeed, I_{xx}. For the right-hand side of the theorem we have

$$\oint_C M dy + N dx = -\frac{1}{3} \oint_C y^3 dx$$

The integral on the second segment yields

$$-\frac{1}{3} \int_{B/2}^{0} H^3 \left(1 - \frac{2x}{B} \right) dx = -\frac{H^3}{3} \left[x - \frac{3B^2}{4B} + \frac{4x^3}{B^2} - \frac{2x^4}{16B^3} \right]_{B/2}^{0} = \frac{BH^3}{24}$$

For the third segment we get

$$-\frac{1}{3} \int_{0}^{-B/2} H^3 \left(1 + \frac{2x}{B} \right) dx = -\frac{H^3}{3} \left[x + \frac{3B^2}{4B} + \frac{4x^3}{B^2} + \frac{2x^4}{16B^3} \right]_{0}^{-B/2} = \frac{BH^3}{24}$$

Adding the two results we obtain

$$I_{xx} = \frac{BH^3}{12}$$

Second moment with respect to the y-axis. The functions to use are $M = x^3/3$ and $N = 0$. For the left-hand side of Green's theorem we obtain

$$\int\int_A \left(\frac{\partial M}{\partial x} - \frac{\partial N}{\partial y} \right) dx dy = \oint_C x^2 dx dy$$

that is, indeed, I_{xx}. For the right-hand side of the theorem we have

$$\oint_C M dy + N dx = -\frac{1}{3} \oint_C x^3 dy$$

The integral on the second segment yields

$$-\frac{1}{3} \int_{B/2}^{0} x^3 \left(-\frac{2H}{B} B \right) dx = \frac{B^3 H}{96}$$

For the third segment we get

$$-\frac{1}{3} \int_{0}^{-B/2} x^3 \left(\frac{2H}{B} \right) dx = \frac{B^3}{96}$$

Adding the two results we obtain

$$I_{yy} = \frac{B^3 H}{48}$$

Product of inertia. The proposed functions are $M = 0$ and $N = -xy^2/2$. For the left-hand side of Green's theorem we get

$$\int\int_A \left(\frac{\partial M}{\partial x} - \frac{\partial N}{\partial y} \right) dx dy = \oint_C xy dx dy$$

which is, as expected, I_{xy}. The right-hand side of the theorem becomes

$$\oint_C M dy + N dx = -\frac{1}{2} \oint_C xy^2 dy$$

The integration along the second segment yields

$$-\frac{1}{2} \int_{B/2}^{0} xH^2 \left(1 - \frac{2x}{B} \right)^2 dx = \frac{B^2 H2}{96}$$

and along the third segment

$$-\frac{1}{2} \int_{0}^{B/2} xH^2 \left(1 + \frac{2x}{B} \right)^2 dx = -\frac{B^2 H2}{96}$$

Adding the above results we obtain $I_{xx} = 0$, as expected since the y-axis is an axis of symmetry.

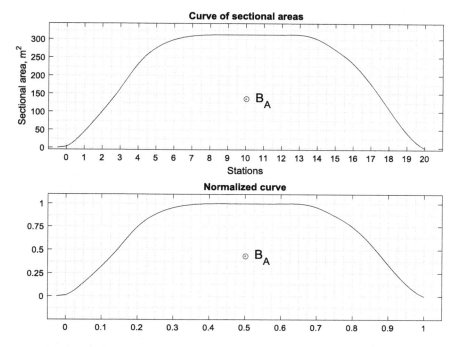

Figure 3.26 A curve of sectional areas

3.9 HULL TRANSFORMATIONS

In Section 3.3.3 we defined the *sectional area curve* as the curve of the areas of transverse hull sections versus the longitudinal coordinate x. This concept is used in some methods of transforming the ship lines with the aim of modifying certain geometrical properties of the hull. One typical application is the development of systematic series of ship forms. In this section we explain the basics of these methods. A seminal, comprehensive paper on the subject is that of Lackenby (1950). A shorter treatment is given in Chapter 9 of Molland (2008). Significant worked examples can be found in Sessa (2008). For more details than shown here we refer the reader to Lackenby (1950) and to Sessa (2008). As an example we show in Fig. 3.26 a curve based on the data of the ship model 903 described in INSEAN (1963). The main dimensions of the full-scale ship are

length between perpendiculars, L_{pp}	197.00 m
moulded beam, B	28.50 "
mean draught, T_m	11.05 "

In the upper curve of the figure we represent the sectional area as a function of the coordinate x, but on the x-axis we mark station numbers instead of x-values. We distinguish three zones. From Station 8 to Station 13 the transverse sections of the hull are identical and so are their areas. This zone is the *parallel body*. Ahead of Station 13 we have the *entrance*, and aft of Station 8 the *run*. The corresponding Italian terms are no less expressive. Going from aft toward bow they are *corpo di uscita*, *corpo cilindrico*, and *corpo di entrata* (Sessa, 2008). Watson (1998) writes, 'The sectional area curve is one of the principal factors which determines the resistance of a ship and careful attention should be paid to its form.'

For some transformations it is convenient to normalize the curve. To do this the x-coordinates are divided by L_{pp} or L_{wl}, and the ordinates are divided the midship-section area, A_m. In Fig. 3.26 the lower curve is the normalized variant of the upper curve.

3.9.1 Numerical Calculations

The area under the upper curve in Fig. 3.26 is proportional to the volume of displacement, ∇. This is an example of one of the three orders of integration mentioned in Example 3.8. The centroid of this area is the point B_A whose x-coordinate is, in fact, the x-coordinate of the centre of buoyancy B, that is LCB. The other coordinate of B_A has no physical meaning, but it is used in the method explained in Section 3.9.3. Let $dL = Lpp/20$ be the station spacing, A_x the sectional area corresponding to x, and let us note the properties of station i by A_i for the area, j_i for the station number, and x_i for its x-coordinate measured from the aft perpendicular. The area under the sectional area curve, between the aft and the forward perpendicular, equals

$$\nabla = \int_{x_0}^{x_{20}} A_x dx$$

We calculate this integral numerically using Simpson's rule

$$\nabla \approx \frac{dL}{3} (A_0 + 4A_1 + 2A_3 + 4A_3 + 2A_4 \ldots + 4A_{19} + A_{20})$$

The x-coordinate of B_A is given by

$$LCB = \frac{\int_{x_0}^{x_{20}} x A_x dx}{\nabla} \tag{3.86}$$

Station	Sectional area		Simpson	Functions	j_i	Functions	Functions
No.	measured	actual	multiplier	of area		of M_y	of M_x
	mm	m²		m²		m²	m⁴
1	2	3	4	5 = 3x4	6	7 = 5x6	8 = 3x5
0	0.50	3.154	1	3.154	0	0.000	9.950
1	7.00	44.161	4	176.643	1	176.643	7800.679
2	16.00	100.939	2	201.878	2	403.755	20377.285
3	26.00	164.026	4	656.102	3	1968.307	107617.534
4	36.95	233.106	2	466.211	4	1864.844	108676.396
5	43.72	275.815	4	1103.261	5	5516.306	304296.282
6	47.41	299.094	2	598.189	6	3589.132	178914.795
7	49.00	309.125	4	1236.500	7	8655.502	382233.282
8	49.51	312.406	2	624.811	8	4998.490	195194.518
9	49.52	312.406	4	1249.622	9	11246.602	390389.036
10	49.53	312.406	2	624.811	10	6248.112	195194.518
11	49.52	312.406	4	1249.622	11	13745.846	390389.036
12	49.51	312.406	2	624.811	12	7497.734	195194.518
13	49.50	312.406	4	1249.622	13	16245.091	390389.036
14	47.60	300.293	2	600.586	14	8408.202	180351.704
15	43.00	271.273	4	1085.092	15	16276.382	294356.243
16	37.00	233.421	2	466.842	16	7469.472	108970.713
17	28.00	176.643	4	706.572	17	12011.718	124810.868
18	17.00	107.247	2	214.495	18	3860.909	23004.044
19	6.50	41.006	4	164.026	19	3116.486	6726.096
20	0.00	0.000	1	0.000	20	0.000	0.000
Σ				13299.697		133299.533	3604896.530
Integral				43667.338			

Figure 3.27 The spreadsheet of sectional areas

With $x_i = j_i dL$, we write the numerical approximation of Eq. (3.86) as

$$xB_A = \frac{(dL^2/3)(j_0 A_0 + 4j_1 A_1 + 2j_2 A_2 \ldots + 4j_{19} A_{19} + j_{20} A_{20})}{(dL/3)(A_0 + 4A_1 + 2A_2 + \ldots + 4A_{19} + A_{20})}$$

$$= dL \frac{j_0 A_0 + 4j_1 A_1 + 2j_2 A_2 \ldots + 4j_{19} A_{19} + j_{20} A_{20}}{(A_0 + 4A_1 + 2A_2 + \ldots + 4A_{19} + A_{20})}$$

The moment of the area under the curve, with respect to the x-axis, is

$$\int_{x_0}^{x_{20}} \frac{A_x}{2} A_x dx = \frac{1}{2} \int_{x_0}^{x_{20}} A_x^2 dx$$

Then, the numerical approximation of the height of B_A above the x-axis is

$$y_B = \frac{1}{2} \frac{(dl/3)(A_0^2 + 4A_1^2 + \ldots A_{20}^2)}{(dL/3)(A_0 + 4A_1 + \ldots A_{20})} = \frac{1}{2} \frac{A_0^2 + 4A_1^2 + \ldots A_{20}^2}{A_0 + 4A_1 + \ldots A_{20}} \quad (3.87)$$

Simpson's method belongs to traditional Naval Architecture. Once calculated manually, the procedure is implemented today on computer programs. In our example, we perform the calculations in the Excel spreadsheet shown in Fig. 3.27. Column 1 contains the numbers of the 21 stations that cover the length between perpendiculars. In column 2 we show the ordinates of the sectional area curve as measured on the drawing included in INSEAN (1963). These values are scaled so as to obtain the midship area $A_M = 312.406$ m^2 and the results appear in column 3. Column 4 contains the *multipliers* used in Simpson's rule. The products of the multipliers by the sectional areas are displayed in column 5. The j_i values are entered into column 6; their products by the values in column 5 appear in column 7. Column 8 is produced by multiplying the values in column 3 by those in column 5. The line marked $\sum$ contains the sums of columns 5, 7, and 8. In the last line of the spreadsheet, column 5, we multiply the sum in the cell above by $dL/3$ and display the area under the curve, i.e. the volume of displacement, ∇, in m^3.

Our example is based on a drawing in which the aft perpendicular coincides with the rudder stock. The waterplane, however, extends about half a station interval aft of this perpendicular. This corresponds to an additional area under the curve that we roughly estimate as less than 8 m^3. The corrected area under the curve is then 43675.1 m^3. As to the coordinates of the point B_A, the correction affects only the second digit after the decimal point. The area under the normalized curve has another significance; it is equal to the prismatic coefficient, C_p. To prove this we calculate

$$\int_{x_0}^{x_n} \frac{A_x}{A_M} \cdot \frac{dx}{L_{pp}} = \frac{\int_{x_0}^{x_n} A_x dx}{L_{pp} A_M} = \frac{\nabla}{L_{pp} A_M} = C_p \tag{3.88}$$

3.9.2 The 'One Minus Prismatic' Method

In this section we show how to change the prismatic coefficient of a hull with parallel middle body. To explain the method we show in Fig. 3.28 only the part of the normalized sectional-area curve that corresponds to the forebody. Correctly we should use a special notation for its prismatic coefficient. For the sake of simplicity, however, we note it by C_p. Let us change the prismatic coefficient from C_p to $C_p + \delta C_p$. To do this we extend the parallel middle body by translating each transverse section forward a distance δx. We keep L_{pp} constant, which means that Station 20 is not moved. As formulated by Lackenby, 'the new spacing of the sections from the end of the body is made proportional to the difference between the respective prismatic and unity'. The term 'unity' corresponds to the total area of the figure in the normalized plot. It follows that

$$\frac{1 - (x + \delta x)}{1 - x} = \frac{1 - (C_p + \delta C_p)}{1 - C_p}$$

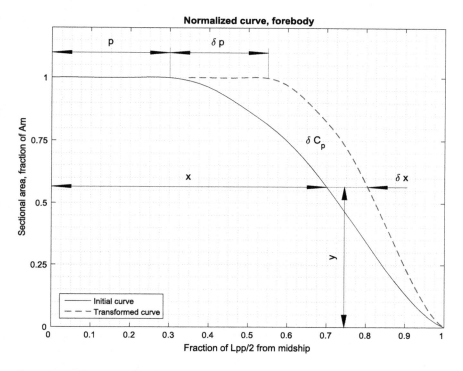

Figure 3.28 The principle of the 'one minus prismatic' method

which yields

$$\delta x = \frac{\delta C_p}{1 - C_p}(1 - x) \tag{3.89}$$

To prove that the additional area equals, indeed, δC_p we calculate

$$\int_0^1 \delta x \, dy = \frac{\delta C_p}{1 - C_p} \int_0^1 (1 - x) \, dy$$

The first term of the integral yields

$$\int_0^1 dy = 1$$

The second term gives

$$\int_0^1 x \, dy = C_p$$

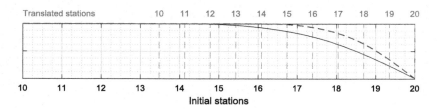

Figure 3.29 The transformed design waterline

Summing up, the additional area is

$$\frac{\delta C_p}{1 - C_p}(1 - C_p) = \delta C_P$$

Fig. 3.28 is based on the same data as Figs 3.26 and 3.27. Integration of the area under the curve yields $C_p = 0.7122$. Let δC_p be 0.1. The transformed curve is shown as a dashed line, blue in the electronic edition of the book. The added area under the curve is noted δC_p. The original extension of the parallel middle body is p; the transformation increases it by δp. In this example the centroid of the area under the curve is translated forward. What happens to the LCB value of the whole hull depends on the transformation applied to the after-body curve. This subject is treated by Lackenby. In Fig. 3.29 we show how the transformation affects the design waterline. The shifted stations are shown in dashed lines, in red in the electronic version of the book, and their numbers appear on top. The transformed waterline appears also as a dashed line, in blue in the electronic edition. After transforming all waterlines, the new half-breadths are read from the waterlines plan at the sections spaced at dL intervals. In Exercise 2.4 we ask the reader to prove that $C_p = C_b/C_m$. In our transformation the midship section does not change and so does the midship coefficient, C_m. It follows that changing the prismatic coefficient, C_p, causes a proportional change of the displacement.

3.9.3 Swinging the Curve

We show now how to change the longitudinal centre of buoyancy, LCB, by transforming the sectional-area curve. This method can be applied also to ships that have no parallel middle body; we explain it, however, using the same data as for the 'one minus prismatic' method. Consider in Fig. 3.30 a strip of width dx located at the distance x from the midship section and having the height A_x; it is coloured in yellow in the electronic version of

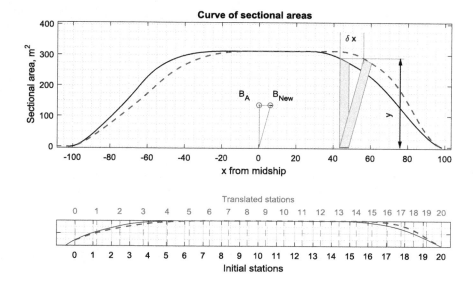

Figure 3.30 Swinging the sectional-areas curve

the book. The transformation 'swings' the strip so that its top is moved a distance $\delta x = A_x \tan \theta$. In this strip we consider an elemental area $dxdy$ at the coordinate y above the x-axis. The transformation moves the elemental area a distance equal to $y \tan \theta$. The change of the moment with respect to the y axis is

$$dM_y = \int_0^{A_x} y \tan \theta \, dxdy = \frac{1}{2} \tan \theta A_x^2 dx$$

The change of moment for the whole area under the curve is

$$M_y = \tan \theta \int_{x_0}^{x_n} \frac{A_x^2}{2} dx$$

The integral represents the moment of the area under the curve with respect to the x-axis (see point B_A in Fig. 3.26). Let us note the height of B_A above x-axis by y_B. Then, $M_y = \nabla y_B$. The change of moment causes a change δLCB of the longitudinal centre of buoyancy. We can write that

$$\nabla \delta LCB = \nabla \tan \theta y_B$$

which yields

$$\tan \theta = \frac{\delta LCB}{y_B} \tag{3.90}$$

We conclude that a station situated at the coordinate x from the midship is moved a distance

$$\delta x = A_x \tan \theta = \frac{\delta LCB}{\gamma_B} A_x \tag{3.91}$$

In Fig. 3.30 we assumed $\delta LCB = 6$ m. The transformed sectional-area curve appears in the upper plot as a dashed line, blue in the electronic edition of the book. The lower plot shows the effect of the transformation on the design waterline. All calculations and the plotting were carried on in a MATLAB script Swinging.m that can be found on the Mathworks site of the book. Integrating with the help of the MATLAB function trapz we obtain

	Initial	Swinged	
Volume of displacement	43692.38	43692.38	m^3
Longitudinal centre of buoyancy, LCB ..	0.19	6.20	m
Waterplane area	2385.45	2390.27	m^2
Longitudinal centre of floatation, LCF.	-2.26	1.18	m

Obviously, in this method it is not possible to specify the variations of the parallel middlebody length and neither the variations of the waterplane area and of the longitudinal centre of buoyancy, LCB.

3.9.4 Lackenby's General Method

In this section we derive the equation of a method for changing both the prismatic coefficient and the extension of the parallel middlebody. Other procedures described by Lackenby, including the 'one minus prismatic' method, are particular cases of the general method. We consider again the normalized curve of the forebody as in Section 3.9.2. The proposed form of equation is

$$\delta x = c(1 - x)(x + d) \tag{3.92}$$

where x is the normalized abscissa of the sectional-area curve measured from midship, and c and d are constants whose expressions we are going to derive. The proposed equation keeps the hull length constant since $\delta x = 0$ at $x = 1$. The length of the parallel body from midship is p. According to Eq. (3.92) this length is increased by

$$\delta p = c(1 - p)(p + d)$$

which yields

$$d = \frac{\delta p}{c(1-p)} - p \tag{3.93}$$

Substituting Eq. (3.93) into Eq. (3.92) we obtain

$$\delta x = c(1-x)\left(x + \frac{\delta p}{c(1-p)} - p\right) = c(1-x)(x-p) + \frac{\delta p}{1-p}(1-x)$$

In the normalized representation of the sectional-area curve the change of the prismatic coefficient equals the areas between the initial and the transformed curves

$$\delta C_p = \int_0^1 \delta x\, dy = c\int_0^1 (x - p - x^2 + px)\, dy + \frac{\delta p}{1-p}\int_0^1 (1-x)\, dy$$

$$= c(C_p - p - 2C_p y_B + pC_p) + \frac{\delta p}{1-p}(1 - C_p)$$

where y_B is again the y-coordinate of the point B_A in Fig. 3.28. Let

$$A = C_p(1 - 2y_B) - p(1 - C_p) \tag{3.94}$$

Then

$$\delta C_p = cA + \frac{\delta p}{1-p}(1 - C_p) \tag{3.95}$$

and, finally,

$$c = \frac{1}{A}\left(\delta C_p - \delta p \frac{1 - C_p}{1-p}\right) \tag{3.96}$$

3.10 APPLICATIONS

3.10.1 The planimeter

In the first section of this chapter we mentioned that the calculations of the geometrical properties of areas and volumes required in ship design involve many integrations. At this point we can add that the amount of these calculations is tremendous. Today the work is done with the aid of digital computers. At the time of this writing there are still veteran Naval Architects who remember other techniques, but young engineers and students may think that it always was like today. The computer era is only a thin, superficial layer in the long history of Naval Architecture. This does not mean

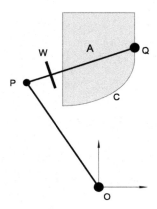

Figure 3.31 The principle of the planimeter

that before the advent of digital computers the calculations were carried on only manually. Digital computers were preceded by analog computers and the first among these were mechanical instruments. The oldest known today is the *Antikytera mechanism*. The dates estimated by various researchers range between 200 BC and 80 BC. However, the level of sophistication of this instrument is such that it must have evolved from previous devices. A milestone was the calculator built in 1642 by Blaise Pascal (French, 1623–1662). The idea of a mechanism that calculates the area enclosed by a given curve appeared at the beginning of the 19th century. James Clerk Maxwell (Scottish, 1831–1879) designed an appropriate mechanism, but did not build it. More *planimeters* were invented and built. To the best of our knowledge the most popular instrument in Naval Architecture was the **polar planimeter** invented by Jakob Amsler-Laffon (Swiss, 1823–1912) in 1854. A sketch of the planimeter is shown in Fig. 3.31. The instrument has two arms, *PO* and *PQ*, connected at *P* by an articulation. The point *O* is kept in a fixed position, in most cases by a weight. The point *Q* is a cursor. *W* is a wheel that rotates around an axis parallel to the arm *PQ*. In the figure we show, as an example, a ship section whose area *A* is enclosed by the curve *C*. While the pointer *Q* follows the contour *C* and returns to the starting point, the wheel *W* turns a number of times that is proportional to the area *A*. The motion of the wheel is transmitted to a gear that turns a counter.

   Amsler has given a proof that his instrument yields a number proportional to the area that must be calculated. Various other proofs were published in the decades that followed (for example, Barbarin, 1880). Some of these proofs are rather long and tedious, others are only partial and do not cover all possible cases. Examples of the latter appear in older books on Naval Architecture, such as Vrijlandt (1948) and Scheltema de Heere (1970). A complete treatment is given by Johow in Krieger (2010). It seems, however, that an essential aspect was missed during nearly a whole century. On one hand the result required from the planimeter is an area; we defined this concept as a double integral. On the other hand the cursor of the planimeter runs along a closed curve; its wheel calculates a line integral. Guido Ascoli (Italian 1887–1957) was the first to relate the planimeter to Green's theorem. He did it in a report presented in 1947, a short time after surviving the racial persecutions of the fascist regime. Ascoli treats the subject at a high mathematical level, within a general theory of integrating devices. This approach is reflected in the title of his communication: '*Vedute sintetiche sugli strumenti integratori*'. We give here a proof suitable for the polar planimeter; it is based on a document of Rabelo and Manso (2004).

In Fig. 3.31 we assume a system of coordinates with origin in O. Let the coordinates of P and Q be

$$P = \begin{vmatrix} \xi \\ \eta \end{vmatrix}, \quad Q = \begin{vmatrix} x \\ y \end{vmatrix} \tag{3.97}$$

We assume that the vectors $\overrightarrow{OP}$ and $\overrightarrow{PQ}$ have the same length r. In addition we observe that the wheel at w rotates only when the local motion has a component in its plane, that is perpendicular to $\overrightarrow{OP}$. As this vector has the components

$$\overrightarrow{PQ} = \begin{vmatrix} x - \xi \\ y - \eta \end{vmatrix}$$

the unit vector perpendicular to it is

$$\overrightarrow{W} = \frac{1}{r} \begin{vmatrix} -y + \eta \\ x - \xi \end{vmatrix} \tag{3.98}$$

To apply Green's theorem we choose as functions M and N the components of the vector $\overrightarrow{W}$

$$M = \frac{1}{r}(x - \xi), \quad N = -\frac{1}{r}(y - \eta) \tag{3.99}$$

If we can prove that, with this choice, the quantity between parentheses in the left-hand side of Eq. (3.80) is equal to a constant k, then the left-hand side is equal to k times the area enclosed by the curve C. To start the proof we write first that the point P moves on a circle with centre at O and radius r, and Q on a circle with centre at P and radius r

$$\xi^2 + \eta^2 = r^2$$
$$(x - \eta)^2 + (y - \eta)^2 = r^2 \qquad (3.100)$$

The solution of this system of equations is

$$\xi = \frac{x}{2} + \frac{y}{2}\sqrt{\frac{4r^2}{x^2 + y^2} - 1}$$

$$\eta = \frac{y}{2} - \frac{x}{2}\sqrt{\frac{4r^2}{x^2 + y^2} - 1} \qquad (3.101)$$

The functions to be used with Green's theorem are

$$M = \frac{1}{r}(x - \xi) = \frac{1}{r}\left(\frac{x}{2} - \frac{y}{2}\sqrt{\frac{4r^2}{x^2 + y^2} - 1} \right)$$

$$N = -\frac{1}{r}(y - \eta) = \frac{1}{r}\left(-\frac{y}{2} + \frac{x}{2}\sqrt{\frac{4r^2}{x^2 + y^2} - 1} \right) \qquad (3.102)$$

and the partial derivatives that appear in the left–hand side of Eq. (3.80)

$$\frac{\partial M}{\partial x} = \frac{1}{r}\left[\frac{1}{2} + 4xyr^2(x^2 + y^2)\left(\frac{4r^2}{x^2 + y^2} - 1 \right)^{-1/2} \right] \qquad (3.103)$$

$$\frac{\partial N}{\partial y} = -\frac{1}{r}\left[\frac{1}{2} + 4xyr^2(x^2 + y^2)\left(\frac{4r^2}{x^2 + y^2} - 1 \right)^{-1/2} \right] \qquad (3.104)$$

Thus, the left–hand side of Eq. (3.80) becomes

$$\frac{\partial M}{\partial x} - \frac{\partial N}{\partial y} = \frac{1}{r} \qquad (3.105)$$

which means that the line integral yields the area enclosed by the curve divided by r. We find in the literature various explanations on how the planimeter treats the line integral, but it seems to us that the following approach is simpler. Let us return to the integrand in the left–hand side

$$M\,dy + N\,dx$$

This is the scalar product of the vector $\overrightarrow{W}$ (see Eq. (3.98)) by the vector

$$dS = \left| \begin{array}{c} dx \\ dy \end{array} \right|$$

The latter is the tangent vector to the curve C. It follows that the scalar product represents the projection of dS on the perpendicular to the axis of wheel, or, in other words, that component of dS that causes the wheel to rotate. The integral of this component is the length run by the wheel. Let D be the wheel diameter, and n the number of rotations achieved while the pointer of the planimeter traces the curve C and returns **exactly** to the staring point. We can write that

$$S = n\pi D$$

The area measured by the planimeter is

$$n\left(\frac{\pi D}{r}\right) \tag{3.106}$$

The quantity between parentheses is the *constant of the planimeter*. It remains to take into account the scale of the drawing on which the planimeter is used. The advent of the digital era did not end the use of the planimeter. An electronic part was added, the user could enter the scale and the result was displayed automatically.

It is remarkable how such a simple and easy-to-use instrument can calculate the areas of irregularly-shaped figures. It can do even more. As shown in Section 3.4.3, by measuring with the planimeter the area under a Bonjean curve yields a static moment. Continuing this line of reasoning it is possible to draw a curve of static moment; the area under it gives a moment of inertia. Several methods of hydrostatic calculations have been developed specifically for use with the planimeter. Jakob Amsler-Laffon continued his work and developed an **integrator** that used gears to yield also the first and second moments of the area enclosed by the curve on which the tracer arm runs. According to McGee (1998), 'the first properly working instrument was imported to Britain' in 1878 and 'the first extended notice of the instrument in naval circles took place in 1880'. McGee explains how the stage was set to receive the integrator, without detailing the fact that the theory of ship stability already existed as developed in France. Drawings of ship lines had been in use for several decades; they contained all the information necessary for hydrostatic and weight calculations. However, these calculations implied an enormous amount of work that could be carried only by mathematically educated and well paid staff. If the results of calculations imposed changes in the design, new calculations were necessary and this could lead to unacceptable delays and expenses. Therefore, there were shipbuilders who refrained from adequately checking the designs of the ships they built. Amsler's integrator shortened considerably the duration of calculations. Moreover, the use of the instrument was so simple that it could be operated by apprentices. The impact was dramatic. Communications presented to meetings of Naval Architects in the years 1882 and 1884 resemble a competition in which each participant tries to calculate more data in shorter times and employing less skilled teams. McGee cites the famous Naval Architect Sir Edward Reed as writing in 1984: 'we happen to be living at a time when stability calculations have become a necessity on an enormous scale, and to supply them would have been at once difficult and expensive beyond limit if it had not been for the advent ... of this mechanical assistant'.

3.10.2 A MATLAB Digitizer

The advent of CAD brought a new instrument, the *digitizing tablet*. As the software evolved, it is now possible to digitize also without such a tablet. For example, we have developed a MATLAB digitizer–integrator that is stored on the Elsevier and Mathworks sites of Biran (2005), and Biran and López-Pulido (2014). See also Biran (2006). Let us suppose that a user is interested in finding the geometrical properties of a ship station. The task is carried on in the steps described below.

1. The drawing of the section is scanned and stored in the JPEG format.
2. The user calls the program with the name of the graphic file as input argument.
3. The user *calibrates* the digitizer by choosing three non–collinear points and enters their real–world coordinates. By solving a system of linear equations the software finds the axes and the scale of the drawing.
4. Going in a counterclockwise direction, the user picks up a sufficient number of points to approximate the shape of the station by straight–line segments and ends by picking a second time the starting point.
5. The software defines the equations of the straight–line segments and uses Green's theorem to calculate the geometrical properties of the digitized station.

Table 3.3 Geometric properties of areas

Term	Definition
Area	$A = \int \int_A dxdy$
Moment with respect to the x-axis	$M_x = \int \int_A y dxdy$
Moment with respect to the y-axis	$M_y = \int \int_A x dxdy$
x-coordinate of centroid C	$x_c = M_y/A$
y-coordinate of centroid C	$y_c = M_x/A$
Second moment with respect to the x-axis	$I_{xx} = \int \int_A y^2 dxdy$
Second moment with respect to the y-axis	$I_{yy} = \int \int_A x^2 dxdy$
Product of inertia with respect to the x- and y-axes	$I_{xy} = \int \int_A xy dxdy$
Polar moment of inertia	$I_{xy} = \int \int_A r^2 dxdy = I_{xx} + I_{yy}$
Radius of gyration with respect to the x-axis	$i_{xx} = \sqrt{I_{xx}/A}$
Radius of gyration with respect to the y-axis	$i_{yy} = \sqrt{I_{yy}/A}$

3.11 SUMMARY

The geometrical properties of a plane area A are defined in Table 3.3.

When the second moments are calculated with respect to centroidal axes, and the axes of coordinates are translated a distance d_1 to the left, and a distance d_2 downwards, the moments of inertia of the area A with respect to the new axes are given by the *parallel translation theorem* (Huyghens, Steiner)

$$I_{x'x'} = I_{xx} + d_1^2 A, \ I_{y'y'} = I_{yy} + d_2^2 A, \ I_{x'y'} = I_{xy} + d_1 d_2 A$$

If the axes of coordinates are rotated counterclockwise by an angle α, the second moments of the area A with respect to the new axes are

$$I_{\xi\xi} = \frac{I_{xx} + I_{yy}}{2} + \frac{I_{xx} - I_{yy}}{2} \cos 2\alpha - I_{xy} \sin 2\alpha$$

$$I_{\eta\eta} = \frac{I_{xx} + I_{yy}}{2} - \frac{I_{xx} - I_{yy}}{2} \cos 2\alpha + I_{xy} \sin 2\alpha$$

$$I_{\xi\eta} = \frac{I_{xx} - I_{yy}}{2} \sin 2\alpha + I_{xy} \cos 2\alpha$$

The *axial* moments, $I_{\xi\xi}$ and $I_{\eta\eta}$, reach extremal values when the angle of rotation fulfills the equation

$$\tan 2\alpha = -\frac{2I_{xy}}{I_{xx} - I_{yy}}$$

At this angle $I_{\xi\eta} = 0$ and the maximum and minimum values of the moments of inertia are given by

$$I_{max,\,min} = \frac{I_{xx} + I_{yy}}{2} \pm \sqrt{\left(\frac{I_{xx} - I_{yy}}{2}\right)^2 + I_{xy}^2}$$

The axes rotated as above are called *principal axes* and the moments of inertia with respect to them are the *principal moments of inertia*. The second moments of an area are the components of the *tensor of inertia*

$$J = \begin{vmatrix} I_{xx} & -I_{xy} \\ -I_{xy} & I_{yy} \end{vmatrix}$$

The eigenvalues of this tensor are the principal moments of inertia, and the eigenvectors indicate the directions of the principal axes.

In many cases the calculations of area properties may be long and tedious. Then, it is difficult to verify the results and it is wise to assess their plausibility by quick means. The following properties can help in performing the job.

- The axial and the polar moments of inertia cannot be zero or negative.
- If one of the axes of coordinates is an axis of symmetry of the figure, the product of inertia equals zero.
- The trace and the determinant of the inertia tensor are invariants of rotation.
- If we can draw a simple figure that circumscribes the given figure, the area and the centroidal axial moments of inertia of the circumscribing figure are greater than those of the given figure.

Areas and first and second moments of areas are defined as double integrals over a region A. Using *Green's theorem* it is possible to turn an area integral into a line integral along the curve C that encloses the domain A. Given two functions M and N defined over the domain A, Green's theorem states that

$$\int\int_A \left(\frac{\partial M}{\partial x} - \frac{\partial N}{\partial y}\right) dx\,dy = \oint_C [M(x, y)dy + N(x, y)dx]$$

The functions M and N to be used for the geometrical properties of areas are listed in Table 3.4.

The geometric properties of a volume V are listed in Table 3.5.

Greens theorem can be used to explain how Amsler's *polar planimeter* measures an area. Other applications of this theorem are digital integrators.

Table 3.4 Functions to be used with Green's theorem

Property	M	N	Equation
Area	$x/2$	$-y/2$	$A = \frac{1}{2}\oint_c x\,dy - y\,dx$
Moment	$-y^2/2$	0	$M_x = -\frac{1}{2}\oint_c y^2\,dx$
Moment	$x^2/2$	0	$M_y = \frac{1}{2}\oint_c x^2\,dx$
Second moment	0	$-y^3/3$	$I_{xx} = -\frac{1}{3}\oint_c y^3\,dx$
Second moment	$x^3/3$	0	$I_{yy} = \frac{1}{3}\oint_c x^3\,dy$
Product of inertia	0	$-xy^2/2$	$I_{xy} = -\frac{1}{2}\oint_C xy^2\,dx$

Table 3.5 Geometric properties of volumes

Term	Definition
Volume	$V = \int\int\int_V dx\,dy\,dz$
Moment with respect to the yOz-plane	$M_{yOz} = \int\int\int_V x\,dx\,dy\,dz$
Moment with respect to the xOz-plane	$M_{xOz} = \int\int\int_V y\,dx\,dy\,dz$
Moment with respect to the xOy-plane	$M_{xOy} = \int\int\int_V z\,dx\,dy\,dz$
x-coordinate of centroid C	$x_c = M_{yOz}/V$
y-coordinate of centroid C	$y_c = M_{xOz}/V$
z-coordinate of centroid C	$z_c = M_{xOy}/V$

An example of the latter is a MATLAB function that allows the user to pick up points on a closed curve displayed on the computer screen and calculates the geometrical properties of the area enclosed by the curve.

The *sectional area curve* is a representation of the areas of transverse ship sections versus the ship length. The area under the curve equals the volume of displacement, ∇, and the longitudinal position of its centroid coincides with the longitudinal centre of buoyancy, *LCB*. The sectional area curve can be used to transform the hull and modify thus some of its properties. In such applications it is convenient to normalize the curve by dividing the longitudinal coordinates by the waterline length, L_{wl}, and the ordinates by the midship area, A_m. The area under the normalized curve is equal to the prismatic coefficient, C_p. Below we list three methods based on the sectional-area curve.

The 'one minus prismatic' method — useful for changing the prismatic coefficient, C_p. Can be used only for hulls with parallel middlebody. Causes uncontrollable changes of C_p and *LCF*.

Swinging the curve — Useful for changing the longitudinal centre of buoyancy, *LCB*. Can be used also for hulls without parallel middlebody. Causes uncontrollable changes of *LCF*.

Lackenby's general method — It allows changes of the parallel middlebody length and of the prismatic coefficient, C_p. All other methods are particular cases of this one.

Example 3.12 (Sectional modulus). In Section 3.5.9 we mentioned the calculation of the *sectional modulus* as an application of concepts and methods discussed in this chapter. The subject is related to the *longitudinal strength* of ships. The ship is treated as a beam loaded by its weights and supported by the buoyancy force. As the ship masses have another distribution along the ship than the buoyancy force, the local differences produce a shear force and a bending moment. In the classic theory of bending the maximum stress in a transverse section of the beam is equal to the bending moment in that section divided by the local sectional modulus, while the latter is obtained by dividing the moment of inertia of the section by the largest distance from the neutral axis. The usual notations in English literature are

$$\sigma = \frac{M}{Z}, \quad Z = \frac{I}{c} \qquad (3.107)$$

Instead of Z the German literature uses the notation W, the initial of the term *Widerstansmoment*. This approach is common in fields like mechanical or civil engineering. In Naval Architecture, however, it is necessary to calculate two sectional moduli, one, Z_B, equal to the moment of inertia divided by the distance between the neutral axis and the keel, the other, Z_D, equal to the moment of inertia divided by the distance between the neutral axis and the deck. To explain why it is necessary to do so let us assume first that $Z_D < Z_B$. Then, the bending stress in the deck plating is greater than that in the bottom. It may happen, however, that because of the structural design the bending stress in the bottom plating can cause local buckling. Thus, the knowledge of Z_B is also necessary. We can reason in the same way for the case $Z_D > Z_B$. Classification societies require, therefore, the calculation of both moduli taking into consideration only those structural members that effectively contribute to the longitudinal strength of the ship. Some details can be found, for example in DNV-GL (2015), Part 3, Chapter 5, or in Lloyd's Register (2016), Part 3, Section 3.4.

To illustrate a procedure for calculating the sectional modulus we show in Fig. 3.32 a structure that is simpler than the actual midship section of a real ship. The dimensions of the structural members have not been calculated as required by classification societies, but assumed only for this example and do not represent the current practice. As ships are symmetric about the centreplane xz, only half of the section is shown. Also, only

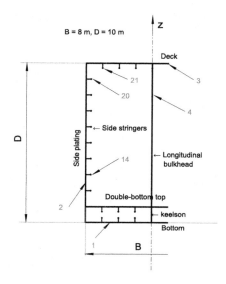

Figure 3.32 A hypothetical midship section

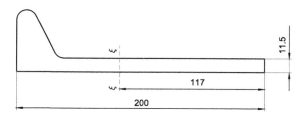

Figure 3.33 Hot rolled bulb flat 200 × 11.5

structural members contributing to the longitudinal strength are shown and taken into account. In this example the longitudinal reinforcements noted *stringers* are made of *bulb flats* (in German *Wulstflachstahl*, also *Hollandpro-fil*). For this example we have chosen the 220 × 11.5 profile illustrated in Fig. 3.33. The $\xi - \xi$ axis is centroidal and perpendicular to the main side of the profile. It is the right-hand end of the profile that is welded to plates. According to an older DIN standard the geometrical properties of the above profile are

Area	28.6 cm^2
Moment of inertia about $\xi - \xi$	1126 cm^4

The drawing submitted for approval should contain all dimensions and each structural member should be marked by an identifying number.

No.	Item	Dimensions	Area	z above BL	Static moment	Own I	Steiner's term	Total I
			cm^2	cm	cm^3	cm^4	cm^4	cm^4
1	2	3	4	5	6 = 5x4	7	8 = 5x6	9 = 7 + 8
1	Bottom	0.6x400 cm^2	240.0	0.3	72	7	22	29
2	Side	0.6x1000 cm^2	600.0	500.0	300000	50000000	150000000	200000000
3	Deck	0.6x400 cm2	240.0	999.7	239928	7	239856022	239856029
4	Longitudinal bulkhead	0.25x900 cm^2	225.0	550.0	123750	15187500	68062500	83250000
5	Double bottom top	0.5*400 cm^2	200.0	99.8	19950	4	1990013	1990017
6	Keelson	0.5*100 cm^2	50.0	50.0	2500	40429	125000	165429
7	Bottom stringer	HP 200x11.5	28.6	13.7	392	1126	5368	6494
8	Bottom stringer	HP 200x11.5	28.6	13.7	392	1126	5368	6494
9	Bottom stringer	HP 200x11.5	28.6	13.7	392	1126	5368	6494
10	Double bottom stringer	HP 200x11.5	28.6	87.8	2511	1126	220473	221599
11	Double bottom stringer	HP 200x11.5	28.6	87.8	2511	1126	220473	221599
12	Double bottom stringer	HP 200x11.5	28.6	87.8	2511	1126	220473	221599
13	Side stringer	HP 200x11.5	28.6	200.0	5720	0	1144000	1144000
14	Side stringer	HP 200x11.5	28.6	300.0	8580	0	2574000	2574000
15	Side stringer	HP 200x11.5	28.6	400.0	11440	0	4576000	4576000
16	Side stringer	HP 200x11.5	28.6	500.0	14300	0	7150000	7150000
17	Side stringer	HP 200x11.5	28.6	600.0	17160	0	10296000	10296000
18	Side stringer	HP 200x11.5	28.6	700.0	20020	0	14014000	14014000
19	Side stringer	HP 200x11.5	28.6	800.0	22880	0	18304000	18304000
20	Side stringer	HP 200x11.5	28.6	900.0	25740	0	23166000	23166000
21	Deck stringer	HP 200x11.5	28.6	987.7	28248	1126	27900767	27901893
22	Deck stringer	HP 200x11.5	28.6	987.7	28248	1126	27900767	27901893
23	Deck stringer	HP 200x11.5	28.6	987.7	28248	1126	27900767	27901893
	Totals		2041.2	443.6	905493	65238082	625637379	690875460

$$I_{BL} \quad 1381750921 \quad cm^4 \qquad I_{\xi\xi} = \quad 578382094 \quad cm^4$$
$$Z_B = \quad 1303812 \quad cm^3 \qquad Z_D = \quad 1039523 \quad cm^3$$

Figure 3.34 Spreadsheet for calculating the sectional modulus

Fig. 3.32 is rather small; to keep it readable we do not show dimensions, but assume that the stringers are equally spaced. Also, we show only a few identifying numbers and trust the reader for making the right connection between Figs 3.32 and 3.34. We calculate the sectional modulus in the spreadsheet shown in Fig. 3.34. Like the drawing, it corresponds to a half section. Therefore, the longitudinal bulkhead marked 4 is entered with half of its thickness. The height of the neutral axis above BL is calculated in the last line of column as $905449/2041.2 = 443.6$ cm.

3.12 EXERCISES

Exercise 3.1 (Subdividing a triangle). In Section 3.3.1 we have shown that the area of the triangle AMN in Fig. 3.3 equals four times the area of the triangle ABC. Subdivide the triangle AMN into suitable elementary figures and prove the above result by using the second axiom of areas.

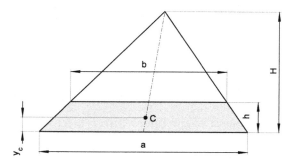

Figure 3.35 A trapezoid as the difference of two triangles

Exercise 3.2 (Scaled rectangle). An engineer measures the dimensions of a rectangular cabin in the general arrangement of a ship drawn at the scale 1:200 and finds that the length is 15 mm and the width 10 mm. What is the actual area of the cabin in m²?

Exercise 3.3 (Area and centroid of trapezoid). This exercise refers to the trapezoid with base a and height h and the triangle with the same base and height H shown in Fig. 3.35. Considering that the area of the trapezoid is the difference between the area of the triangle with base a and height H, and the area of the triangle with base b and height $H - h$, you are asked to:
1. write the equation that gives the triangle height H;
2. write the equation that yields the area of the trapezoid as the difference between the area of the triangle with base a and height H and the area of the triangle with base b and height $H - h$;
3. derive the equation that gives the coordinate y_c of the centroid of the trapezoid;
4. calculate H, the area and the coordinate y_c of the trapezoid for $a = 100$ cm, $b = 75$ cm, and $h = 15$ cm. Redraw at scale Fig. 3.35 for the above values and mark on it the centroid C.

Exercise 3.4 (Isosceles triangle). In Example 3.5 we calculated the second moments of an isosceles triangle. In this exercise you are asked to calculate the second moment with respect to the y-axis of the same triangle by using the formula for the second moment with respect to the x-axis, specifically Eq. (3.54). To do this consider that the given isosceles triangle is composed of two right-angled triangles with bases on the y-axis. Calculate the second moment of a right-angled triangle with respect to its basis and multiply it by 2. Check your results by comparing with Eq. (3.56).

Exercise 3.5 (Parallel translation of the product of inertia). Prove that the parallel-translation theorem for the product of inertia of a plane figure with area A_F is

$$I_{xy} = I_{x'y'} + x_c y_c A_F \qquad (3.108)$$

where

- $I_{x'x'}$ is the product of inertia of the figure with respect to its own centroidal axes;
- I_{xy} is the product of inertia with respect to arbitrary axes O_x and O_y;
- x_c and y_c are the coordinates of the centroid of the area with respect to the axes O_x and O_y.

Exercise 3.6 (Product of inertia with respect to principal axes). Prove that the product of inertia with respect to principal axes is equal to zero. **Hint:** use the identity

$$\tan 2\alpha = \frac{\sin 2\alpha}{\cos 2\alpha}$$

Exercise 3.7 (A right-angled isosceles triangle). Assume that the dimensions in Fig. 3.15 are $B = H = 10$ cm. You are asked to:

1. calculate the moments and product of inertia with respect to the given axes;
2. calculate the angle of the principal axes;
3. calculate the principal moments of inertia;
4. draw the triangle at the scale 2:1 and on it the centroid and the principal axes;
5. explain the results.

Exercise 3.8 (A simplified twin-hull waterplane). In Fig. 3.36 we consider the simplified waterplane area of a twin-hill vessel, let's say a *catamaran*. We assume the data

L_{OA}	110.00 m	B	24.80 m
a	6.88 "	b	17.64 "

1) You are asked to

1. calculate the waterplane area A_w;
2. calculate the waterplane coefficient, C_w, of a single hull and of the twin-hull configuration;
3. calculate the coordinates x_F and y_F of the centre of floatation (waterplane centroid);

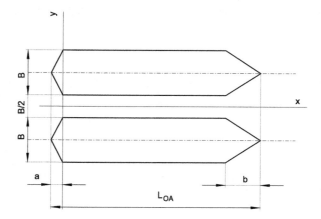

Figure 3.36 Simplified twin-hull — Intact condition

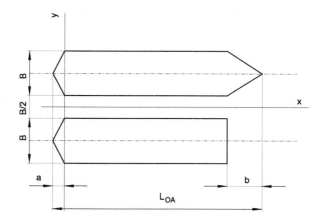

Figure 3.37 Simplified twin-hull — Damage condition

4. calculate the moments of inertia of the waterplane with respect to the given axes;
5. calculate the centroidal moments of inertia;
6. draw at standard scale the waterplane and mark on it the centroid F.

   2) Assume that the forward part of the starboard hull (it can be the *forepeak*) is damaged and contributes no more to the buoyancy and stability of the vessel. For the damaged waterplane shown in Fig. 3.37:
1. calculate the waterplane area A_w;
2. calculate the waterplane coefficient, C_w, of a single hull and of the twin–hull configuration;

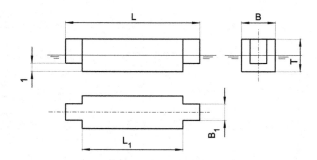

Figure 3.38 A hypothetical barge

3. calculate the coordinates x_F and y_F of the centre of floatation (water-plane centroid);
4. calculate the moments of inertia of the waterplane with respect to the given axes;
5. calculate the centroidal moments of inertia;
6. calculate the angle of the principal axes;
7. calculate the principal moments of inertia;
8. on the drawing of the damaged waterplane mark the centroid F and add the principal axes.

Exercise 3.9 (A hypothetical barge, 1). Fig. 3.38 shows the sketch of a hypothetical barge with very simple forms. Your data are:

L	78.80 m	B_1	9.95 m
L_1	58.50 m	Displacement mass, Δ	13444.000 t
B	19.90 m	Water density, ρ	1.025 t·m^{-3}

You are asked to:
1. calculate the volume of displacement, ∇;
2. calculate the draught, T;
3. calculate the longitudinal centre of floatation, LCF;
4. calculate the vertical centre of buoyancy, $\overline{KB}$;
5. calculate the second moment with respect to the x-axis, I_{xx};
6. calculate the transverse metacentric radius, $\overline{BM}$;
7. calculate the second moment with respect to the transverse centroidal axis, I_L;
8. calculate the longitudinal metacentric radius, $\overline{BM_L}$;

9. draw the waterplane at standard scale and mark on it the centre of floatation, F.

Hint. Begin with Archimedes' principle according to which $\Delta = \rho \nabla$.

Exercise 3.10 (A hypothetical barge, 2). Like Exercise 3.9, but using the data

L	66.60 m	B_1	9.65 m
L_1	49.50 m	Displacement mass, Δ	10118.80 t
B	19.30 m	Water density, ρ	1.025 t·m^{-3}

Exercise 3.11 (Area by Green's theorem). Another set of functions for calculating an area by means of Green's theorem is

$$M = 0, \quad N = y$$

1. Show that the use of these functions yields, indeed, an area.
2. Show that the resulting equation is

$$A = -\oint_C y\,dx$$

3. Use the above equation to calculate the area of the circle with parametric equations

$$x = r\cos 2\pi t, \quad y = r\sin 2\pi t, \quad t = [0\ 1]$$

Exercise 3.12 (Waterplane properties by Green's theorem). Use Green's theorem to calculate the properties of the waterplane shown in Fig. 3.17 of Example 3.7.

Exercise 3.13 (Section area curve). This exercise refers to the data used in Section 3.9.1. You are asked to
1. calculate the prismatic coefficient, C_p;
2. repeat the calculations using this time the normalized data;
3. check that the area under the normalized section area curve yields, indeed, the prismatic coefficient.

Exercise 3.14 (Sectional modulus, 1). In an old version of the regulations of the German Federal Navy we find the following equation for the calculation of the moment of inertia that defines the sectional modulus

$$J = \sum J' + \sum x^2 \cdot A - a \sum x \cdot A$$

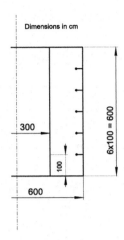

Figure 3.39 A hypothetical midship section

where J is the moment of inertia of the section with respect to the neutral axis, $\sum J'$ the sum of the own moments of inertia of the structural components, $\sum x^2 \cdot A$ the sum of the squared distances of the centroids of structural components above the base line, BL, multiplied by the areas of their sections, a the distance of the neutral axis above BL, and $\sum x \cdot A$ the first moments of the structural components above BL. Show that this formula corresponds to the equations developed in this chapter.

Exercise 3.15 (Sectional modulus, 2). Fig. 3.39 shows a hypothetical section. Assume that the thickness of the plates is 0.7 cm and the stringers are made of bulb flat profiles like that shown in Fig. 3.33.

PART 2

Differential Geometry

CHAPTER 4

Parametric Curves

Contents

4.1 INTRODUCTION

Curves can be represented by implicit, explicit, or parametric equations. Implicit equations cannot be used in computer graphics, explicit can, but not always in a convenient way. Parametric equations suit naturally the work with computers. All splines used in CAD are in parametric form. Parametric equations can be derived from geometric definitions or from laws of mechanics. In this chapter we show examples of parametric equations, explain how to draw them, and discuss their geometrical properties. As an example in Naval Architecture, we derive the parametric equations of the locus of centres of buoyancy of a floating body that inclines at constant displacement around axes parallel to a given direction.

Geometry for Naval Architects
https://doi.org/10.1016/B978-0-08-100328-2.00014-6

Copyright © 2019 Elsevier Ltd.
All rights reserved.
197

4.2 PARAMETRIC REPRESENTATION

In geometry the circle is defined as the locus of the points, in plane, that lie at a constant distance, r, from a fixed point called the centre of the circle. Let x_c, y_c be the coordinates of the centre. There are three possibilities of writing equations that represent the circle as a curve, and they are:

1. **Implicit equation**
2. **Explicit equation**
3. **Parametric equations**

The *implicit* equation is obtained directly from the definition of the circle as a geometric locus

$$(x - x_c)^2 + (y - y_c)^2 = r^2 \qquad (4.1)$$

This equation cannot be used directly to plot the circle by means of a computer language. However, we can obtain from it the *explicit* representation

$$y = \pm\sqrt{r^2 - (x - x_c)^2} + y_c \qquad (4.2)$$

We can use Eq. (4.2) to plot the circle. For example, in MATLAB, given the values of x_c, y_c, r, we can use the commands

```
xc = 2;                            % x-coordinate of centre
yc = 3;                            % y-coordinate of centre
r  = 1.5;                          % circle radius
x  = (xc - r): r/50: (xc + r);     % relevant x interval
y1 = yc + (r^2 - (x - xc).^2).^(1/2);    % upper half of circle
y2 = yc - (r^2 - (x - xc).^2).^(1/2);    % lower half
plot(x, y1, 'k-', x, y2,'k-'), axis equal
xlabel('x'), ylabel('y')
```

We had to use two statements, one to generate points on the upper half-circle, the other for the lower one. In our view this is somewhat awkward as we treated one curve only, not two. The third possibility avoids this nuisance. The *parametric equations* of the circle are

$$x = x_c + r\cos\theta$$
$$y = y_c + r\sin\theta \qquad (4.3)$$

where θ is the angle at centre as shown in Fig. 4.1; it runs from 0 to 2π radians. The variable θ is called **parameter**. Often it is convenient to

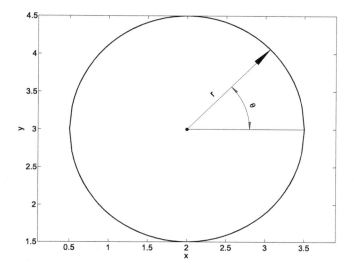

Figure 4.1 Plotting a circle

parametrize the curve so that the parameter belongs to the closed interval [0, 1]. Thus, we **reparametrize** Eqs (4.3) as

$$x = x_c + r\cos 2\pi t$$
$$y = y_c + r\sin 2\pi t, \quad t \in [0,\ 1] \tag{4.4}$$

These equations can be written in vector form as

$$\left| \begin{matrix} x \\ y \end{matrix} \right| = \left| \begin{matrix} x_c \\ y_c \end{matrix} \right| + r \left| \begin{matrix} \cos 2\pi t \\ \sin 2\pi t \end{matrix} \right| \tag{4.5}$$

or, concisely,

$$X = X_c + r \left| \begin{matrix} \cos 2\pi t \\ \sin 2\pi t \end{matrix} \right|, \quad t \in [0,\ 1] \tag{4.6}$$

Mathematicians like to say that X is a *vector-valued function* defined on the closed interval [0, 1] and write

$$X : [0,\ 1] \to \mathbb{R}^2 \tag{4.7}$$

where $\mathbb{R}$ is the set of real numbers. This mapping is illustrated in Fig. 4.2.

Using parametric equations we define a sense for describing the curve. Thus, with Eqs (4.4), when t runs from 0 to 1 the curve is described in a

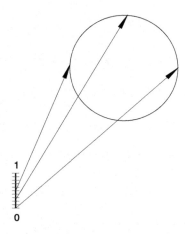

Figure 4.2 The mapping described by Eq. (4.6)

counterclockwise sense. We can better appreciate and exploit this attribute of sense if we consider that the parameter t represents time.

We have seen that for a given curve there may be more than one parametrization. As another example, we can write the following alternative parametrization of a half-circle

$$x = \frac{1 - u^2}{1 + u^2}$$
$$y = \frac{2u}{1 + u^2}, \quad u \in [-1, 1] \tag{4.8}$$

Example 4.1 (MATLAB plot from vector equation). The following code implements Eq. (4.6)

```
function CircleVplot(Xc, r, w, c, s)
%CIRCLEVPLOT plots circle using vectorial formulation
% Calls the function pline
%    Input arguments:
%        Xc, coordinates of centre given as column vector(2x1)
%        r, radius of circle
%        c, line colour
%        w, line width
%        s, line style
% Written by Adrian Biran 2013, last version March 2016

t     = 0: 0.01: 1;          % parameter
```

```
theta = 2*pi*t;              % angle at centre, radian
XC    = repmat(Xc, 1, length(theta));
X     = XC + [ r* cos(theta); r*sin(theta) ];
% plot
pline(X, w, c, s)
axis equal
end
```

We used the function `pline` developed in Biran (2011); it can be found on the MATLAB site of Biran and López-Pulido (2014). To plot the same circle as in Fig. 4.1 call the function with

```
CircleVplot([ 2; 3 ], 1.5, 1.5, 'k', '-')
```

4.3 PARAMETRIC EQUATION OF STRAIGHT LINE

The parametric equation of a straight line passing through a given point, $\mathbf{P_1}$, and having its direction defined as a vector $\mathbf{D}$, is

$$\mathbf{P} = \mathbf{P_1} + \mathbf{D}t \qquad (4.9)$$

In plane this vectorial equation stands for two scalar equations

$$x = x_1 + td_x$$
$$y = y_1 + td_y$$

If the straight line is defined by two points, $\mathbf{P_1}$, $\mathbf{P_2}$, then its direction is that of the vector $\mathbf{P_2} - \mathbf{P_1}$ and the corresponding parametric equation, in vectorial form, is

$$\mathbf{P} = \mathbf{P_1} + (\mathbf{P_2} - \mathbf{P_1})t = (1 - t)\mathbf{P_1} + t\mathbf{P_2} \qquad (4.10)$$

or, in expanded form

$$\begin{vmatrix} P_x \\ P_y \end{vmatrix} = (1 - t) \begin{vmatrix} P_{1x} \\ P_{1y} \end{vmatrix} + t \begin{vmatrix} P_{2x} \\ P_{2y} \end{vmatrix} \qquad (4.11)$$

As we will see in Section 9.2, this is the Bézier curve of first-degree. For values of t in the interval $[0, 1]$ we obtain points that lie between P_1 and P_2. In MATLAB it is possible to work with Eq. (4.10) and there is no need to write two equations. Here is an example

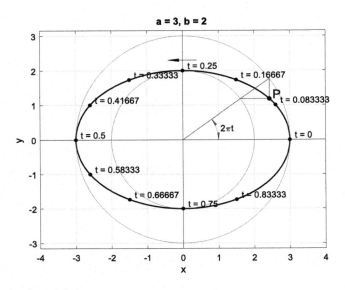

Figure 4.3 Ellipse plot based on parametric equations

```
P1 = [ 2; 3 ]; P2 = [ 5; 8 ];
t = 0.7;
P = (1 - t)*P1 + t*P2
P =
   4.1000
   6.5000
```

Example 4.2 (Parametric equations of ellipse). Let us consider an ellipse with the centre in the origin of coordinates, semi-axis major a and semi-axis minor b. The parametric equations of this curve are

$$x = a\cos 2\pi t$$
$$y = b\sin 2\pi t, \quad t \in [0\ 1] \tag{4.12}$$

Squaring the above equations and adding them side by side yields, indeed, Eq. (1.7). As an example, we show in Fig. 4.3 an ellipse drawn with the above equations. A simple graphic method for finding points on the ellipse is based on these equations and it is applied here for the point P. On the figure an arrow shows the sense in which the curve is generated and we marked several points with the corresponding values of the parameter t.

Example 4.3 (Parametric equations of hyperbola). The parametric equations of the hyperbola with centre in the origin of coordinates, and axes a

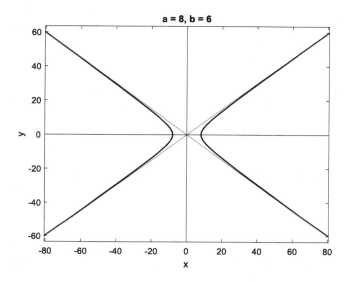

Figure 4.4 Hyperbola plot based on parametric equations

and b are

$$x = \pm a \cosh t, \quad y = \pm b \sinh t \qquad (4.13)$$

The following MATLAB lines plot the curve and its asymptotes as in Fig. 4.4

```
% axes and parameter
a = 8; b = 6;
t = -3: 0.1: 3;
% plot curve
x = a*cosh(t); y = b*sinh(t);
hp = plot(x, y, 'k-', -x, y, 'k-');
set(hp, 'LineWidth', 1.5)
title([ 'a = ' num2str(a) ', b = ' num2str(b) ])
xlabel('x'), ylabel('y')
axis equal
hold on
% show axes
plot([ -80 80 ], [ 0 0 ], 'k-', [ 0 0 ], [ -62 62 ], 'k-')
% show assymptotes
plot([ -80 80 ], [ -80*(b/a) 80*(b/a) ], 'r-')
plot([ -80 80 ], [ 80*(b/a) -80*(b/a) ], 'r-')
hold off
```

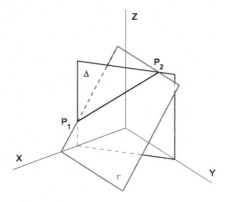

Figure 4.5 The geometrical significance of Eqs (4.14)

4.4 CURVES IN 3D SPACE

4.4.1 The Straight Line

In the plane, that is in 2D space, we could represent a curve by one implicit or explicit equation. This is not possible in 3D space. As a simple example, in plane the equation

$$ax + by + c = 0$$

defines a straight line. In 3D space, however,

$$Ax + By + Cz + D = 0$$

represents a plane. To define a straight line we need two equations, for example the following explicit equations

$$y = -\frac{4}{3}x + 4$$

$$z = -\frac{5}{3.5}x + \frac{20}{3.5} \tag{4.14}$$

Fig. 4.5 explains the geometrical significance of Eqs (4.14). The first equation represents the plane Δ, perpendicular to the coordinates plane xOy. The latter plane was defined in Chapter 1 as the horizontal projection plane π_1. The second equation represents the plane Γ perpendicular to the frontal plane xOz (π_2.) The straight line defined by Eqs (4.14) is the intersection of the two planes. In the figure we see a segment, $\overline{P_1 P_2}$, of this line. To obtain the coordinates of the point $\mathbf{P_1}$ set $y = 0$, extract x

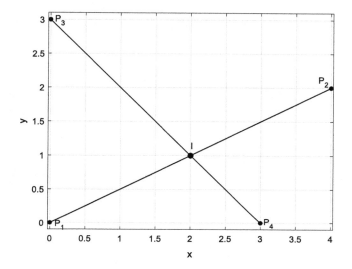

Figure 4.6 Intersecting lines defined by parametric equations

from the first equation and substitute it in the second one. To calculate the coordinates of the point $\mathbf{P_2}$ start by setting $z = 5$.

$$\mathbf{P_1} = \begin{vmatrix} 3 \\ 0 \\ 5/3.5 \end{vmatrix}, \; \mathbf{P_2} = \begin{vmatrix} 0.5 \\ 2 \\ 5 \end{vmatrix}. \tag{4.15}$$

Knowing two points of the line we are interested in, we can write the parametric equation

$$\mathbf{P} = \mathbf{P_1} + (\mathbf{P_2} - \mathbf{P_1})t = (1 - t)\mathbf{P_1} + t\mathbf{P_2} \tag{4.16}$$

This equation is based on the same reasoning as in Section 4.3. The vectorial equation stands for

$$\begin{vmatrix} P_x \\ P_y \\ P_z \end{vmatrix} = (1 - t) \begin{vmatrix} P_{1x} \\ P_{1y} \\ P_{1z} \end{vmatrix} + t \begin{vmatrix} P_{2x} \\ P_{2y} \\ P_{2z} \end{vmatrix} \tag{4.17}$$

4.4.2 Working With Parametric Equations

Fig. 4.6 shows two segments of straight lines defined by the points P_1, P_2, P_3, and P_4. We learned above how to write the parametric equations of the lines. We show now how to use the parametric equations to draw the lines and find the point of intersection, I. In this point the equations of the two lines should yield the same result, that is

$$(1 - t_1)P_1 + t_1 P_2 = (1 - t_2)P_3 + t_2 P_4$$

We rearrange this equation in matrix form

$$\left| \begin{array}{cc} (P_2 - P_1) & -(P_4 - P_3) \end{array} \right| \left| \begin{array}{c} t_1 \\ t_2 \end{array} \right| = |P_3 - P_1|$$

We plot the lines, solve the equation, and calculate the coordinates of the intersection point with the MATLAB commands

```
% define points
P1 = [ 0; 0 ]; P2 = [ 4; 2 ]; P3 = [ 0; 3 ]; P4 = [ 3; 0 ];
pline([ P1 P2 ], 1, 'k', '-'), grid
axis equal
xlabel('x', 'FontSize', 14);
ht = ylabel('y', 'FontSize', 14);
hold on
% show given points
point(P1, 0.03)
text((P1(1) + 0.07), -0.03, 'P_1')
point(P2, 0.03)
text((P2(1) - 0.18), (P2(2) + 0.1), 'P_2')
point(P3, 0.03)
text((P3(1) + 0.06), P3(2), 'P_3')
point(P4, 0.03)
text((P4(1) + 0.05), 0.0, 'P_4')
pline([ P3 P4 ], 1, 'k', '-')
% build matricial equations and solve system
A = [ (P2 - P1) -(P4 - P3) ];
B = P3 - P1;
T = A\B;
% calculate intersection point
t1 = T(1); t2 = T(2);
I1 = (1 - t1)*P1 + t1*P2;
I2 = (1 - t2)*P3 + t2*P4;
point(I1, 0.04)
ht = text(I1(1), (I1(2) + 0.15), 'I')
% print results
```

```
fid = fopen('IntersectLines.txt', 'w')
fprintf(fid, 't1 ..................... %5.3f \n', t1)
fprintf(fid, 'I calculated with t1 ... %5.3f %5.3f \n', I1(1), I1(2))
fprintf(fid, 't2 ..................... %5.3f \n', t2)
fprintf(fid, 'I calculated with t2 ... %5.3f %5.3f \n', I2(1), I2(2))
fclose(fid)
hold off
```

We used the functions `pline` and `point` explained in Biran (2011) and downloadable from the Mathworks site of Biran and López–Pulido (2014). The resulting printout is

```
t1 ..................... 0.500
I calculated with t1 ... 2.000 1.000
t2 ..................... 0.667
I calculated with t2 ... 2.000 1.000
```

4.4.3 The Helix

As a classical example of curve in three-dimensional space we return to the **helix** defined in Section 1.13 and remind that the helix is the curve described by a point that turns at a constant distance, r, from a straight-line axis and, at the same time, advances at constant speed in parallel to the given axis. The distance covered by the point in parallel with the axis of rotation, for one complete turn, is called *pitch* and we note it by p. The parametric equations corresponding to one complete turn around the z-axis are

$$x = r\cos 2\pi t$$
$$y = r\sin 2\pi t$$
$$z = pt, \quad t \in [0\ 1] \tag{4.18}$$

This arc of helix is shown in Fig. 4.7.

4.5 DERIVATIVES OF PARAMETRIC FUNCTIONS

In this chapter and the following we calculate various derivatives of functions related to curves in parametric representation. When we use the Leibniz notation for derivatives we accept as self-evident equalities like

$$\frac{dy}{dx} = \frac{1}{\frac{dx}{dy}}, \quad \frac{dy}{dx} = \frac{\frac{dy}{dt}}{\frac{dx}{dt}}, \quad \text{a.s.o.}$$

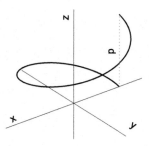

Figure 4.7 The helix

In this section we show that expressions like those above are not just a result of symbolic manipulations, but that they can be derived in a rigorous manner. We follow here examples from an excellent book by Piskunov (1965). We begin by considering the explicit representation

$$y = f(x) \tag{4.19}$$

and its *inverse* representation

$$x = g(y) = g(y(x)) \tag{4.20}$$

If $y(x)$ is a monotonically increasing or decreasing function its inverse is unique. Otherwise, y can have more than one inverse. If the derivative of y with respect to x is not zero, we can differentiate both sides of Eq. (4.20) with respect to x and write

$$1 = \frac{dg}{dy}\frac{df}{dx} = \frac{dx}{dy}\frac{dy}{dx} \tag{4.21}$$

We conclude that

$$\boxed{\frac{dy}{dx} = \frac{1}{\frac{dx}{dy}}} \tag{4.22}$$

Let a planar curve be represented by

$$x = f_1(t)$$
$$y = f_2(t), \quad t_0 \leq t \leq T \tag{4.23}$$

where $f_1(t)$ has an inverse, $t = g(x)$, and at least a first derivative. Then, treating y as a composite function, $y = f_2(g(x))$, we can write

$$\frac{dy}{dx} = \frac{dy}{dt}\frac{dt}{dx} \tag{4.24}$$

Applying Eq. (4.22) we rewrite Eq. (4.24) as

$$\boxed{\frac{dy}{dx} = \frac{\frac{dy}{dt}}{\frac{dx}{dt}}} \tag{4.25}$$

Let us find also the expression of the second derivative of y, with respect to x, in terms of derivatives with respect to the parameter t. We start with

$$\frac{d^2y}{dx^2} = \frac{d}{dx}\left(\frac{dy}{dx}\right) = \frac{d}{dt}\left(\frac{\frac{dy}{dt}}{\frac{dx}{dt}}\right)\frac{dt}{dx}$$

and continue with

$$\frac{d^2y}{dx^2} = \frac{\frac{dx}{dt}\frac{d^2y}{dt^2} - \frac{dy}{dt}\frac{d^2x}{dt^2}}{\left(\frac{dx}{dt}\right)^2}\frac{1}{\frac{dx}{dt}}$$

The final result is

$$\boxed{\frac{d^2y}{dx^2} = \frac{\frac{dx}{dt}\frac{d^2y}{dt^2} - \frac{dy}{dt}\frac{d^2x}{dt^2}}{\left(\frac{dx}{dt}\right)^3}} \tag{4.26}$$

4.6 NOTATION OF DERIVATIVES

We will continue to use the Leibniz notation for derivatives wherever necessary and practical, but, as a shorter notation, we are going to use the dot notation of Newton for derivatives with respect to a general parameter. Thus, $\dot{x}$ will stand for dx/dt and $\ddot{x}$ for d^2x/dt^2. For derivatives with respect to an arc-length parameter (see Section 4.9 below), or to a general independent variable, we are going to use Legendre's prime-mark notation. Thus, x' will stand for dx/ds, while y' may mean dy/dx and y'' may stand for d^2y/dx^2. With these conventions Eq. (4.26) becomes

$$\frac{d^2y}{dx^2} = \frac{\dot{x}\ddot{y} - \dot{y}\ddot{x}}{\dot{x}^3} \tag{4.27}$$

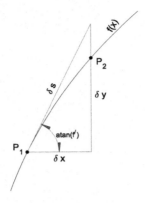

Figure 4.8 ArcLength

4.7 TANGENTS

For a curve defined by $x(t)$, $y(t)$, the *tangent* vector has the components $\dot{x}(t)$, $\dot{y}(t)$, and the *normal* vector, $-\dot{y}(t)$, $\dot{x}(t)$. The magnitude of both vectors, called also the *speed*, is $\sqrt{\dot{x}^2 + \dot{y}^2}$. The *unit* tangent and normal vectors are

$$\mathbf{t} = \left| \begin{array}{c} \frac{\dot{x}}{\sqrt{\dot{x}^2+\dot{y}^2}} \\ \frac{\dot{y}}{\sqrt{\dot{x}^2+\dot{y}^2}} \end{array} \right|, \quad \mathbf{n} = \left| \begin{array}{c} \frac{\dot{y}}{\sqrt{\dot{x}^2+\dot{y}^2}} \\ \frac{\dot{x}}{\sqrt{\dot{x}^2+\dot{y}^2}} \end{array} \right| \tag{4.28}$$

4.8 ARC LENGTH

In textbooks of calculus we can find rigorous treatments of the notion of **arc length**; they are based on Riemannian sums. We present here a shorter, more intuitive approach. In Fig. 4.8 we consider an arc of the curve $f(x)$ extending from P_1 to P_2. The difference of the x-coordinates is δx, and the difference of y-coordinates is δy. We write

$$\delta s = \sqrt{\delta x^2 + \delta y^2} = \sqrt{1 + (\delta y/\delta x)^2}\delta x$$

This is an approximation of the arc length between the two points. The expression is exact when δx approaches 0, hence

$$ds = \sqrt{1 + (dy/dx)^2}dx \tag{4.29}$$

It remains to integrate from the first to the second point. For a curve given in parametric form we rewrite Eq. (4.29) as

$$ds = \sqrt{\dot{x}^2 + \dot{y}^2}\, dt \tag{4.30}$$

and the integral is

$$s = \int_{t_1}^{t_2} \sqrt{\dot{x}^2 + \dot{y}^2}\, dt \tag{4.31}$$

The extension to 3D is obvious.

Example 4.4 (The length of a helix arc). To calculate the length of an arc of helix corresponding to one rotation (or, in other words, for one pitch) we begin with the derivatives

$$\frac{dx}{dt} = -2\pi r \sin 2\pi t$$

$$\frac{dy}{dt} = 2\pi r \cos 2\pi t$$

$$\frac{dz}{dt} = p \tag{4.32}$$

Then, the arc length corresponding to one pitch is

$$s = \int_0^1 \sqrt{4\pi^2 r^2 \sin^2 2\pi t + 4\pi^2 r^2 \cos^2 2\pi t + p^2} = \sqrt{4\pi^2 r^2 + p^2} \tag{4.33}$$

The above result can be obtained in a simpler way by applying Pitagora's theorem to the developed arc. We illustrated this approach in Chapter 1, Fig. 1.47.

4.9 ARC-LENGTH PARAMETRIZATION

As a simple example, let us consider again the circle defined by Eq. (4.4). The length of the arc corresponding to the angle $\theta = 2\pi t$ is $s = 2\pi r t$. Extracting t and substituting into the parametric equations we obtain

$$\left| \begin{array}{c} x \\ y \end{array} \right| = \left| \begin{array}{c} x_c \\ y_c \end{array} \right| + r \left| \begin{array}{c} \cos \frac{s}{r} \\ \sin \frac{s}{r} \end{array} \right| \tag{4.34}$$

The tangent vector in the point corresponding to $s = s_0$ is

$$\mathbf{t} = \begin{vmatrix} \frac{dx}{ds} \\ \frac{dy}{ds} \end{vmatrix}_{s=s_0} = r \begin{vmatrix} -\frac{1}{r} \sin \frac{s_0}{r} \\ \frac{1}{r} \cos \frac{s_0}{r} \end{vmatrix}$$

and its magnitude equals

$$\sqrt{\mathbf{t} \cdot \mathbf{t}} = r \sqrt{\frac{1}{r^2} \sin^2 \frac{s_0}{r} + \frac{1}{r^2} \cos^2 \frac{s_0}{r}} = 1$$

where the dot under the square-root symbol stays for the dot (or scalar) product.

An example of calculation in MATLAB is

```
r = 2; s = 1.5;
t = r*[ -sin(s/r)/r; cos(s/r)/r ]; % tangent vector
Vmagn = norm(t)
Vmagn =
     1
```

The tangent vector of a curve parametrized by arc-length is a unit vector.

Example 4.5 (Helix — Arc-length parametrization). We return to the helix in Example 4.4; its speed is constant and equal to

$$v = \sqrt{4\pi^2 r^2 + p^2} \tag{4.35}$$

The arc length as a function of the parameter t is

$$s(t) = \int_0^t v\, du = vt \tag{4.36}$$

After reparametrizing Eq. (4.18) and putting it in vectorial form we obtain

$$H = \begin{vmatrix} r \cos \frac{2\pi}{v} s \\ r \sin \frac{2\pi}{v} s \\ \frac{p}{v} s \end{vmatrix} \tag{4.37}$$

In Exercise 4.3 we ask the reader to prove that in this parametrization the speed is, indeed, equal to 1.

Example 4.6 (A mechanical example — Ballistics without air resistance). Galileo Galilei (Italian, 1564–1642) showed that the trajectory of a shell fired in a space void of air, that is without air resistance, is a parabola. The proof of this is easy and it provides a good example of parametric equations in kinematics. The parameter t is here exactly the time. Assuming that the initial velocity of the shell is V_0, and that it is fired at an angle α, the distances travelled along the horizontal and vertical axes x, z are

$$x = (V_0 \cos \alpha)t$$
$$z = (V_0 \sin \alpha)t - \frac{g}{2}t^2 \tag{4.38}$$

Eliminating the parameter t we obtain the explicit, second-degree equation of a parabola.

4.10 THE CURVE OF CENTRES OF BUOYANCY

4.10.1 Parametric Equations

In this section we derive the parametric equations of the curve of centres of buoyancy of a floating body that inclines freely around axes of inclination with a constant direction. As the mass of the ship does not change during heeling, the volume of displacement, ∇, is constant. For the set of positions that comply with this condition the French literature uses the term '*carènes isoclines*', the Italian literature the term '*carene isocline*'. We remember having learned the equations from Pierrotet (1942); they are attributed to Scribanti (Angelo, Italian, 1869–1926). Our treatment is based on Ilie (1974), but we present it in a simpler form and in our notation.

Fig. 4.9 shows a vertical section of a floating body. In upright condition the waterline is $W_0 L_0$ and the corresponding centre of buoyancy is B_0. We assume that the plane of the section contains B_0 and is perpendicular to the axes of inclination. As usual in books in Romanic languages, we call it *plane of inclination*. Further, we assume that all waterlines are tangent to a cylinder whose section is the curve C. For the sake of generality we do not start the derivation from the upright condition, but from an arbitrary waterline $W_1 L_1$ inclined by an angle ϕ with respect to the horizontal waterplane $W_0 L_0$, and tangent in P to the curve C. The corresponding centre of buoyancy is B_ϕ. Finally, we adopt a system of coordinates with the origin in B_0, the axes y and z in the shown plane, and the x-axis pointing towards us.

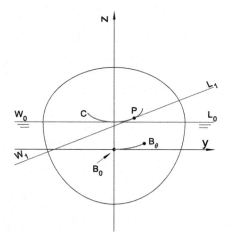

Figure 4.9 Floating body — Centres of buoyancy, 1

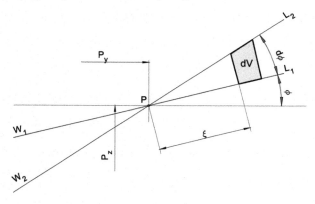

Figure 4.10 Floating body — Centres of buoyancy, 2

Fig. 4.10 is a 'zoom' of Fig. 4.9. The coordinates of the point P are P_y, P_z. Let the body heel from the angle ϕ by an infinitesimal angle $d\phi$, to a new waterline W_2L_2. Two wedges are formed between the waterlines W_1L_1 and W_2L_2. The right-hand wedge is a volume that submerges and adds a contribution to buoyancy, the left-hand wedge is a volume that emerges and its contribution to buoyancy is lost. The first moments of the wedges cause changes of the first moments of the floating body, hence a change in the position of the centre of buoyancy. To calculate the change of volume and its moments we consider an elemental volume dV with base dA situated at the distance ξ from the axis of inclination that passes through P, and at the distance x from the coordinate plane yOz. The height of the elemental volume is $\xi\,d\phi$ and the volume

$$\xi\,d\phi\,dA \tag{4.39}$$

The moment with respect to the xOy plane is

$$(P_z + \xi\sin\phi)\xi\,d\phi\,dA \tag{4.40}$$

The moment with respect to the yOz plane is

$$x\xi\,d\phi\,dA \tag{4.41}$$

The moment with respect to the xOz plane is

$$(P_y + \xi\cos\phi)\xi\,d\phi\,dA \tag{4.42}$$

Integrating over the whole waterplane area $W_\phi L_\phi$, from here on noted $WL(\phi)$, we obtain

$$dv = d\phi \int\int_{WL(\phi)} \xi\,dA$$
$$dM_{xOy} = P_z\,d\phi \int\int_{WL(\phi)} \xi\,dA + \sin\phi\,d\phi \int\int_{WL(\phi)} \xi^2\,dA$$
$$dM_{yOz} = d\phi \int\int_{WL(\phi)} x\xi\,dA$$
$$dM_{yOz} = P_y\,d\phi \int\int_{WL(\phi)} \xi\,dA + \cos\phi \int\int_{WL(\phi)} x^2\,dA \tag{4.43}$$

The integral in the first equation is the first moment of the waterplane $WL(\phi)$ with respect to the axis of inclination

$$\int\int_{WL(\phi)} \xi\,dA = M_x(\phi) \tag{4.44}$$

The second integral in the second equation represents the second moment of the waterplane $WL(\theta)$ with respect to the axis of inclination

$$\int\int_{WL(\phi)} \xi^2\,dA = I(\phi) \tag{4.45}$$

Finally, the integral in the third equation is the product of inertia of the area $WL(\theta)$ with respect to the axis of inclination and an axis perpendicular to it in the plane yOz

$$\int\int_{WL(\phi)} x\xi\,dA = I_{x\xi}(\phi) \qquad (4.46)$$

Integrating from 0 to ϕ we obtain

$$v = \int_0^\phi M_x d\phi \qquad (4.47)$$

We assumed that the heeling occurs at constant volume. Then, the net volume of the wedges should be zero for any angle ϕ. This means that the moment M_x should be zero for any angle ϕ and this happens when the axis of inclination is a centroidal axis of the waterplane. We integrate now Eqs (4.43) from 0 to ϕ, take into consideration that $M_X = 0$, and divide by the volume of displacement

$$x = \frac{1}{\nabla}\int_0^\phi I_{x\xi}(\phi)d\phi$$

$$y = \frac{1}{\nabla}\int_0^\phi I(\phi)\cos\phi\,d\phi$$

$$z = \frac{1}{\nabla}\int_0^\phi I(\phi)\sin\phi\,d\phi \qquad (4.48)$$

If the floating body is symmetric about the yOz plane, as ships are, $Ix\xi$ is zero for small angles of inclination and the curve starts in the yOz plane. As the inclination increases, the waterplane is usually no more symmetric and the centre of buoyancy leaves the yOz plane.

4.10.2 A Theorem on the Axis of Inclination

In the preceding section we proved that for small angles of inclination the axis of inclination is a centroidal axis of the waterplane. We conclude that *the cylinder with section C is the locus of all centroidal axes parallel to a given direction*. The curve C, the trace of the cylinder on the plane of inclination, is the locus of the projections on that plane of the centres of floatation of a floating body that inclines at constant volume around axes perpendicular to the same plane. We can state the important theorem

If a floating body is inclined by a small angle, at constant volume of displacement, the intersection of the new waterplane with the initial one is a centroidal axis of the initial waterplane.

4.10.3 The Tangent and the Normal to the B-Curve

We consider the projection of the curve of buoyancy on the plane of inclination and call it *B-curve*. Let us calculate its slope starting from Eqs (4.48)

$$\frac{dz}{dy} = \frac{\frac{dz}{d\phi}}{\frac{dy}{d\phi}} = \frac{I(\phi)\sin\phi}{I(\phi)\cos\phi} \cdot \frac{\nabla}{\nabla} = \tan\phi \qquad (4.49)$$

We conclude that *the tangent in a point of the B-curve is parallel to the corresponding waterplane.* It follows that, the force of buoyancy acts along the normal to the B-curve. The tangent vector has the components

$$\frac{I(\phi)}{\nabla}\cos\phi, \quad \frac{I(\phi)}{\nabla}\sin\phi \qquad (4.50)$$

and the magnitude $I(\phi)/\nabla$. As shown in Section 5.6, and in Biran and López–Pulido (2014), Section 2.8.2 this is the value of the *metacentric radius*, the normal to the B-curve has the components

$$-\frac{I(\phi)}{\nabla}\sin\phi, \quad \frac{I(\phi)}{\nabla}\cos\phi \qquad (4.51)$$

and the same magnitude as the tangent vector.

4.10.4 Parametric Equations for Small Angles of Inclination

For small angles of inclination we can assume that $I(\phi)$ and $I_{x\xi}$ are constant, $\cos\phi \approx 1$, and $\sin\phi \approx \phi$. Then Eqs (4.48) become

$$x = \frac{1}{\nabla}I_{x\xi}\phi$$
$$y = \frac{1}{\nabla}I\phi$$
$$z = \frac{1}{2\nabla}I\phi^2 \qquad (4.52)$$

The projection on the yOz plane is a parabola. As written above, as ship hulls are symmetric with respect to the centreplane, $I_{x\xi}$ is zero for small angles and the curve lies in the yOz plane.

In Eqs (4.48) we noted $I(\phi)$, $I_{x\xi}(\phi)$ because these quantities are functions of the angle of inclination. Let us suppose that we want to calculate points on the curve; we can do it only numerically. We work in steps of ϕ, sufficiently small to achieve the desired accuracy. At each step we calculate $I(\phi)$, $I_{x\xi}(\phi)$, and the position of the centroidal axis perpendicular to

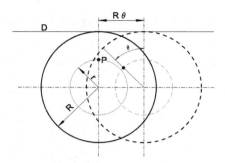

Figure 4.11 For the equations of the trochoid

the plane of inclination. This axis is the next axis of inclination. Preceding Scribanti's equations, this procedure was implemented in the method of Le Parmantier. Hervieu (1985) describes the method and mentions that it was used by the French Navy from 1910 until the introduction of digital computers. The indicated ϕ step is $10°$.

Example 4.7 (Trochoidal waves). According to the oldest theory of water waves, their surface is a curve called trochoidal. This theory was used in Naval Architecture during many decades. Classification societies required the calculation of longitudinal bending in trochoidal waves, and the regulations of the German Federal Navy specified that the stability of ships should be checked assuming trochoidal waves perpendicular to the vessel. Therefore, we treated the subject in the first edition of Biran (2003) and the first reprint, Biran (2005), Section 10.2.3. The theory used today in such calculations and in seakeeping is that of sinusoidal waves. The trochoid is the curve traced by a point on a circle that rolls without sliding under a straight line. We use Fig. 4.11 to derive the parametric equations of the curve. On a circle with radius R we consider a point P situated at radius r. As the circle rolls under the line D turning by an angle θ, it advances to the right a distance $R\theta$. At the same time the point P moves horizontally to the left by $r\sin\theta$. The resulting equations are

$$x = R\theta - r\sin\theta$$
$$y = r\cos\theta \qquad (4.53)$$

If the cycloid represents a wave of length λ and height H, we have

$$\lambda = 2\pi R$$
$$H = 2r$$

Then we rewrite Eqs (4.53) as

$$x = \frac{\lambda}{2\pi}\theta - \frac{H}{2}\sin\theta$$
$$y = \frac{H}{2}\cos\theta \qquad\qquad (4.54)$$

4.11 SUMMARY

A curve can be represented by an implicit equation, $f(x, y) = 0$, $f(x, y, z) = 0$, by an explicit equation, $y = f(x)$, $z = f(x, y)$, or by parametric equations, $x = f_1(t)$, $y = f_2(t)$, $z = f_3(t)$. Implicit equations are not suitable for computer graphics. Explicit equations can be used for drawing curves, but sometimes not in a convenient way. Parametric equations suit naturally the drawing of curves by computer software. The equations representing a curve can be seen as a vector function defined over the closed interval of the parameter. It is convenient to consider that the parametric equations are the components of a vector, and MATLAB can treat them as such, that is as a single vectorial equation. In continuation we refer to plane curves, but the extension to space is straightforward. It is only necessary to add a third element. Given two points

$$\mathbf{P_1} = \begin{vmatrix} x_1 \\ y_1 \end{vmatrix}, \quad \mathbf{P_2} = \begin{vmatrix} x_2 \\ y_2 \end{vmatrix}$$

the equation of the straight line that passes through them is

$$\mathbf{P} = (t - 1)\mathbf{P_1} + t\mathbf{P_2}$$

For points lying between $\mathbf{P_1}$ and $\mathbf{P_2}$ $t \in [0\ 1]$.

Our notation for derivatives with respect to the parameter t is

$$\frac{dx}{dt} = \dot{x}, \quad \frac{dy}{dt} = \dot{y}$$
$$\frac{d^2x}{dt^2} = \ddot{x}, \quad \frac{d^2y}{dt^2} = \ddot{y}$$

The first and the second derivatives with respect to x, in terms of the parameter t, are

$$\frac{dy}{dx} = \frac{\dot{y}}{\dot{x}}$$

$$\frac{d^2y}{dx^2} = \frac{\dot{x}\ddot{y} - \dot{y}\ddot{x}}{\dot{x}^3}$$

A plane curve represented by $x(t)$, $y(t)$, has the tangent and normal vectors

$$\mathbf{t} = \begin{vmatrix} \dot{x} \\ \dot{y} \end{vmatrix}, \quad \mathbf{n} = \begin{vmatrix} -\dot{y} \\ \dot{x} \end{vmatrix}$$

with magnitude $\sqrt{\dot{x}^2 + \dot{y}^2}$. This magnitude is also called the *speed* of the curve. The length of an arc of curve between $t = 0$ and $t = t_0$ is

$$\int_0^{t_0} = \sqrt{\dot{x}^2 + \dot{y}^2}\, dt$$

An example derived from mechanical principles, and of importance in Naval Architecture, are the parametric equations of the curve of centres of buoyancy of a floating body that inclines at constant displacement around axes parallel to a given direction,

$$x = \frac{1}{\nabla} \int_0^\phi I_{x\xi}(\phi)\,d\phi$$

$$y = \frac{1}{\nabla} \int_0^\phi I(\phi) \cos\phi\, d\phi$$

$$z = \frac{1}{\nabla} \int_0^\phi I(\phi) \sin\phi\, d\phi$$

We call *B-curve* the projection of the above curve on the plane of inclination. The tangent in a point of this curve is parallel to the corresponding waterplane.

4.12 EXERCISES

Exercise 4.1 (Implicit equations of a half-circle). **1)** Starting from Eqs (4.8) derive the implicit equation of a circle with centre in the origin of the coordinate axes and with radius 1.

2) Plot the curve described by Eqs (4.8).

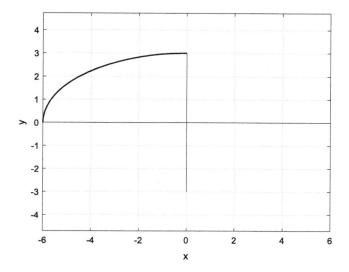

Figure 4.12 An arc of ellipse

Exercise 4.2 (Arc of ellipse). Write the parametric equations that define the arc of ellipse shown in Fig. 4.12 and indicate the corresponding range of the parameter.

Exercise 4.3 (The helix). In continuation of Examples 4.4 and 4.5 you are asked to prove the following statements:
1. the circle is a limit case of the helix;
2. in the arc–length parametrization, Eqs (4.37), the speed is equal to 1.

Exercise 4.4 (Ballistics in void). Referring to Example 4.6 find the components and the magnitude (here exactly the *speed*) of the velocity vector.

Exercise 4.5 (Trochoid compared to sine). Plot the trochoidal wave with $\lambda = 70$ m, $H = \lambda/20$ and the sine wave with the same length and amplitude. To compare, take care of the phase.

Exercise 4.6 (Parametric equations of hyperbola). Recover the canonic equation of the hyperbola from the parametric equations shown in Example 4.3 (see Section 1.16.5).

Exercise 4.7 (Left-handed helix). The helix in Fig. 4.7 is *right-handed*. This means that if the helix would represent a screw thread, and we would turn it in the counterclockwise sense, the screw would advance in the positive sense of the z-axis. With the same rotation a left-handed helix would advance in the negative sense of the z-axis. What should be set in Eqs (4.18) to represent a left-handed helix?

Exercise 4.8 (Tangent to the B-curve). Starting from Eqs (4.52) show that the tangent in a point of the B-curve for small angles of inclination is parallel to the corresponding waterplane.

CHAPTER 5

Curvature

Contents

5.1 INTRODUCTION

English	curvature	osculating circle
French	courbure	cercle osculateur
German	Krümmung	Krümmungskreis, Schmiegekreis
Italian	curvatura	cerchio osculatore

The term **curvature** suggests deviation from a straight line or a plane. The term **radius of curvature** suggests the radius of a circle that approximates a small arc of a given curve. In this chapter we give the mathematical definitions of these notions and show how to calculate their values for various

Geometry for Naval Architects
https://doi.org/10.1016/B978-0-08-100328-2.00015-8
Copyright © 2019 Elsevier Ltd.
All rights reserved.

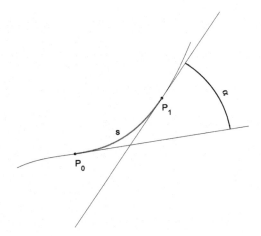

Figure 5.1 The definition of curvature

mathematical representations of curves. The notions of curvature and radius of curvature have important applications in the hydrostatics of floating bodies, computer graphics, dynamics and elasticity. A few applications are shown in this chapter, more in Chapters 6 and 7. The notion of *fair curve* introduced in Chapter 2 is related to curvature and work has been done on how to use this mathematical concept for more rigorous definitions and for *fairing* ship lines. Therefore, there are Naval Architectural programs that produce displays of the curvature of ship lines. We give an example in MultiSurf.

5.2 THE DEFINITION OF CURVATURE

The term **curvature** suggests deviation from a straight line. To define a measure of this deviation we use Fig. 5.1. On a given curve we consider an initial point, P_0, and another point, P_1. Let the arc length between the two points be s, and the angle between the tangents to the curve in the two points be α. This angle is a measure of the deviation from a straight line. Obviously, as the length of the arc increase, so may the deviation. Therefore, it makes sense to consider the ratio α/s; we may call it *mean curvature*. The **curvature** is defined as the limit of this ratio when P_1 tends to P_0, that is

$$\kappa = \lim_{s \to 0} \frac{\alpha}{s} = \frac{d\alpha}{ds} \qquad (5.1)$$

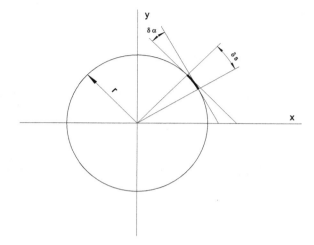

Figure 5.2 The curvature of the circle

We use the Greek letter *kappa* to note the curvature. The inverse of the curvature is the **radius of curvature** that we note by the Greek letter *rho*

$$\rho = \frac{1}{s} = \frac{ds}{d\alpha} \tag{5.2}$$

Let us calculate this radius for the circle shown in Fig. 5.2. We consider the tangents in two points separated by an arc of length δs. Let the angle between the two tangents be $\delta \alpha$. The tangent in a point of a circle is perpendicular to the radius corresponding to that point. Therefore, the angle subtended by the arc δs is equal to $\delta \alpha$ and we can write

$$\delta s = r \delta \alpha \tag{5.3}$$

Hence the radius of curvature of the circle is

$$\rho = \lim_{s \to 0} \frac{\delta s}{\delta \alpha} = \frac{ds}{d\alpha} = r \tag{5.4}$$

This is a justification for calling *radius of curvature* the inverse of the curvature. The generalization of this notion is developed in Section 5.3.

5.2.1 Curvature in Explicit Representation

We rewrite Eq. (5.2) as

$$\rho = \frac{ds}{d\alpha} = \frac{ds}{dx} / \frac{d\alpha}{dx} \tag{5.5}$$

Noting that $\alpha = \arctan y'$ we calculate

$$\frac{d\alpha}{dx} = \frac{d\alpha}{dy'} \cdot \frac{dy'}{dx} = \frac{y''}{1 + (y')^2} \tag{5.6}$$

From Section 4.8 we know that

$$\frac{ds}{dx} = \sqrt{1 + (y')^2} \tag{5.7}$$

Combining these results yields

$$\rho = \frac{(1 + (y')^2)^{3/2}}{y''} \tag{5.8}$$

5.2.2 Curvature in Parametric Representation

Starting from the definition of the radius of curvature we write

$$\rho = \frac{ds}{d\alpha} = \frac{ds}{dt} \bigg/ \frac{d\alpha}{dt} \tag{5.9}$$

From

$$\tan \alpha = \frac{\dot{y}}{\dot{x}} \tag{5.10}$$

we get

$$\alpha = \arctan \frac{\dot{y}}{\dot{x}} \tag{5.11}$$

Differentiating with respect to t yields

$$\frac{d\alpha}{dt} = \frac{1}{1 + (\dot{y}/\dot{x})^2} \frac{\dot{x}\ddot{y} - \dot{y}\ddot{x}}{\dot{x}^2} = \frac{\dot{x}\ddot{y} - \dot{y}\ddot{x}}{\dot{x}^2 + \dot{y}^2} \tag{5.12}$$

We also have

$$\frac{ds}{dt} = (\dot{x}^2 + \dot{y}^2)^{1/2} \tag{5.13}$$

In conclusion

$$\rho = \frac{ds}{dt} \bigg/ \frac{d\alpha}{dt} = (\dot{x}^2 + \dot{y}^2)^{1/2} \frac{\dot{x}^2 + \dot{y}^2}{\dot{x}\ddot{y} - \dot{y}\ddot{x}} = \frac{(\dot{x}^2 + \dot{y}^2)^{3/2}}{\dot{x}\ddot{y} - \dot{y}\ddot{x}} \tag{5.14}$$

5.3 OSCULATING CIRCLE

5.3.1 Definition

The following definitions are equivalent.

Definition 1   Given a curve C and three points, P_1, P, and P_2 lying in this order on C, we consider the circle passing through these points. When the points P_1 and P_2 tend to P, the circle tends to what we call the **osculating circle** in the point P. The radius of the osculating circle is the *radius of curvature* of C in P, and the centre of this circle is the *centre of curvature* of C in P.

Definition 2   Given a curve C and two points, P, P_1, on C, we consider the intersection, C_c, of the normals to C in P and P_1. When P_1 tends to P, the intersection tends to the centre of the osculating circle in P. The distance between P and C_c is the *radius of curvature* of C in P, and C_c becomes the *centre of curvature* of C in P.

Definition 3   Given a curve C and a point P on it, the *osculating circle* in P is the circle tangent to C, in P, that best approximates the curve in that point.

The notion of osculating circle is due to Leibniz (Gottfried Wilhelm, German, 1646–1716) who coined the Latin term *circulus osculans*, i.e. *kissing circle*.

5.3.2 Definition 1 detailed

This is the most natural definition. However, putting it into equations and deriving results involves a lot of tedious algebra. We start by defining three points

$$P_1 = \begin{vmatrix} x - h \\ f(h - h) \end{vmatrix} \tag{5.15}$$

$$P_0 = \begin{vmatrix} x \\ f(h) \end{vmatrix} \tag{5.16}$$

$$P_2 = \begin{vmatrix} x + h \\ f(h + h) \end{vmatrix} \tag{5.17}$$

Next, we write that the circle of radius r, and centre at (x_c, y_c), passes through these three points

$$(x - h - x_c)^2 + (f(x - h) - y_c)^2 = r^2 \tag{5.18}$$

$$(x - x_c)^2 + (f(x) - y_c)^2 = r^2$$
$$(x + h - x_c)^2 + (f(x + h) - y_c)^2 = r^2$$

In continuation we must solve the system, rearrange the results in a suitable form and let $h \to 0$.

5.3.3 Definition 2 Detailed

This definition reminds that of the metacentre (see Biran and López–Pulido, 2014, Section 2.5):

*'Let us assume that from an initial angle of heel, ϕ_0, the ship heels by an additional small angle $\delta\phi$. We consider the lines of action of the buoyancy force in the two positions. When $\delta\phi \to 0$, the intersection of the two lines tends to a point we call **metacentre**.'*

With this definition we also state that the metacentre is the centre of curvature of the *curve of centres of buoyancy*, i.e. the *B-curve*.

The mathematical approach is based on the fact that the direction of the normal in a point P of the curve C coincides with the direction of the normal to any circle tangent in that point to the curve C (Fornero, 2011). Let the coordinates of the point P be $(x_0, f(x_0))$, and those of the point P_1, $(x_0 + h)$, $f(x_0 + h)$. The slope of the curve at P is $f'(x_0)$, and at P_1 is $f'(x_0 + h)$. The slopes of the normals at those points are obtained by taking the inverses of the above-mentioned slopes and changing their signs. Then, the equations of the normals to C in the two points are

$$y - f(x_0) = -\frac{1}{f'(x_0)}(x - x_0) \tag{5.19}$$

$$y - f(x_0 + h) = -\frac{1}{f'(x_0 + h)}(x - (x_0 + h))$$

where

$$f'(x_0) = \left.\frac{df}{dx}\right|_{x=x_0} , \, f'(x_0 + h) = \left.\frac{df}{dx}\right|_{x=x_0+h}$$

We rearrange the system as

$$x + f'(x_0)y = f'(x_0)f(x_0) + x_0 \tag{5.20}$$

$$x + f'(x_0 + h)y = f'(x_0 + h)f(x_0 + h) + (x_0 + h)$$

The x-coordinate of the intersection of the two normals is given by

$$x_c = \frac{\begin{vmatrix} f'(x_0)f(x_0) + x_0 & f'(x_0) \\ f'(x_0 + h)f(x_0 + h) + (x_0 + h) & f'(x_0 + h) \end{vmatrix}}{\begin{vmatrix} 1 & f'(x_0) \\ 1 & f'(x_0 + h) \end{vmatrix}} \tag{5.21}$$

Expanding the determinants yields

$$x_c = \frac{f'(x_0)f'(x_0 + h)(f(x_0) - f(x_0 + h)) + f'(x_0 + h)x_0 - f'(x_0)(x_0 + h)}{f'(x_0 + h) - f'(x_0)} \tag{5.22}$$

We divide both nominator and denominator by h and calculate their limits as $P_1 \to P$. For the denominator we obtain

$$\lim_{h \to 0} \frac{f'(x_0 + h) - f'(x_0)}{h} = f'' x_0$$

For the nominator we look for the limit of

$$-f'(x_0)f'(x_0 + h)\frac{f(x_0 + h) - f(x_0)}{h}$$
$$+ \frac{f'(x_0 + h)x_0 - f'(x_0)x_0}{h} - \frac{f'(x_0)h}{h}$$

which yields

$$-(f'(x_0))^3 + f''(x_0)x_0 - f'(x_0) = -f'(x_0)(1 + (f'(x_0))^2) + f''(x_0)$$

Summing up

$$\lim_{h \to 0} x_c = \frac{f''(x_0)x_0 - f'(x_0)(1 + (f'(x_0))^2)}{f''(x_0)} = x_0 - \dot{f}'(x_0)(1 + (f'(x_0))^2)f''(x_0) \tag{5.23}$$

Substituting this value for x in the first of Eqs (5.20) we obtain

$$y_c = f(x_0) + \frac{(1 + f'(x_0))^2}{f''(x_0)} \tag{5.24}$$

5.3.4 Definition 3 Detailed

Let us assume that a given plane curve, C, is described by the explicit function $f(x)$. By *'the circle that best approximates the curve C in the point corresponding to x_0'* we understand the circle described by $g(x)$ that satisfies the following conditions:

$$g(x_0) = f(x_0),\ g'(x_0) = f'(x_0),\ g''(x_0) = f''(x_0) \tag{5.25}$$

We remind the reader that by $f'(x_0)$ we mean in this section the derivative of f with respect to x at the point x_o, a.s.o. Noting with x_c, y_c the coordinates of the centre of the circle $g(x)$, and with r its radius, we can write the equation

$$(x - x_c)^2 + (g(x) - y_c)^2 = r^2 \tag{5.26}$$

Differentiating this equation with respect to x, and dividing both sides of the equation by 2, we obtain

$$(x - x_c) + (g(x) - y_c)g'(x) = 0 \tag{5.27}$$

A second differentiation with respect to x yields

$$1 + (g'(x))^2 + (g(x) - y_c)g''(x) = 0 \tag{5.28}$$

Substituting the values given by Eq. (5.25) into Eqs (5.27) and (5.28) we get

$$(x_0 - x_c) + (f(x_0) - y_c)f'(x_0) = 0 \tag{5.29}$$
$$1 + (f'(x_0))^2 + (f(x_0) - y_c)f''(x_0) = 0 \tag{5.30}$$

From Eq. (5.30) we obtain

$$y_c = f(x_0) + \frac{1 + (f'(x_0))^2}{f''(x_0)} \tag{5.31}$$

Substituting Eq. (5.31) into Eq. (5.29) yields

$$x_c = x_0 + f(x_0) - f(x_0) - \left(\frac{1 + (f'(x_0))^2}{f''(x_0)}\right)f'(x_0)$$
$$= x_0 - \frac{1 + (f'(x_0))^2}{f''(x_0)}f'(x_0) \tag{5.32}$$

Substituting the coordinates of the centre into Eq. (5.26) lets us find the radius of the osculating circle

$$r^2 = (x_0 - x(c))^2 + (f(x_0) - y_c)^2 \tag{5.33}$$

$$= \left(\frac{1 + (f'(x_0))^2}{f''(x_0)} f'(x_0)\right)^2 + \left(\frac{1 + (f'(x_0))^2}{f''(x_0)}\right)^2 \tag{5.34}$$

$$= \left(\frac{1 + (f'(x_0))^2}{f''(x_0)}\right)^2 (1 + f'(x_0))^2) \tag{5.35}$$

hence

$$r = \left|\frac{(1 + (f'(x_0))^2)^{3/2}}{f''(x_0)}\right| \tag{5.36}$$

This is the radius of curvature that we have already found in Eq. (5.8). Using the absolute value we avoid the necessity to give a sign to the radius of curvature.

5.3.5 Centre of Curvature in Parametric Representation

Substituting Eqs (4.24) and (4.27) into Eqs (5.32) and (5.31) we obtain

$$x_c = x_0 - \frac{\dot{y}(\dot{x}^2 + \dot{y}^2)}{\dot{x}\ddot{y} - \ddot{x}\dot{y}}$$

$$y_c = y_0 + \frac{\dot{x}(\dot{x}^2 + \dot{y}^2)}{\dot{x}\ddot{y} - \ddot{x}\dot{y}} \tag{5.37}$$

Obviously, the derivatives should be calculated at the point with coordinates x_0, y_0.

Example 5.1 (The curvature of the ellipse). In this example we consider the ellipse with parametric equations

$$x = a\cos t \tag{5.38}$$

$$y = b\sin t \tag{5.39}$$

The radius of curvature is given by

$$\rho = \frac{(\dot{x}^2 + \dot{y}^2)^{3/2}}{\dot{x}\ddot{y} - \ddot{x}\dot{y}} \tag{5.40}$$

$$= \frac{(a^2 \sin^2 t + b^2 \cos^2 t)^{3/2}}{ab\sin^2 t + ab\cos^2 t} = \frac{(a^2 \sin^2 t + b^2 \cos^2 t)^{3/2}}{ab} \tag{5.41}$$

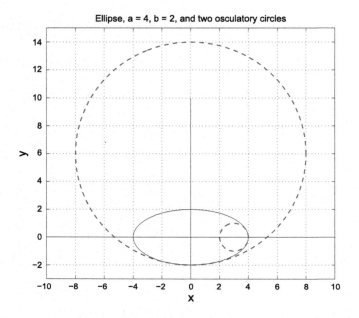

Figure 5.3 An ellipse and two of its osculating circles

For $t = 0$ we obtain

$$\rho_1 = \frac{b^3}{ab} = \frac{b^2}{a} \tag{5.42}$$

For $t = 3\pi/2$ we obtain

$$\rho_2 = \frac{a^3}{ab} = \frac{a^2}{b} \tag{5.43}$$

Fig. 5.3 shows the ellipse with axes $x = 4$, $b = 2$ and the osculating circles calculated above. As we can see, in the chosen points the ellipse is well approximated by the corresponding osculating circles, a fact that justifies a popular geometric construction that approximates the ellipse by four arcs of circle (see, for example, Thuillier et al., 1991, p. 258).

Example 5.2 (The curvature of the sine curve). The curvature of $y = \sin x$ is

$$\kappa = \frac{\frac{d^2 y}{dx^2}}{\left(1 + \left(\frac{dy}{dx}\right)^2\right)^{3/2}} = \frac{-\sin x}{(1 + \cos^2 x)^{3/2}} \tag{5.44}$$

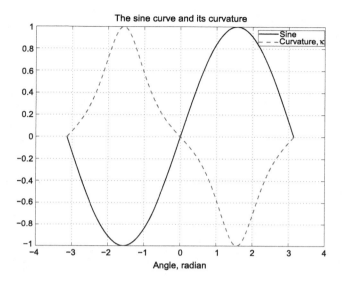

Figure 5.4 The curvature of the sine curve

In Fig. 5.4 we see the graph of the curvature plotted over that of the sine function. In the origin the curvature is zero. Therefore, in the neighbourhood of the origin the sine is well approximated by the tangent to the curve. The curvature reaches maximum values for $x = -\pi/2$ and $x = \pi/2$. As can be seen in Fig. 5.5, in the neighbourhood of these points the sine is well approximated by the corresponding osculating circles. The radii of the latter are calculated with Eq. (5.8).

5.4 AN APPLICATION IN KINEMATICS — THE CENTRIFUGAL ACCELERATION

5.4.1 Position

In this section we consider a point that moves along a plane curve and derive the equations giving its velocity and acceleration. We obtain a general expression for the normal acceleration on a curved trajectory, rather than the traditional, well-known expression for a point moving on a circular trajectory. We assume that the trajectory of the point is parametrized by arc length and the *position* vector is

$$\mathbf{R} = \begin{vmatrix} x(s) \\ y(s) \end{vmatrix}$$

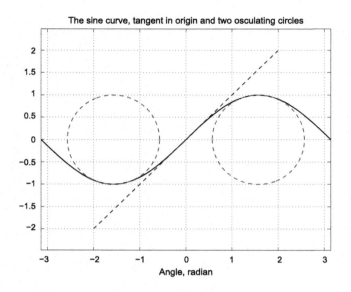

Figure 5.5 The sine curve — Tangent in origin and two osculating circles

We note the time by t and assume that s is a function of t.

5.4.2 Velocity

The *velocity* of the point is the vector

$$\mathbf{V} = \frac{d\mathbf{R}}{dt} = \frac{d\mathbf{R}}{ds} \cdot \frac{ds}{dt}$$

As the curve is parametrized by arc length, $d\mathbf{R}/ds$ is a unit vector that we note by $\mathbf{T}$. The derivative ds/dt is the *speed* that we note by v, and the velocity vector can be written as

$$\mathbf{V} = v\mathbf{T} \tag{5.45}$$

In terms of the parameter t the unit tangent vector is

$$\mathbf{T} = \frac{1}{\sqrt{\dot{x}^2 + \dot{y}^2}} \begin{vmatrix} \dot{x} \\ \dot{y} \end{vmatrix} \tag{5.46}$$

and the unit *normal* vector

$$\mathbf{N} = \frac{1}{\sqrt{\dot{x}^2 + \dot{y}^2}} \begin{vmatrix} -\dot{y} \\ \dot{x} \end{vmatrix} \tag{5.47}$$

5.4.3 Acceleration

The *acceleration* vector is defined by

$$\begin{aligned}
\frac{d^2\mathbf{R}}{dt^2} = \frac{d\mathbf{V}}{dt} &= \frac{d}{dt}v\mathbf{T} \\
&= \frac{dv}{dt}\mathbf{T} + v\frac{d\mathbf{T}}{dt} = \frac{dv}{dt}\mathbf{T} + v\frac{d\mathbf{T}}{ds}\cdot\frac{ds}{dt} \\
&= \frac{dv}{dt}\mathbf{T} + v^2\frac{d\mathbf{T}}{ds}
\end{aligned} \tag{5.48}$$

As $\mathbf{T}$ is a unit vector, its magnitude is $\sqrt{\mathbf{T}\cdot\mathbf{T}} = 1$, equivalent to $\mathbf{T}\cdot\mathbf{T} = 1$. Differentiating this scalar product with respect to s we get

$$2\mathbf{T}\cdot\frac{d\mathbf{T}}{ds} = 0$$

We conclude that the vector $d\mathbf{T}/ds$ is perpendicular to the vector $\mathbf{T}$, or, in other words, lies along the normal unit vector $\mathbf{N}$. Then, we can write

$$\frac{d\mathbf{T}}{ds} = c\mathbf{N} \tag{5.49}$$

and, following an idea of Walser (2003), we are going to show that the factor c is the curvature $\kappa = 1/\rho$. We start with

$$\frac{d\mathbf{T}}{ds} = \frac{d\mathbf{T}}{dt}\frac{dt}{ds}$$

From (4.29) we know that

$$\frac{dt}{ds} = \frac{1}{(\dot{x}^2 + \dot{y}^2)^{1/2}}$$

We begin with the component of $\mathbf{T}$ that is parallel to the x-axis in Eq. (5.46) and noting it $\mathbf{T}_1$

$$\begin{aligned}
\frac{d\mathbf{T}_1}{ds} &= \frac{1}{(\dot{x}^2 + \dot{y}^2)^{1/2}}\frac{d}{dt}\frac{\dot{x}}{(\dot{x}^2 + \dot{y}^2)^{1/2}} \\
&= \frac{1}{(\dot{x}^2 + \dot{y}^2)^{1/2}}\frac{\ddot{x}(\dot{x}^2 + \dot{y}^2)^{1/2} - \dot{x}\frac{2(\dot{x}\ddot{x}+\dot{y}\ddot{y})}{2(\dot{x}^2+\dot{y}^2)^{1/2}}}{\dot{x}^2 + \dot{y}^2} \\
&= \frac{1}{(\dot{x}^2 + \dot{y}^2)^{1/2}}\frac{1}{(\dot{x}^2 + \dot{y}^2)^{3/2}}[\ddot{x}(\dot{x}^2 + \dot{y}^2) - \dot{x}(\dot{x}\ddot{x} + \dot{y}\ddot{y})]
\end{aligned} \tag{5.50}$$

After simplifying the expression between square parentheses we have

$$\frac{d\mathbf{T}_1}{ds} = \frac{\dot{x}\ddot{y} - \ddot{x}\dot{y}}{(\dot{x}^2 + \dot{y}^2)^{3/2}} \cdot \frac{-\dot{y}}{(\dot{x}^2 + \dot{y}^2)^{1/2}} \tag{5.51}$$

The first factor is the curvature, $\kappa = 1/\rho$, as given by Eq. (5.14). The second factor is the first component of the unit normal vector, as seen in Eq. (5.47). After a similar treatment of the component of $\mathbf{T}$ that is parallel to the y-axis in Eq. (5.46) and noting it $\mathbf{T}_2$ we obtain

$$\frac{d\mathbf{T}_2}{ds} = \frac{\dot{x}\ddot{y} - \ddot{x}\dot{y}}{(\dot{x}^2 + \dot{y}^2)^{3/2}} \cdot \frac{\dot{x}}{(\dot{x}^2 + \dot{y}^2)^{1/2}} \tag{5.52}$$

The first factor is again the curvature, the second factor is the second component of the unit normal vector. Summing up, the acceleration is

$$\mathbf{A} = \frac{dv}{dt}\mathbf{T} + \frac{v^2}{\rho}\mathbf{N} \tag{5.53}$$

If the trajectory is a circle with radius r Eq. (5.53) reduces to the well-known expression

$$\mathbf{A} = \frac{dv}{dt}\mathbf{T} + \frac{v^2}{r}\mathbf{N} \tag{5.54}$$

5.5 ANOTHER APPLICATION IN MECHANICS — THE ELASTIC LINE

In this example we derive the differential equation of the *elastic line*, that is the shape of a beam deflected in bending. We have here an application of the notion of radius of curvature. The result has been used in a traditional method for calculating the deflection of ships under a bending moment.

Fig. 5.6 shows schematically a segment of a beam deflected in bending. In the absence of a bending moment or any other loading, the points A and B belong to the same transversal section of the beam. Similarly, the points C and D belong to another transversal section of the beam. Before deflection the distance between the points A and C was equal to that between the points B and D. In *simple beam theory* it is assumed that plane, transversal sections remain plane after the application of a bending moment. After bending the upper part of the beam is in tension and the distance between the points A and C increases. The lower part of the beam is compressed and the distance between the points B and D is shortened. Somewhere within

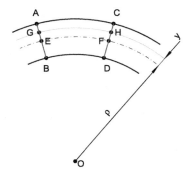

Figure 5.6 For the derivation of the equation of the elastic line

the beam there is a longitudinal line that keeps its length. For a ship this is the *neutral axis* defined in Example 3.12. As the transversal sections remain plane and normal to the beam, the continuations of their lines intersect in the centre of curvature of the curve assumed by the beam. This centre is marked O in the figure, the radius of curvature is ρ, and let the angle between the normals be θ. By definition, the *strain* is

$$\epsilon = \frac{(\rho + y)\theta - \rho\theta}{\rho\theta} = \frac{y}{\rho} \tag{5.55}$$

On the other hand we have

$$\epsilon = \frac{\sigma}{E}$$

where σ is the *stress* and E the *Young modulus* known also as *modulus of elasticity*. From beam theory we know that

$$\sigma = \frac{My}{I}$$

where M is the *bending moment*, y the distance of the measurement point from the neutral axis, and I the *moment of inertia* of the transversal section under consideration. Combining the above equations we obtain

$$\frac{1}{\rho} = \frac{M}{EI}$$

If the concavity is such that the curve turns to the right, as in our figure, it is usual to consider the curvature as negative and we write

$$\frac{1}{\rho} = -\frac{d^2y/dx^2}{(1 + (dy/dx)^2)^{3/2}} = \frac{M}{EI} \tag{5.56}$$

Assuming that the slope of the elastic line is small, we rewrite the above equation as

$$\frac{d^2 y}{dx^2} = -\frac{M}{EI} \tag{5.57}$$

5.6 AN APPLICATION IN NAVAL ARCHITECTURE — THE METACENTRIC RADIUS

From Eqs (4.48) we choose those that give the projection on the plane of inclination. We called this projection *B-curve* and its parametric equations are

$$y = \frac{1}{\nabla} \int_0^\phi I(\phi) \cos \phi \, d\phi$$

$$z = \frac{1}{\nabla} \int_0^\phi I(\phi) \sin \phi \, d\phi$$

To calculate the radius of curvature with Eq. (5.14) we need the derivatives

$$\dot{y} = \frac{I(\phi)}{\nabla} \cos \phi$$

$$\dot{z} = \frac{I(\phi)}{\nabla} \sin \phi$$

$$\ddot{y} = \frac{1}{\nabla} \left[\frac{dI(\phi)}{d\phi} \cos \phi - I(\phi) \sin \phi \right]$$

$$\ddot{z} = \frac{1}{\nabla} \left[\frac{dI(\phi)}{d\phi} \sin \phi - I(\phi) \cos \phi \right]$$

The radius of curvature, known in this case as *metacentric radius*, is given by

$$\overline{BM} = \frac{(\dot{y}^2 + \dot{z}^2)^{2/3}}{\dot{y}\ddot{z} - \dot{z}\ddot{y}}$$

Substituting the derivatives into the numerator we obtain

$$\left(\frac{I(\phi)}{\nabla} \right)^3 (\cos^2 \phi + \sin^2 \phi)^{2/3} = \left(\frac{I(\phi)}{\nabla} \right)^3$$

Substituting the derivatives into the denominator yields

$$\frac{I(\phi)}{\nabla^2}\left[\cos\phi\left(\frac{dI(\phi)}{d\phi}\sin\phi + I(\phi)\cos\phi\right)\right.$$
$$\left. - \sin\phi\left(\frac{dI(\phi)}{d\phi}\cos\phi - I(\phi)\sin\phi\right)\right] = \left(\frac{I(\phi)}{\nabla}\right)^2$$

Dividing the numerator by the denominator we get

$$\overline{BM} = \frac{I(\phi)}{\nabla}$$

This is the result obtained in a different way in Biran and López-Pulido (2014), Section 2.8.2.

5.7 DIFFERENTIAL METACENTRIC RADIUS

In Fig. 4.9 the waterlines for various heel angles are tangent to the curve C. As we are going to learn in Section 5.11, the curve C is the *envelope* of the waterlines. The radius of curvature of C is given by

$$\rho_C = \frac{dI}{d\nabla}$$

and is called *differential metacentric radius*. Here, the moment of inertia of the waterplane with respect to the axis of inclination, I, and the volume of displacement, ∇, are considered functions of the draught, T. We remind this notion only as an example of radius of curvature in Naval Architecture. The reader interested in proofs of the formula can find them in Semyonov-Tyan-Shansky (2004), pp. 99–102, or Hervieu (1985), pp. 120–124. As the centre of floatation, F, lies on the cylinder with trace C, the curve is often noted by F.

5.8 CURVES IN SPACE

We consider now a curve in space and assume that it is parametrized by arc length

$$\mathbf{R} = \begin{vmatrix} x(s) \\ y(s) \\ z(s) \end{vmatrix}$$

The *tangent* vector at a given point is defined by

$$\mathbf{T} = \frac{d\mathbf{R}}{ds} \tag{5.58}$$

Reasoning like in Section 5.4 we can write

$$\frac{d\mathbf{T}}{ds} = \kappa \mathbf{N} \tag{5.59}$$

where $\mathbf{N}$ is the unit vector normal to the curve. While at a given point there is only one tangent to the curve, at the same point there are infinitely many normal vectors. However, the vector defined by Eq. (5.59) is unique and we call it *principal normal vector*. The vectors $\mathbf{T}$ and $\mathbf{N}$ define the *osculating plane* at the given point. The vector normal to this plane at the given point is called *binormal* and is given by the cross-product

$$\mathbf{B} = \mathbf{T} \times \mathbf{N} \tag{5.60}$$

We are going to show that the vector $d\mathbf{B}/ds$ is parallel to the principal normal, $\mathbf{N}$. As defined, the three vectors $\mathbf{T}$, $\mathbf{N}$ and $\mathbf{B}$ are unit vectors, each one perpendicular to the other two, so that

$$\mathbf{B} \cdot \mathbf{T} = 0$$

Differentiating this relationship we obtain

$$\mathbf{B}' \cdot \mathbf{T} + \mathbf{B} \cdot \mathbf{T}' = \mathbf{B}' \cdot \mathbf{T} + \mathbf{B} \cdot \kappa \mathbf{N} = 0$$

Being perpendicular one to another, $\mathbf{B} \cdot \mathbf{N} = 0$ and then $\mathbf{B}' \cdot \mathbf{T} = 0$, meaning that $\mathbf{B}'$ is perpendicular to $\mathbf{T}$. We have also $\mathbf{B} \cdot \mathbf{B} = 1$, which yields $\mathbf{B}' \cdot \mathbf{B} = 0$. Thus $\mathbf{B}'$ is perpendicular also to $\mathbf{B}$ and we conclude that it is parallel to $\mathbf{N}$. Therefore, we can write

$$\mathbf{B}' = -\tau \mathbf{N} \tag{5.61}$$

We call τ the *torsion* of the curve. While the curvature κ is a measure of the deviation of the curve from a straight line, the torsion τ is a measure of the twisting of the curve out of the osculating plane. The plane defined by $\mathbf{N}$ and $\mathbf{B}$ is called *normal plane*, and the one defined by $\mathbf{T}$ and $\mathbf{B}$ *rectifying plane*.

5.9 EVOLUTES

English	evolute	involute
French	développée	développante
German	Evolute	Evolvente
Italian	evoluta	evolventa
Spanish	evoluta	evolvente

Given a curve, C, the locus of its centres of curvature is called **evolute** of C. Let us note the evolute by E_C. Then, C is called the **involute** of E_C. A relevant example for this book is the relationship between the curve of centres of buoyancy, B, and the locus of the corresponding metacentres, M. As shown, for example, in Biran and López-Pulido (2014), Section 2.10, the B-curve is the involute of the curve M, and M the **metacentric evolute**. In the next section we are going to show another important relationship between a curve and its evolute. Given the equation of the curve as $f(x_c)$, the equation of its evolute is obtained from Eqs (5.31) and (5.32) if for x_0 and y_0 we substitute x_c and $f(x_c)$:

$$x = x_c - \frac{1 + (f'(x_c))^2}{f''(x_c)} f'(x_c)$$

$$y = f(x_c) + \frac{1 + (f'(x_c))^2}{f''(x_c)}$$

or, in parametric form

$$x = x_c - \frac{\dot{y}(\dot{x}^2 + \dot{y}^2)}{\dot{x}\ddot{y} - \ddot{x}\dot{y}}$$

$$y = y_c + \frac{\dot{x}(\dot{x}^2 + \dot{y}^2)}{\dot{x}\ddot{y} - \ddot{x}\dot{y}} \tag{5.62}$$

As an example let us calculate and plot the evolute of an ellipse with parametric equations

$$x_e = a\cos\theta$$

$$y_e = b\sin\theta, \quad \theta = [0\ 2\pi]$$

The first derivatives are

$$\dot{x} = -a\sin\theta$$

$$\dot{y} = b\cos\theta$$

and the second derivatives

$$\ddot{x} = -a\cos\theta$$

$$\ddot{y} = -b\sin\theta$$

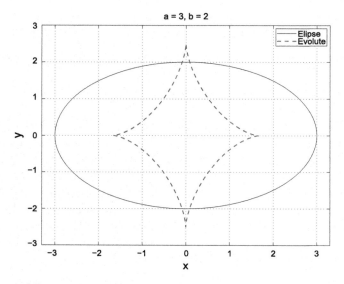

Figure 5.7 An ellipse and its evolute

Then, the equations of the evolute are

$$x = x_e + \frac{\cos\theta(a^2\sin^2\theta + b^2\cos^2\theta)}{a}$$

$$y = y_e + \frac{\sin\theta(a^2\sin^2\theta + b^2\cos^2\theta)}{b} \tag{5.63}$$

Fig. 5.7 shows the ellipse with $a = 3$, $b = 2$, and its evolute. The evolute of the ellipse is called **astroid**. Fig. 5.8 shows several normals of the ellipse in the third quadrant. It is easy to see that the normals of the ellipse are tangent to the evolute. This evolute has four *cusps*, that is points in which two branches with curvature of different sign join. At this point the tangents of the two branches lie on the same straight line, but change orientation. Mind the difference from an inflection point.

5.10 A LEMMA ON THE NORMAL TO A CURVE IN IMPLICIT FORM

We consider the implicit representation of a curve

$$f(x, y) = 0 \tag{5.64}$$

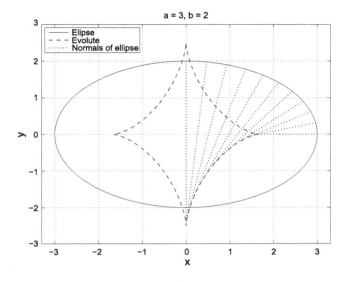

Figure 5.8 An ellipse, its normals and its evolute

where y can be represented as a function $\phi(x)$. Differentiating Eq. (5.64) with respect to x yields

$$\frac{\partial f}{\partial x} + \frac{\partial f}{\partial y} \cdot \frac{d\phi}{dt} = 0 \tag{5.65}$$

This is the scalar product of the vectors

$$\mathbf{T} = \begin{bmatrix} 1 \\ \frac{d\phi}{dx} \end{bmatrix}, \ \mathbf{N} = \begin{bmatrix} \frac{\partial f}{\partial x} \\ \frac{\partial f}{\partial y} \end{bmatrix} \tag{5.66}$$

$\mathbf{T}$ is the tangent vector. Then, Eq. (5.65) simply says that $\mathbf{N}$ is normal to the curve f and has the components indicated above. As an example let us consider the following equation of a straight line

$$f = ax - y + b = 0$$

The partial derivatives are

$$\frac{\partial f}{\partial x} = a, \ \frac{\partial f}{\partial y} = -1$$

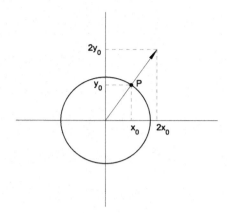

Figure 5.9 A normal of circle

The normal vector is

$$\mathbf{N} = \begin{bmatrix} a \\ -1 \end{bmatrix}$$

obviously perpendicular to the tangent vector

$$\mathbf{T} = \begin{bmatrix} 1 \\ a \end{bmatrix}$$

Example 5.3 (The normals of the circle). Let the implicit equation of the circle in Fig. 5.9 be

$$f = x^2 + y^2 - r^2 = 0$$

The partial derivatives are

$$\frac{\partial f}{\partial x} = 2x, \quad \frac{\partial f}{\partial y} = 2y$$

As we can see in Fig. 5.9, these are, indeed, the components of the normal vector.

5.11 ENVELOPES

Let us consider a family of planar curves depending on a parameter λ, and let their general equation be $f(x, y, \lambda)$. We say that this family has an **envelope** if

1. for any value of λ the corresponding curve f is tangent to the envelope;

2. any point of the envelope belongs to one of the curves f.

Let us consider a point P of the envelope; it is also a point on one of the curves that corresponds to a certain value of the parameter λ so that we can write its coordinates as $x(\lambda)$, $y(\lambda)$. This point satisfies the equation

$$f(x(\lambda),\ y(\lambda),\ \lambda) = 0 \tag{5.67}$$

The tangent of the envelope in a given point coincides with the tangent to the f curve that passes through that point. This means that there the tangent to the envelope is perpendicular to the normal to the corresponding f curve so that their scalar product is zero:

$$\frac{\partial f}{\partial x} \cdot \frac{dx}{d\lambda} + \frac{\partial f}{\partial y} \cdot \frac{dy}{d\lambda} = 0 \tag{5.68}$$

Differentiating Eq. (5.67) with respect to the parameter λ we obtain

$$\frac{\partial f}{\partial x} \cdot \frac{dx}{d\lambda} + \frac{\partial f}{\partial y} \cdot \frac{dy}{d\lambda} + \frac{\partial f}{\partial \lambda} = 0 \tag{5.69}$$

From Eqs (5.69) and (5.68) we deduce

$$\frac{\partial f}{\partial \lambda} = 0$$

In conclusion, the envelope must satisfy the system

$$f(x(\lambda),\ y(\lambda),\ \lambda) = 0$$
$$\frac{\partial f(x,\ y,\ \lambda)}{\partial \lambda} = 0 \tag{5.70}$$

To obtain the equation of the envelope it remains to eliminate λ. This is not always possible, and not every family of curves has an envelope. As an example that can be easily solved let us consider the family of circles

$$x^2 + (y - \lambda)^2 = \left(\frac{\lambda}{2}\right)^2 \tag{5.71}$$

In Fig. 5.10 we see the circles corresponding to the values $\lambda = -5, -4, \ldots 0, \ldots 5$. The origin of coordinates is a member of the family. To find the equation of the envelope we write the system

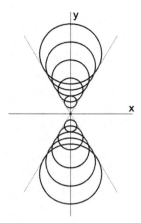

Figure 5.10 A family of circles and its envelopes

$$f(x, y, \lambda) = x^2 + (y - \lambda)^2 - \left(\frac{\lambda}{2}\right)^2 = 0$$

$$f_\lambda(x, y, \lambda) = -2(y - \lambda) - \frac{\lambda}{2} = 0 \qquad (5.72)$$

Retrieving λ from the second equation and substituting it in the first yields

$$x^2 - \frac{1}{3}y^2 = 0$$

which can be rewritten as

$$\left(x - \frac{1}{\sqrt{3}}y\right)\left(x + \frac{1}{\sqrt{3}}y\right) = 0$$

The envelopes of the given family of circles are the straight lines

$$x + \frac{1}{\sqrt{3}}y = 0$$

$$x - \frac{1}{\sqrt{3}}y = 0 \qquad (5.73)$$

5.12 THE METACENTRIC EVOLUTE

In this section we discus an important application in Naval Architecture. The evolute of the B-curve is known as **metacentric evolute**. In detail,

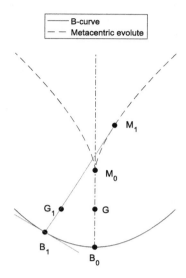

Figure 5.11 Metacentric evolute — Translation of GM

it is the projection on the plane of inclination of the locus of metacentres of a floating body that inclines at constant volume of displacement around axes parallel to a given direction. For ships the above notions refer usually to inclinations around longitudinal axes, namely *heeling*. Assuming that the floating body can incline 360°, the B-curve is a closed curve and so is the metacentric evolute. One of the features of this evolute is that, like the evolute of the ellipse and the evolutes of other curves, it has several cusps. The shape of the evolute and the number of cusps depend on the shape of the floating body. One of the cusps corresponds to the upright condition and the tangent in that point is the vertical for that position. Metacentric evolutes have been used during a long time as a useful tool for finding the floating position and for explaining certain phenomena. With the advent of digital computer and Naval-Architectural software this evolute has been gradually forgotten. We believe, however, that it can still be used for learning as in some situations it gives immediate and clear answers. As a first example let us consider Fig. 5.11; it shows a typical B-curve and the corresponding metacentric evolute. In this case the evolute has ascending branches, a feature sometimes important. We'll see later what may happen when the branches are descending. In upright condition the centre of gravity, G, is located somewhere between the centre of buoyancy, B, and the *initial metacentre*, M_0. The *metacentric height*, $\overline{GM}$, is positive, as required

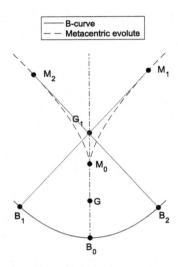

Figure 5.12 Negative $\overline{GM}$ analysis on metacentric evolute

for statical stability. Let us suppose that because of some displacement of weights the centre of gravity moves transversally to G_1. For equilibrium the ship must heel until the corresponding centre of buoyancy, B_1, and the centre of gravity, G_1, lie on the same vertical. This condition is known as *Stevin's law*. Then, all we have to do is to draw from G_1 the tangent to the metacentric evolute and to continue it until it intersects the B-curve. The tangent to the B-curve, in B_1, is parallel to the new waterline. We see immediately that the new position is stable as $\overline{GM} > 0$. No calculations were needed!

Let us refer now to Fig. 5.12. The definitions for the initial conditions are the same as in the preceding case, but we assume now that due to the loading of the ship the centre of gravity moved upwards to G_1. The metacentric height is now negative and the position is unstable. To find a new position of equilibrium we try to draw a tangent from G_1 to a branch of the metacentric evolute. We find two possible positions that are symmetric with respect to the upright position. The ship will heel to one of them, at an angle of *loll*. Usually small perturbations, such as waves or gusts of wind, can throw the ship to the other position. From the figure we can also understand that the higher the centre of gravity above the initial metacentre M_0, the larger the angle of loll. Moreover, we cannot reduce the loll to zero by adding ballast on one side of the ship because such an action will only move transversely the centre of gravity, leaving it still above M_0. We can also see that while the angle of loll is reduced on one side, it may increase on the other. As shown in Biran and López-Pulido (2014), Section 6.11, the dynamical stability is reduced and this can result in capsizing. What should be done is to lower the centre of gravity. We leave to the reader to see what happens if the metacentric evolute has descending branches, as happens with the evolute of the ellipse. Unless there is another cusp in the vicinity, the ship will find no angle of loll and will capsize.

As remarked by Hervieu (1985), the metacentric radius depends on the moment of inertia of the waterplane with respect to the axis of inclination. Therefore, the largest sections influence most the shape of the metacentric evolute. Hervieu (1985), pp. 106–113, approximates the hull shape by that of a cylinder with transverse sections congruent to the midship section and shows how the shape of the latter section determines the metacentric evolute. A table that shows visually the influence of the hull form on the shape of the metacentric evolute can be found in Henschke (1957), Table 1.7. A good analysis of the relationship between forms of the hull and the shape of the metacentric evolute is given in Ilie (1974).

5.13 CURVATURE AND FAIR LINES

To what extent a curve may be considered fair depends very much on the distribution of curvature along the curve. Therefore, various attempts have been made to quantify the notion of fairness in terms of curvature, and to use curvature as a criterion for fairing ship lines. An interesting review is given in Narli and Sariöz (2004) who also describe a computer procedure. Their approach is to fair the ship lines by working on plane sections, such as stations. Among the criteria cited in the paper we remark the one attributed to Roulis and Rando: 'A C^2 curve is considered fair if it minimizes the integral of the squared curvature with respect to arc length'. Other researchers worked with the curvatures of surfaces, but this subject belongs to Chapter 6.

5.14 EXAMPLES

Example 5.4 (Connecting two railway segments). A common problem in the design of railway networks is the connection of straight line segments. To avoid centrifugal acceleration, when passing from a straight-line segment to the connecting curve, or *vice-versa*, the curvature of the connecting segment in the transition points should be zero. Let us show in the

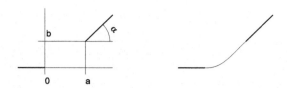

Figure 5.13 Connecting two railway segments

following example how we can solve such a problem. The left–hand part of Fig. 5.13 shows two segments of straight lines, one horizontal, the other with slope α. To simplify the mathematics, we assume that the origin of coordinates is the end of the horizontal segment. Let the coordinates of the first point of the second segment be $(a; b)$. We impose the following conditions:

1. the y-coordinate of the first point of C is zero;
2. the slope of C in the first point is zero;
3. the second derivative of y with respect to x, in the first point of C, is zero;
4. the x-coordinate of the last point of C equals that of the first point of the second line;
5. the slope of C at its last point equals that of the second straight line;
6. the second derivative of y with respect to x in the last point of C is zero.

As we have to fulfill six conditions, we assume a fifth-degree polynomial of the form

$$y = C_1 x^5 + C_2 x^4 + C_3 x^3 + C_4 x^2 + C_5 x + C_6$$

From conditions 1, 2, and 3 we immediately find that $C_4 = 0$, $C_5 = 0$, $C_6 = 0$. Using the remaining conditions we can write the equation

$$\begin{vmatrix} a^5 & a^4 & a^3 \\ 5a^4 & 4a^3 & 3a^2 \\ 20a^3 & 12a^2 & 6a \end{vmatrix} \begin{vmatrix} C_1 \\ C_2 \\ C_3 \end{vmatrix} = \begin{vmatrix} b \\ \tan\alpha \\ 0 \end{vmatrix} \tag{5.74}$$

A MATLAB function that solves this equation is shown below.

```
function    Raccordement1(a, b, alpha)
%RACCORDEMENT1 Connects two straight lines using an explicit function
% y = f(x). The first line is assumed as lying on the x-axis and its
```

```
% end is taken as origin of coordinates.
% Input arguments:
%   a = x-coordinate of the first point of the second line, m
%   b = y-coordinate of the first point of the second line, m
%   alpha = angle of the second line measured from the horizontal,
%               degrees

% build the coefficient matrix
A = [    a^5        a^4        a^3
       5*a^4      4*a^3      3*a^2
      20*a^3     12*a^2      6*a   ];
% build the free vector
B = [ b; tand(alpha); 0 ];
% solve the system
C = A\B
% build x-axis
x = 0: a/50: a;
% evaluate fifth-degree polynomial by Horner's scheme
y = x.^3.*((C(1)*x + C(2)).*x + C(3));
plot(x, y, 'k-')
end
```

Let us run this function for $a = 3$, $b = 5$, and $\alpha = 45°$.

```
%TESTRACCORD1 Illustrates the example that uses the function
%   Raccordement1. Uses the function ArcDim.

subplot(1, 2, 1)
    hp = plot([ -2 0 ], [ 0 0 ], 'k-', [ 3 5 ], [ 2 4 ], 'k-');
    set(hp, 'LineWidth', 1.5)
    axis([ -3 6 -1 5 ]), axis equal, axis off
    ht = text(0, -1, '0'); set(ht, 'FontSize', 14)
    ht = text(3, -1, 'a'); set(ht, 'FontSize', 14)
    hold on
    % add axes
    plot([ 0 5 ], [ 0 0 ], 'k-', [ 0 0 ], [ -0.5 4.5 ], 'k-')
    plot([ -0.5 0.5 ], [ 2 2 ], 'k-')
    plot([ 3 3 ], [-0.5 2 ], 'k-')
    ht = text(0.1, 2.6, 'b'); set(ht, 'FontSize', 14)
    plot([ 0 5 ], [ 2 2 ], 'k-')
    ArcDim([ 3; 2 ], 1.8, 0.1, 0, 45, 'k')
    ht = text(5, 3.1, '\alpha', 'Rotation', -50);
```

```
    set(ht, 'FontSize', 14)
    hold off
subplot(1, 2, 2)
    hp = plot([ -2 0 ], [ 0 0 ], 'k-', [ 3 5 ], [ 2 4 ], 'k-');
    set(hp, 'LineWidth', 1.5)
    axis([ -3 6 -1 5 ]), axis equal, axis off
    hold on
    Raccordement1(3, 2, 45)
    hold off
```

Our function `Raccordement1` worked well in the above example and the resulting curve is plotted in the right-hand side of Fig. 5.13. Running the function for higher slopes of the second segment yields unsatisfactory results. This is due to what is called *polynomial inflexibility*, a behaviour studied by Carl Runge (German, 1856–1927). See also Biran and Breiner (2002), Example 9.3.

5.15 SUMMARY

The notion of curvature refers on how much a curve deviates from a straight line. For a formal definition we consider two points along the curve separated by an arc length δs, and the tangents to the curve in the two points. Let the angle between the two tangents be $\delta \alpha$. The curvature in the first point is by definition

$$\kappa = \lim_{\delta s \to 0} \frac{d\alpha}{\delta s} = \frac{d\alpha}{ds}$$

The inverse of curvature is the *radius of curvature*

$$\rho = \frac{1}{\kappa} = \frac{ds}{d\alpha}$$

For a curve represented as $y = f(x)$, the radius of curvature is given by

$$\rho = \frac{(1 + (y')^2)^{3/2}}{y''}$$

For a curve in parametric representation the radius of curvature is

$$\rho = \frac{(\dot{x}^2 + \dot{y}^2)^{3/2}}{\dot{x}\ddot{y} - \dot{y}\ddot{x}}$$

The notion of curvature and radius of curvature have applications in many fields of science and technology and this chapter includes examples from elasticity, kinematics and Naval Architecture. In particular it is shown that a material point moving with speed v along a curved trajectory is subjected to an acceleration that can be decomposed into a tangential and a normal components. The latter has the magnitude v^2/ρ where ρ is the radius of curvature. This generalizes the well-known theorem according to which, if the trajectory is a circle, ρ equals the radius of the circle. In Naval Architecture a crucial application in the study of the stability of a floating body is based on the curve of centres of buoyancy. The projection of this curve on the plane of inclination is the *B-curve*. The radius of curvature of the B–curve is the *metacentric radius*

$$\overline{BM} = \frac{I}{\nabla}$$

where I is the second moment of the waterplane with respect to the axis of inclination, and ∇ the volume of displacement.

The *osculating circle* of a curve, in a given point, is the circle tangent to the curve in that point that 'best' approximates there the curve. There are three equivalent definitions. Formulating them mathematically and solving the resulting equations shows that the radius of the osculating circle is the radius of curvature at the given point. The centre of the osculating circle is the *centre of curvature* of the curve at the given point. The locus of the centres of curvature of a given curve, C, is the *evolute*, E, of the curve C. Conversely, C is the *involute* of E. The normals of C are tangent to the evolute E. In Naval Architecture the evolute of the B-curve is called *metacentric evolute*. The metacentric evolute was used during a long period as a convenient and quick tool for assessing stability. With the advent of digital computers and Naval-Architectural software, the metacentric evolute was forgotten. However, we think that it may still be useful in explaining quickly what happens when a mass is moved. One important case that can be understood in this way is that of a ship with negative initial metacentric height.

Given a family of planar curves depending on a parameter λ, and general equation $f(x, y, \lambda)$, we say that this family has an **envelope** if

1. for any value of λ the corresponding curve f is tangent to the envelope;
2. any point of the envelope belongs to one of the curves f.

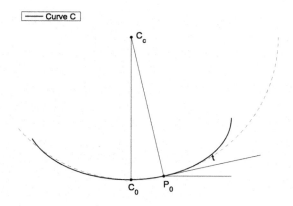

Figure 5.14 Exercise 5.1

5.16 EXERCISES

Exercise 5.1 (The place of the centre of curvature). Let us consider the curve shown in Fig. 5.14. C_c is the centre of curvature corresponding to the point P_0, and C_0 is the intersection of the vertical passing through C_c and the given curve. The curve is shown as a solid line, the osculating circle as a dashed line. The tangent in P_0 is marked t. Calculate the angle $\widehat{C_0 C_c P_0}$ using Eqs (5.31) and (5.32) and show that $\overline{C_c P_0}$ is normal to the given curve.

Exercise 5.2 (The osculating circle of the parabola). The exercise refers to the parabola $x^2 = 4py$, where $p = 3$. You are asked to:
1. draw the parabola in a suitable interval;
2. derive the general equation of the radius of curvature and calculate the value corresponding to the point of the parabola in the origin of coordinates;
3. on the graph add the osculating circle corresponding to the point in the origin of coordinates;
4. try to draw a few other circles tangent to the parabola in the origin of coordinates and check that the osculating circle is the best approximation of the given curve.

Exercise 5.3 (The envelope of ballistic parabolas). This is another classic exercise on envelopes. Refer to the equations of ballistics developed in Example 4.6 and find the envelope of the trajectories for fixed initial speed, V_0, and angles varying from 0 to 90 degrees. Points situated above this envelope are *safe*.

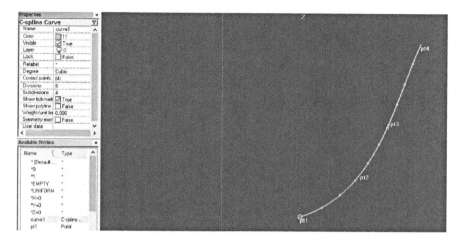

Figure 5.15 C-spline interpolated between points on a parabola

Exercise 5.4 (The evolute of the parabola). Consider the parabola

$$y = 0.5x^2$$

1. Write the equation of the evolute.
2. Plot the parabola and its evolute in the interval $[-1, 1]$.
3. Do you find similar curves in one of the figures of this chapter?

APPENDIX 5.A CURVATURE IN MULTISURF

Some computer programs used in the design of ship hulls provide a possibility for visualizing the curvature of lines. As an example, we show here how to do it in MultiSurf. The following points belong to the parabola $y = 0.5x^2$:

y	0.000	1.500	2.250	3.000
z	0.000	1.125	2.531	4.500

We entered these points in MultiSurf, at $x = 0$, and interpolated a C-spline, that is a curve that passes through all given points. The resulting projection on the YZ plane is shown in Fig. 5.15. Suppose you are sitting before the screen. Select the curve by clicking on it. The curve appears as a thick line. The origin of the parameter t is marked by an 'O' and the orientation is indicated by an arrow. If you want to see points corresponding to the values $t = 0, 0.1, 0.2, \ldots, 1$, turn to the Properties manager and

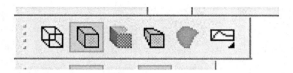

Figure 5.16 The rightmost icon opens the menu for curvature

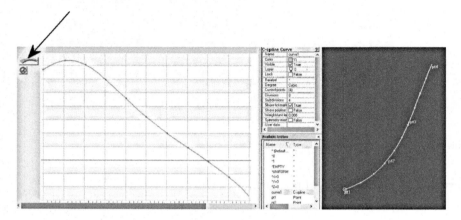

Figure 5.17 Curvature plot for curve in Fig. 5.15

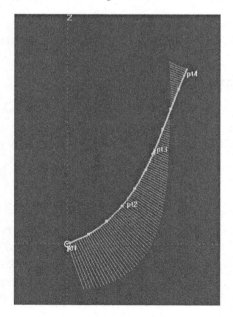

Figure 5.18 Another curvature plot for the curve in Fig. 5.15

set `show tickmark` to `true`. The curve looks like a ship station, but careful inspection reveals that it is not a parabola. The property manager informs us that the curve is of degree 3 and we can see that the tangent in the first point is not horizontal. We get more evidence if we check the curvature. With the curve selected click on the rightmost icon shown in Fig. 5.16. In the menu that opens choose `curvature`. The curvature plot that appears is shown in Fig. 5.17; it is a plot of curvature versus arc length. The maximum of the curvature is not at the first point. Moreover, as the 'curve of curvature' crosses zero, there is a point of inflection immediately after $t = 0.7$. Try to drag a point of the given curve (Fig. 5.17) and see how this affects the curvature. If you click on the icon indicated by an arrow in Fig. 5.17 you obtain the nice plot of curvature shown in Fig. 5.18. The point where the curvature changes sign is obvious.

CHAPTER 6

Surfaces

Contents

6.1 INTRODUCTION

English	surface
French	surface
German	Fläche
Italian	superficie

In the preceding chapter we discussed curves as one-dimensional mathematical objects; we could define any point on a curve as a function of one parameter. In this chapter we deal with *surfaces* as two-dimensional objects; we need two parameters to define a point on a surface. We studied the curves either as embedded in a two-dimensional space, namely the plane, or as embedded in the three-dimensional space. We are going to study some properties of surfaces as objects embedded in the three-dimensional space. A curve that moves in the 3D space generates a surface. Conversely, the intersection of two surfaces is a curve.

In Naval Architecture we meet the concept of surface first when we define the *hull surface*. The traditional treatment of this surface was the

Geometry for Naval Architects
https://doi.org/10.1016/B978-0-08-100328-2.00016-X
Copyright © 2019 Elsevier Ltd.
All rights reserved.

subject of Chapter 2. The analysis of the geometrical properties of the hull surface leads to the abstract concept of surface of the centres of buoyancy, a subject well analyzed in the past in French literature, but forgotten today.

In this chapter we explain the parametric representation of surfaces as functions of two parameters, curves on surfaces, the calculation of distances, angles between curves, and of areas on surfaces. Several kinds of curvatures are defined on surfaces and we show their relationships. The curvature of surfaces plays an essential role in the design of ship hulls as it influences the fairing of ship lines and the possibility of developing the plates of the ship shell. Among the curves lying on a surface we distinguish the geodesics. The shortest path on a surface between two of its points lies on a geodesic of the surface. Geodesics have the important property that they are mapped onto straight lines on the developed surface. Several procedures for developing ship plates are based on the use of geodesics.

By the end of this chapter we mention that there is one property of surfaces, the Gaussian curvature, that does not depend upon the embedding in space and can be detected and measured in two dimensions only.

6.2 PARAMETRIC REPRESENTATION

Like curves, surfaces can be represented mathematically in three forms:
1. — implicit equations;
2. — explicit equations;
3. — parametric equations.

Implicit equations are of the form $f(x, y, z) = 0$. For example, the following equation represents a plane

$$Ax + By + ZCz + D = 0$$

and the equation of a sphere with centre in the origin of coordinates and radius R is

$$x^2 + y^2 + z^2 = R^2$$

Explicit equations can be, for example, of the form $z = f(x, y)$. Such an expression is known as *Monge form*. For example, the following equation defines a hyperbolic paraboloid:

$$z = \frac{1}{2p}\left[-\frac{x^2}{a^2} + \frac{y^2}{b^2}\right]$$

Parametric equations are of the form

$$x = f_1(u, v)$$
$$y = f_2(u, v)$$
$$z = f_3(u, v)$$

This representation is due to Gauss (Johann Carl Friedrich, German, 1777–1855, written in German as Gauß). As a simple example, the following equations represent a circular cylinder with axis along the z-axis, radius r, and height h

$$x = r\cos 2\pi u$$
$$y = r\sin 2\pi u$$
$$z = hv, \quad u \in [0,\ 1], \quad v \in [0,\ 1] \tag{6.1}$$

A line of constant u is obtained by setting u equal to a constant value and letting v run through its domain of values, while a line of constant v is obtained by setting v equal to a constant value and letting u to be a variable. These lines form a net of coordinate lines on the surface. A point on the surface is defined by the intersection of a line of constant u with a line of constant v. Returning to our example of cylinder, the lines of constant u are straight lines parallel to the axis, and the lines of constant v are circles in planes perpendicular to the axis.

Let us write the parametric equations as components of a vector

$$X = \begin{vmatrix} f_1(u, v) \\ f_2(u, v) \\ f_3(u, v) \end{vmatrix} \tag{6.2}$$

Then, $\partial X / \partial u$ calculated at a point P is the tangent vector at P to the constant v curve that passes through P. We are going to use for this derivative the symbol X_u. Similarly, $\partial X / \partial v$ calculated at a point P is the tangent vector at P to the constant u curve that passes through P. We'll use for this derivative the notation X_v. It is usual to assume that the surfaces we are dealing with fulfill the condition

$$X_u \times X_v \neq 0 \tag{6.3}$$

The '$\times$' sign stands for the cross, or, in other words, vector product. Eq. (6.3) says that at a given point of the surface the tangents to the constant

u and constant v curves that pass through that point have distinguished, non-zero tangents. These tangents are also tangent to the surface in the directions of the constant-u and constant-v lines through the given point.

Example 6.1 (Cylinder — Plotting lines of constant u). We refer to Eqs (6.1). The following MATLAB script uses them to plot the curves corresponding to $v = 0, 0.1, \ldots, 1$.

```
%PARAMCYLINDER Plots the curves of constant v of a parametric cylinder

% constants
r = 6;                                      % base radius
h = 10;                                     % cylinder height
% define frame
plot3([ -1.5*r 1.5*r ], [ 0 0 ], [ 0 0 ], 'k-',...
       [ 0 0 ], [ -1.5*r 1.5*r ], [ 0 0 ], 'k-',...
       [ 0 0 ], [ 0 0 ], [ -1.1*h 1.5*h ], 'k-')
set(gca, 'Xdir', 'reverse', 'Ydir', 'reverse')        % first angle view
axis equal, axis off
hold on
ht = text(1.4*r, 0, 0.18*h, 'x'); set(ht, 'FontSize', 14)
ht = text(0, 1.4*r, 0.15*h, 'y'); set(ht, 'FontSize', 14)
ht = text(0, 0.1*r, 1.42*h, 'z'); set(ht, 'FontSize', 14)
% plot curves v = ct
u   = 0: 0.02: 1; v   = 0: 0.1: 1;
x   = r*cos(2*pi*u); y   = r*sin(2*pi*u);
z   = v*h;
l   = length(x);
for k = 1:11
    plot3(x, y, z(k)*ones(l));
end
hold off
```

The result is shown in Fig. 6.1. We invite the reader to write a similar script that plots the generators corresponding to $u = 0, 0.1, \ldots, 1$, as seen in Fig. 6.2.

Fig. 6.3 is a complete view of the cylinder. The lines of constant u and constant v can be easily distinguished; they form a net of coordinate lines of the surface. The MATLAB code that produces the figure is shown below and it uses the parametric equations introduced at the beginning of this section.

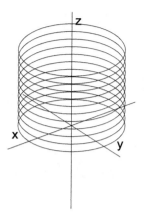

Figure 6.1 Constant-*v* curves

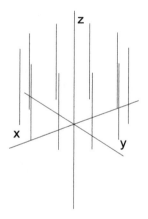

Figure 6.2 Constant-*u* curves

```
%CYLSURF Plots a cylindric surface using parametric equations.

% constants
h = 10;
m = 30;                  % subdivides u
n = 10;                  % subdivides v
r = 5*ones((n + 1), 1); % base radius
% parameters
u = 0: 1/m: 1; v = 0: 1/n: 1;
% coordinates
```

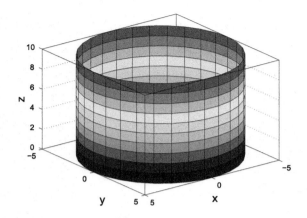

Figure 6.3 The surface of a parametric cylinder

```
x = r*cos(2*pi*u);
y = r*sin(2*pi*u);
z = h*v;
% form array for plotting
Z = z'*ones(1, (m + 1));
surf(x, y, Z)
set(gca, 'Xdir', 'reverse', 'Ydir', 'reverse')
ht = xlabel('x'); set(ht, 'FontSize', 14)
ht = ylabel('y'); set(ht, 'FontSize', 14)
ht = zlabel('z'); set(ht, 'FontSize', 14)
box
```

Example 6.2 (The Earth sphere). Perhaps the most popular example of coordinate lines on a surface is that of meridians and parallels on the spherical model of the Earth. Here, instead of u and v, the parameters are called longitude, λ, and latitude, ϕ. These parameters are collectively known as *geographic coordinates*. We use Fig. 6.4 to derive the parametric equations of the sphere. Let r be the radius of the sphere and P a point on its surface. P_h is the projection of P on the xOy plane, P_x the projection of P_h on the axis x, and P_y the projection of P_h on the axis y. The parameter λ is the angle between $\overline{OP_x}$ and $\overline{OP_h}$, and the parameter ϕ the angle between $\overline{OP_h}$ and $\overline{OP}$. The parametric equations of the sphere are

$$\mathbf{P} = \begin{vmatrix} r\cos\phi\cos\lambda \\ r\cos\phi\sin\lambda \\ r\sin\phi \end{vmatrix} \tag{6.4}$$

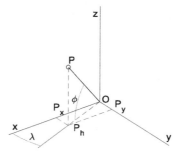

Figure 6.4 Spherical coordinates

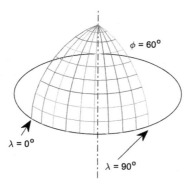

Figure 6.5 The net of geographic coordinates on a sphere

The lines of constant λ are called *meridians*, and the lines of constant ϕ are the *parallels*. In Fig. 6.5 we show part of the net of coordinate lines. The tangent vectors to the constant-ϕ lines, that is to the parallels, are given by

$$\mathbf{P}_\lambda = \begin{vmatrix} -r\cos\phi\sin\lambda \\ r\cos\phi\cos\lambda \\ 0 \end{vmatrix} \tag{6.5}$$

As expected, these tangents lie in the planes of the parallels. The tangent vectors to the constant-λ lines, that is the meridians, are

$$\mathbf{P}_\phi = \begin{vmatrix} -r\sin\phi\cos\lambda \\ -r\sin\phi\sin\lambda \\ r\cos\phi \end{vmatrix} \tag{6.6}$$

The normal vector is given by the cross product

$$\mathbf{P}_\lambda \times \mathbf{P}_\phi = \begin{vmatrix} \mathbf{i} & \mathbf{j} & \mathbf{k} \\ -r\cos\phi\sin\lambda & r\cos\phi\cos\lambda & 0 \\ -r\sin\phi\cos\lambda & -r\sin\phi\sin\lambda & r\cos\phi \end{vmatrix}$$

and the result can be written in the form

$$\mathbf{n} = \begin{vmatrix} r^2\cos^2\phi\cos\lambda \\ r^2\cos^2\phi\sin\lambda \\ r^2\sin\phi\cos\phi \end{vmatrix} = r^2\cos\phi \begin{vmatrix} \cos\phi\cos\lambda \\ \cos\phi\sin\lambda \\ \sin\phi \end{vmatrix}$$

This vector has the direction and sense of $\overrightarrow{OP}$, the position vector that lies along the local radius of the sphere. The factor $r^2\cos\phi$ is the length of the normal as it can be seen from

$$||\mathbf{P}_\lambda \times \mathbf{P}_\phi|| = r^2\cos\phi\sqrt{\cos^2\phi\cos^2\lambda + \cos^2\phi\sin^2\lambda + \sin^2\phi} = r^2\cos\phi$$

6.3 CURVES ON SURFACES

Given the parametric equations of a surface it is possible to derive from them the parametric equations of certain curves on that surface. We have already seen one possibility, namely how to obtain coordinate lines. To do this one has to set a fixed value for one of the parameters and let the other vary. Another way of defining a curve on a surface is to impose a relationship between the two parameters u and v. For example, given Eqs (6.1) of a right, circular cylinder, and the relationship $u = v$, we obtain the equations of a helix. This curve is in fact a function of one parameter and we usually return for it to the notation t. A third possibility is to define both parameters u and v as functions of a parameter t. Then, if the parametric equations of the surface are given as a vector $\mathbf{P}(u, v)$, the tangent vector to the curve, also tangent to the surface in the direction of the curve, is given by

$$\dot{\mathbf{P}} = \mathbf{P}_u\dot{u} + \mathbf{P}_v\dot{v} \tag{6.7}$$

As an example, we return to the cylinder represented by Eqs (6.1) and the helix defined by assuming $u = t$, $v = t$. Eq. (6.7) yields

$$\dot{\mathbf{P}} = \begin{vmatrix} -2\pi r\sin 2\pi t \\ 2\pi r\cos 2\pi t \\ 0 \end{vmatrix} + \begin{vmatrix} 0 \\ 0 \\ h \end{vmatrix} = \begin{vmatrix} -2\pi r\sin 2\pi t \\ 2\pi r\cos 2\pi t \\ h \end{vmatrix} \tag{6.8}$$

For a geometric interpretation of this result we first remark that at any point the horizontal projection of this vector is perpendicular to the radius corresponding to that point. Next, we consider a meridian section of the cylinder. The projection of the tangent vector on the plane of the section has the slope $h/(2\pi r)$. This is exactly the slope of the developed helix (see Fig. 1.47).

6.4 FIRST FUNDAMENTAL FORM

We consider a surface $\mathbf{P}(u, v)$ and a curve on it defined by setting u and v as functions of t. Given two points on the curve corresponding to $t = t_1$ and $t = t_2$, the distance between them is obtained from

$$\int_{t_1}^{t_2} \sqrt{\dot{\mathbf{P}} \cdot \dot{\mathbf{P}}}\, dt = \int_{t_1}^{t_2} \sqrt{(\mathbf{P}_u \dot{u} + \mathbf{P}_v \dot{v}) \cdot (\mathbf{P}_u \dot{u} + \mathbf{P}_v \dot{v})}\, dt \qquad (6.9)$$

Expanding the expression under the square-root sign yields

$$ds^2 = \mathbf{P}_u \cdot \mathbf{P}_u du^2 + 2\mathbf{P}_u \cdot \mathbf{P}_v dudv + \mathbf{P}_v \cdot \mathbf{P}_v dv^2$$

Using the traditional notation

$$E = \mathbf{P}_u \cdot \mathbf{P}_u, \;\; F = \mathbf{P}_u \cdot \mathbf{P}_v, \;\; G = \mathbf{P}_v \cdot \mathbf{P}_v$$

we rewrite ds^2 as

$$ds^2 = Edu^2 + 2Fdudv + Gdv^2 \qquad (6.10)$$

This expression is known as the *first fundamental form* and is noted by I. Davies and Sammuels (1996) define also

$$\mathbf{B} = \begin{vmatrix} E & F \\ F & G \end{vmatrix} \qquad (6.11)$$

and call $\mathbf{B}$ the *first fundamental matrix*. With this notation and

$$\mathbf{u} = \begin{vmatrix} u(t) \\ v(t) \end{vmatrix}$$

we can write the first fundamental form under the elegant form

$$ds^2 = \dot{\mathbf{u}}^T \mathbf{B} \dot{\mathbf{u}} \qquad (6.12)$$

This expression can be treated easily in MATLAB. Let us see a few applications of the first fundamental form. We started from the distance between two points of a curve defined on a given surface. As a very simple example we return to the helix on the surface of a right, circular cylinder, Eqs (6.1), and calculate the distance between the points corresponding to $t = 0$ and $t = 1$. The helix is defined by $u = v = t$. We have

$$E = \mathbf{P}_u \cdot \mathbf{P}_u = \begin{vmatrix} -2\pi r \sin 2\pi t \\ 2\pi r \cos 2\pi t \\ 0 \end{vmatrix} \cdot \begin{vmatrix} -2\pi r \sin 2\pi t \\ 2\pi r \cos 2\pi t \\ 0 \end{vmatrix} = 4\pi^2 r^2$$

$$F = \mathbf{P}_u \cdot \mathbf{P}_v = \begin{vmatrix} -2\pi r \sin 2\pi t \\ 2\pi r \cos 2\pi t \\ 0 \end{vmatrix} \cdot \begin{vmatrix} 0 \\ 0 \\ h \end{vmatrix} = 0$$

$$G = \mathbf{P}_v \cdot \mathbf{P}_v = \begin{vmatrix} 0 \\ 0 \\ h \end{vmatrix} \cdot \begin{vmatrix} 0 \\ 0 \\ h \end{vmatrix} = h^2$$

The first fundamental form is

$$ds^2 = 4\pi^2 r^2 + h^2$$

and the integral yields

$$s = \int_0^1 \sqrt{4\pi^2 r^2 + h^2} \, dt = \sqrt{4\pi^2 r^2 + h^2}$$

We have already obtained this result in two other ways. We are going to show more applications of the first fundamental form.

The angle between two lines. The coefficients of the first fundamental form find an application in the calculation of the angle between the constant-u and the constant-v lines. Noting this angle by α we write the definition of the scalar product of the tangents to the above lines as

$$\mathbf{P}_u \cdot \mathbf{P}_v = \sqrt{\mathbf{P}_u \cdot \mathbf{P}_u}\sqrt{\mathbf{P}_v \cdot \mathbf{P}_v} \cos \alpha$$

Extracting $\cos \alpha$ and using the notation of the first fundamental form we obtain

$$\cos \alpha = \frac{F}{\sqrt{EG}} \tag{6.13}$$

In particular, if $F = 0$ the net of coordinate lines is orthogonal. This is the case of the right, circular cylinder. In Exercise 6.5 we ask the reader to prove that the net of parallels and meridians of a sphere is also orthogonal. We are going now to generalize the results for two arbitrary lines on the same surface $\mathbf{P}$. Let the lines be given by $\mathbf{P}(u_1(t), v_1(t))$ and $\mathbf{P}(u_2(t), v_2(t))$. The differentials involved are

$$d\mathbf{P}_1 = \mathbf{P}_u du_1 + \mathbf{P}_v dv_1, \quad d\mathbf{P}_2 = \mathbf{P}_u du_2 + \mathbf{P}_v dv_2$$

and the angle between the two curves is given by

$$
\begin{aligned}
\cos\alpha &= \frac{(\mathbf{P}_u \cdot \mathbf{P}_u)du_1 du_2 + (\mathbf{P}_u \cdot \mathbf{P}_v)(du_1 dv_2 + du_2 dv_1) + (\mathbf{P}_v \cdot \mathbf{P}_v)dv_1 dv_2}{||d\mathbf{P}_1||\,||d\mathbf{P}_2||} \\
&= \frac{E du_1 du_2 + F(du_1 dv_2 + du_2 dv_1) + G dv_1 dv_2}{||d\mathbf{P}_1||\,||d\mathbf{P}_2||}
\end{aligned}
\tag{6.14}
$$

The condition for the orthogonality of the two lines is

$$E du_1 du_2 + F(du_1 dv_2 + du_2 dv_1) + G dv_1 dv_2 = 0$$

Dividing by $du_1 du_2$ and using the notations

$$\lambda_1 = \frac{dv_1}{du_1}, \quad \lambda_2 = \frac{dv_2}{du_2}$$

we get the condition

$$E + F(\lambda_1 + \lambda_2) + G\lambda_1\lambda_2 = 0 \tag{6.15}$$

Area. In Fig. 6.6 we consider the area of the curvilinear parallelogram delimited by two lines of constant u and two lines of constant v. Let the angle between the constant-u and constants-v lines be α. The distance between the constant-u lines is

$$\mathbf{P}_u du = \sqrt{E} du$$

and the distance between the constant-v lines,

$$\mathbf{P}_v dv = \sqrt{G} du$$

For small increments, du, dv, we can use the formula of the parallelogram and write

$$dA = \sin\alpha \mathbf{P}_u du \mathbf{P}_v dv$$

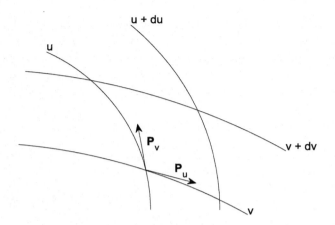

Figure 6.6 Elemental curvilinear parallelogram

From Eq. (6.13)

$$\sin\alpha = \sqrt{1 - \cos^2\alpha} = \sqrt{\frac{EG - F^2}{EG}}$$

Summing up, the elemental area equals

$$dA = \sqrt{\frac{EG - F^2}{EG}}\sqrt{E}\sqrt{G}dudv = \sqrt{EG - F^2}dudv \qquad (6.16)$$

6.5 SECOND FUNDAMENTAL FORM

We consider a parametrized surface $\mathbf{R}(u, v)$ and on it a curve $\mathbf{C}$ defined by $u(t)$, $v(t)$. Without loss of generality, in Figs 6.7 and 6.8 we exemplify our definitions on a sphere. Let the tangent vector to $\mathbf{C}$ in a point P be $\mathbf{T}$, and the principal normal vector $\mathbf{N}$. From Eqs (5.58) and (5.59) we have

$$\frac{d^2\mathbf{R}}{ds^2} = \frac{d\mathbf{T}}{ds} = \kappa\mathbf{N} \qquad (6.17)$$

where κ is the curvature of $\mathbf{C}$ at P. Let the unit normal vector to the surface, at P, be $\mathbf{M}$. As $\mathbf{T}$ is also tangent to $\mathbf{R}$, it is perpendicular to $\mathbf{M}$ and we can write

$$\mathbf{T} \cdot \mathbf{M} = 0$$

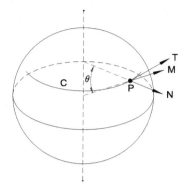

Figure 6.7 Definitions for the second fundamental form

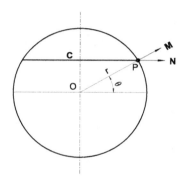

Figure 6.8 Definitions for the second fundamental form

Differentiating with respect to s yields

$$\frac{d\mathbf{T}}{ds} \cdot \mathbf{M} + \mathbf{T} \cdot \frac{d\mathbf{M}}{ds} = \kappa\mathbf{N} \cdot \mathbf{M} + \mathbf{T} \cdot \frac{d\mathbf{M}}{ds} = 0$$

and we rewrite this equation as

$$\kappa\mathbf{N} \cdot \mathbf{M} = -\frac{d\mathbf{R}}{ds} \cdot \frac{d\mathbf{M}}{ds} \tag{6.18}$$

N and **M** are unit vectors; therefore, their scalar product equals the cosine of the angle between these vectors. Noting this angle by θ we have

$$\kappa\cos\theta = -\frac{d\mathbf{R}}{ds} \cdot \frac{d\mathbf{M}}{ds} \tag{6.19}$$

Detailing the numerator in the right-hand side of the equation yields

$$\kappa \cos \theta = -\frac{(\mathbf{R}_u du + \mathbf{R}_v dv) \cdot (\mathbf{M}_u du + \mathbf{M}_v dv)}{ds^2}$$

With the notations

$$e = -\mathbf{R}_u \cdot \mathbf{M}_u, \quad 2f = -(\mathbf{R}_u \cdot \mathbf{M}_v + \mathbf{R}_v \cdot \mathbf{M}_u,) \, g = -\mathbf{R}_v \cdot \mathbf{M}_v$$

the numerator becomes

$$edu^2 + 2fdudv + gdv^2 \tag{6.20}$$

This expression is known as the *second fundamental form* and is noted by II. The denominator is the first fundamental form and the final result is

$$\boxed{\kappa \cos \theta = \frac{\cos \theta}{\rho} = \frac{edu^2 + 2fdudv + gdv^2}{Edu^2 + 2Fdudv + Gdv^2}} \tag{6.21}$$

Various textbooks contain slightly different derivations of the second fundamental form and different notations for its terms. The proof given above follows some Italian lecture notes and is close to that given in Struik (1961). A frequent alternative to the notation e, f, g is L, M, N. As we used two of these letters to note normal vectors, we adopt the notation of Struik. For simplicity we'll ignore the minus sign before e, f, g as this does not influence the explanations that will follow in this chapter.

For a geometrical interpretation of the role of the angle θ we refer to the *meridian section* of a sphere shown in Fig. 6.8. As curve **C** we have chosen the parallel at latitude θ. The normal to the parallel is **N**, the normal to the sphere is **M**, and the angle between these vectors is θ. The vectors **T** and **N** define a plane that cuts the sphere along the curve **C**. The radius of curvature of **C** is $r\cos\theta$. The vectors **T** and **M** define a plane that passes through the centre of the sphere. All planes that contain the centre of the sphere cut the surface along circles with radius r that are called **great circles**. The plane defined by **T** and **M** is normal to the sphere. Therefore, the curvature of the great circle is the *normal curvature of the sphere*. The radius of curvature of the sphere equals the radius r of the sphere for all normal sections. This is not the case for other surfaces.

To interpret Eq. (6.21) let us examine in a point **P** the curvature of the sections obtained by cutting the surface with planes that contain the tangent to the curve under consideration. The largest curvature, i.e. the smallest radius of curvature, is reached when the plane is normal to the surface. A consequence is *Meusnier's theorem* (Jean Baptiste, French, 1754–1793)

Let **C** be a curve on the surface **R**, ρ its radius of curvature, and ρ_n the radius of curvature of the normal section. The radius ρ is the projection of ρ_n on the osculating plane of **C**.

An equivalent statement is

Let **C** be a curve on the surface **R**, and C_c the centre of curvature of the normal section. The centre of curvature of **C** is the projection of C_c on the osculating plane of **C**.

Another consequence is that all curves on **R** that pass through a given point and have the same osculating plane have the same radius of curvature.

Example 6.3 (Sphere — Second fundamental form). On a sphere described by Eqs (6.4) we consider the parallel at latitude ϕ_0; its equation is

$$\mathbf{C} = r \begin{vmatrix} \cos\phi_0\cos\lambda \\ \cos\phi_0\sin\lambda \\ \sin\phi_0 \end{vmatrix}$$

The tangent to the curve is given by

$$\mathbf{C}_\lambda = r \begin{vmatrix} -\cos\phi_0\sin\lambda \\ \cos\phi_0\cos\lambda \\ \sin\phi_0 \end{vmatrix}$$

As the length of this vector equals

$$\|\mathbf{C}_\lambda\| = r\sqrt{\cos^2\phi_0\sin^2\lambda + \cos^2\phi_0\cos^2\lambda + 0} = r\cos\phi_0$$

the unit tangent and normal vectors are

$$\mathbf{T} = \begin{vmatrix} -\sin\lambda \\ \cos\lambda \\ 0 \end{vmatrix}, \quad \mathbf{N} = \begin{vmatrix} \cos\lambda \\ \sin\lambda \\ 0 \end{vmatrix}$$

In Example 6.2 we calculated the tangents to the sphere surface, in the directions of the coordinate lines, and the normal vector to the sphere.

Continuing the calculations we find the unit normal vector to the sphere

$$\mathbf{M} = \begin{vmatrix} \cos\phi\cos\lambda \\ \cos\phi\sin\lambda \\ \sin\phi \end{vmatrix} \tag{6.22}$$

The scalar product of the vectors $\mathbf{N}$ and $\mathbf{M}$ is $\cos\phi_0$, as shown in Fig. 6.8. The expressions for the first normal form are

$$E = \mathbf{P}_\lambda \cdot \mathbf{P}_\lambda = r^2\cos^2\phi\sin^2\lambda + r^2\cos^2\phi\cos^2\lambda = r^2\cos^2\phi$$
$$F = \mathbf{P}_\lambda \cdot \mathbf{P}_\phi = r^2\sin\phi\cos\phi\sin\lambda\cos\lambda - r^2\sin\phi\cos\phi\sin\lambda\cos\lambda = 0$$
$$G = \mathbf{P}_\phi \cdot \mathbf{P}_\phi = r^2\sin^2\phi\cos^2\lambda + r^2\sin^2\phi\sin^2\lambda + r^2\cos^2\phi = r^2 \tag{6.23}$$

The first normal form is

$$I = r^2\cos^2\phi d\lambda^2 + r^2 d\phi^2 \tag{6.24}$$

To experiment with Eq. (6.16) let us calculate the area of the sphere. The elemental area is given by

$$dA = \sqrt{EG - F^2}\,d\phi\,d\lambda = \sqrt{r^4\cos^2\phi}\,d\phi\,d\lambda = r^2\cos\phi\,d\phi\,d\lambda$$

and the area of the Northern hemisphere is

$$A = r^2 \int_0^{2\pi} \int_0^{\pi/2} \cos\phi\,d\phi\,d\lambda = 2\pi r^2$$

For the whole sphere we obtain, as expected,

$$A = 4\pi r^2$$

For the second fundamental form we calculate

$$\mathbf{M}_\lambda = \begin{vmatrix} -\cos\phi\sin\lambda \\ \cos\phi\cos\lambda \\ 0 \end{vmatrix}, \quad \mathbf{M}_\phi = \begin{vmatrix} -\sin\phi\cos\lambda \\ -\sin\phi\sin\lambda \\ \cos\phi \end{vmatrix}$$

and in continuation

$$e = \mathbf{P}_\lambda \cdot \mathbf{M}_\lambda = r\cos^2\phi\sin^2\lambda + r\cos^2\phi\cos^2\lambda = r\cos^2\phi$$
$$2f = \mathbf{P}_\lambda \cdot \mathbf{M}_\phi + \mathbf{P}_\phi \cdot \mathbf{M}_\lambda = r\sin\phi\cos\phi\sin\lambda\cos\lambda - r\sin\phi\cos\phi\sin\lambda\cos\lambda = 0$$
$$g = \mathbf{P}_\phi \cdot \mathbf{M}_\phi = r\sin^2\phi\cos^2\lambda + r\sin^2\phi\sin^2\lambda + r\cos^2\phi = r$$

The second fundamental form of the sphere is

$$II = r\cos^2\phi\, d\lambda^2 + r\, d\phi^2 \qquad (6.25)$$

Summing up, the general equation of the curvature is

$$\kappa\cos\theta = \frac{r\cos^2\phi\, d\lambda^2 + r\, d\phi^2}{r^2\cos^2\phi\, d\lambda^2 + r^2\, d\phi^2} = \frac{1}{r} \qquad (6.26)$$

For $\theta = 0$ we get the curvature of the sphere, $1/r$, and for $\theta = \phi_0$ the curvature $1/(r\cos\phi_0)$ of the parallel at latitude ϕ_0.

6.6 PRINCIPAL, GAUSSIAN, AND MEAN CURVATURES

English	principal curvatures	Gaussian curvature, total curvature	mean curvature
French	courbures principales	courbure de Gauss	courbure moyenne
German	Hauptkrümmungen	Gaussche Krümmung	mittlere Krümmung
Italian	curvature principali	curvatura gaussiana	curvatura media

At a regular point of a surface there is only one normal to the surface, but an infinity of tangents. Each tangent and the normal define a plane normal to the surface. Such a plane cuts the surface along a curve C whose curvature is a *normal curvature* of the surface in the direction of C. Euler (Leonhard, Swiss, 1707–1783) has proved that there are two directions such that for one of them the normal curvature is minimal, while for the other it is maximal. Moreover, these directions are perpendicular one to the other. The directions of the extremal curvatures are called *principal directions*, and the corresponding curvatures are known as *principal curvatures*. In many applications, including the fairing of ship lines and the production of shell plating, it is interesting to analyze the **Gaussian curvature**

$$K = \kappa_{min}\kappa_{max}$$

and the **mean curvature**

$$H = \frac{\kappa_{min} + \kappa_{max}}{2}$$

In textbooks of differential geometry we find various treatments of the principal, Gaussian, and mean curvatures, sometimes at a high mathematical level. In this section we follow the simpler approaches of Bäschlin (1947) and Struik (1961) and try to explain the matters in more detail.

The direction of the normal section is given by $\lambda = dv/du$. With this notation we rewrite Eq. (6.21) as

$$\kappa = \frac{e + 2f\lambda + g\lambda^2}{E + 2F\lambda + G\lambda^2} \tag{6.27}$$

Rearranging the terms we get

$$(\kappa G - g)\lambda^2 + 2(\kappa F - f)\lambda + (\kappa E - e) = 0 \tag{6.28}$$

To obtain the extremal values of the curvature we impose the condition $d\kappa/d\lambda = 0$. Then, the differentiation of Eq. (6.27) with respect to λ yields

$$\frac{(E + 2F\lambda + G\lambda^2)(f + g\lambda) - (e + 2f\lambda + g\lambda^2)(F + G\lambda)}{(E + 2F\lambda + G\lambda^2)^2} = 0 \tag{6.29}$$

Obviously, the numerator must equal zero and we express this condition by the equation

$$(Fg - Gf)\lambda^2 + (Eg - Ge)\lambda + (Ef - Fe) = 0$$

Dividing by the coefficient of the first term we obtain

$$\lambda^2 + \frac{Eg - Ge}{Fg - Gf}\lambda + \frac{Ef - Fe}{Fg - Gf} = 0 \tag{6.30}$$

This equation has two solutions meaning that there are two directions, one for which the curvature reaches a maximum, another one for which the curvature has a minimum. We are going to show that these directions are perpendicular one to the other. To do this we remark that the sum of the roots of Eq. (6.30) is

$$\lambda_1 + \lambda_2 = -\frac{Eg - Ge}{Fg - Gf}$$

and the product

$$\lambda_1\lambda_2 = \frac{Ef - Fe}{Fg - Gf}$$

Substituting the above values into the left-hand side of Eq. (6.15) we obtain, indeed, zero. The condition for orthogonality is fulfilled. We return now to Eqs (6.27) and (6.29) and write

$$\kappa = \frac{e + 2f\lambda + g\lambda^2}{E + 2F\lambda + G\lambda^2} = \frac{f + g\lambda}{F + G\lambda}$$

that is

$$(\kappa G - g)\lambda + (\kappa F - f) = 0 \tag{6.31}$$

Multiplying this equation by λ and subtracting from Eq. (6.28) we obtain a second equation

$$(\kappa F - f)\lambda + (\kappa E - e) = 0 \tag{6.32}$$

We have now a system of two homogeneous equations in two variables

$$(\kappa G - g)du + (\kappa F - f)dv = 0$$
$$(\kappa F - f)du + (\kappa E - e)dv = 0 \tag{6.33}$$

This system has a non-trivial solution only if its determinant is zero

$$\begin{vmatrix} (\kappa G - g) & (\kappa F - f) \\ (\kappa F - f) & (\kappa E - e) \end{vmatrix} = 0$$

Expanding the determinant and simplifying we obtain an equation in κ

$$(EG - F^2)\kappa^2 + (eG - 2fF + gE)\kappa + (eg - f^2) = 0 \tag{6.34}$$

There are two solutions, κ_{min} and κ_{max}, and they can be calculated with the known formula for the second-degree equation. Often, it is more interesting to retrieve the Gaussian curvature that, as the product of the solutions, is given by

$$K = \kappa_{min}\kappa_{max} = \frac{eg - f^2}{EG - F^2} \tag{6.35}$$

and the mean curvature that is derived from the sum of the solutions by

$$H = \frac{\kappa_{min} + \kappa_{max}}{2} = \frac{1}{2}\frac{eG - 2fF + gE}{EG - F^2} \tag{6.36}$$

The normal curvature in a given direction is related to the principal curvatures by an expression due to Euler. To obtain it we rotate the axes of coordinates until they correspond to the principal directions. Let α be the angle that the considered direction makes with the axis that corresponds to κ_{min}. Then the curvature in the direction α is

$$\kappa_\alpha = \kappa_{min} \cos^2 \alpha + \kappa_{max} \sin^2 \alpha \qquad (6.37)$$

Example 6.4 (Sphere, continuation). Using results from Example 6.3 we calculate the Gaussian curvature of the sphere as

$$K = \frac{eg - f^2}{EG - F^2} = \frac{r^2 \cos^2 \phi}{r^4 \cos^2 \phi} = \frac{1}{r^2}$$

The mean curvature is

$$H = \frac{1}{2} \cdot \frac{eG - 2fF + gE}{EG - F^2} = \frac{1}{2} \cdot \frac{r^3 \cos^2 \phi - 0 + r^3 \cos^2 \phi}{r^4 \cos^2 \phi} = \frac{1}{r}$$

These are the expected results as the normal curvature of the sphere is $1/r$ at all points.

6.7 RULED SURFACES

English	ruled surface	developable surface
French	surface réglée	surface développable
German	Regelfläche	abwickelbare Fläche
Italian	superficie rigata	superficie sviluppabile

The following definitions of a **ruled surface** are equivalent:
1. A surface such that through each point of it passes a straight line that is fully contained in the surface.
2. A surface generated by the motion of a straight line.
3. The set of a family of straight lines depending on a parameter that spans a set of real numbers.

The straight lines mentioned in the definitions are the *generators* or *rulings* of the surface. The general equation of a ruled surface is

$$\mathbf{P}(u, v) = \mathbf{C}(u) + v\mathbf{r}(u)$$

Here $\mathbf{C}(u)$ is a curve through which all generators pass; we call it the *directrix* of the surface. The vector $\mathbf{r}(u)$ defines the direction of the generators. Ruled surfaces play a role in the design of hull surfaces and one of the reasons is that all developable surfaces are ruled surfaces. As we will see, the converse is not true. Software for hull design may include the option for generating ruled surfaces. In Chapter 2 we exemplify a tool of MultiSurf. Below we give some representative examples of ruled surfaces.

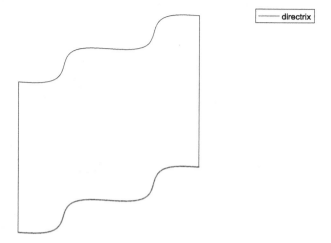

directrix

Figure 6.9 A cylindrical surface patch

6.7.1 Cylindrical Surfaces

As an example we choose as directrix the sine curve

$$\mathbf{C} = \begin{vmatrix} 4\pi u \\ \sin 4\pi u \\ 0 \end{vmatrix}, \quad u \in |0,\ 1|$$

and as direction of generators the vertical vector

$$\mathbf{r} = \begin{vmatrix} 0 \\ 0 \\ H \end{vmatrix}, \quad v \in |0,\ 1|$$

The resulting ruled surface is

$$\mathbf{P} = \begin{vmatrix} 4\pi u \\ \sin 4\pi u \\ vH \end{vmatrix}, \quad u \in |0,\ 1|, \quad v \in |0,\ 1| \tag{6.38}$$

This surface patch is shown in Fig. 6.9.

6.7.2 Conic Surfaces

Another category of ruled surfaces is that of conic surfaces; their generators pass through a given point. As an example we choose now as directrix an ellipse with semi-axes a and b

$$\mathbf{C} = \begin{vmatrix} a\cos((1+u)\pi) \\ b\sin((1+u)\pi) \\ 0 \end{vmatrix}$$

and direction vector

$$\mathbf{r} = \begin{vmatrix} -a\cos((1+u)\pi) \\ -b\sin((1+u)\pi) \\ H \end{vmatrix}$$

The equation of the ruled surface is

$$\mathbf{P} = \mathbf{C} + v\mathbf{r} = \begin{vmatrix} (1-v)a\cos((1+u)\pi) \\ (1-v)a\sin((1+u)\pi) \\ vH \end{vmatrix}, \quad u \in [0, 1], \ v \in [0, 1] \qquad (6.39)$$

We plot several constant-u and constant-v lines using the following MATLAB script.

```
%ELLIPTICONE Right, elliptic-cone coordinate lines
%   plotted from parametric equation

% constants and parameters
a = 6; b = 4; H = 5;
% define axes
plot3([ 0 -1.25*a ], [ 0 0 ], [ 0 0 ], 'k-',...
      [ 0 0 ], [ 0 -1.25*b ], [ 0 0 ], 'k-',...
      [ 0 0 ], [ 0 0 ], [ 0 1.25*H ], 'k-')
axis equal, axis off
hold on
% plot constant-v lines
u  = 0: 0.025: 1; v  = 0: 0.1: 1;
lu = length(u); lv = length(v);
x = zeros(1, lu); y = zeros(1, lu);
for k = 1:lv
    x = (1 - v(k))*a*cos((1 + u)*pi);
    y = (1 - v(k))*b*sin((1 + u)*pi);
```

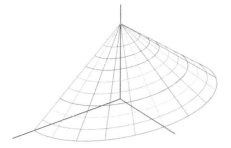

Figure 6.10 A conical surface

```
    z = v(k)*H*ones(1, lu);
    plot3(x, y, z, '-')
end
% plot constant-u lines
u  = 0: 0.1: 1; v  = 0: 0.025: 1;
lu = length(u); lv = length(v);
x = zeros(1, lu); y = zeros(1, lu);
for k = 1:lu
    x = (1 - v)*a*cos((1 + u(k))*pi);
    y = (1 - v)*b*sin((1 + u(k))*pi);
    z = v*H;
    plot3(x, y, z, '-')
end
hold off
```

The resulting plot is shown in Fig. 6.10.

6.7.3 Surfaces of Tangents

A third category of developable surfaces is generated by the tangents to a 3D curve. The general equation of such a surface is

$$\mathbf{P}(u, v) = \mathbf{C}(u) + v\mathbf{C}_u(u)$$

As an example we choose a helix as directrix. The equation of the surface is

$$\mathbf{C} = \begin{vmatrix} r\cos 2\pi u \\ r\sin 2\pi u \\ up \end{vmatrix} + v \begin{vmatrix} -2\pi r\sin 2\pi u \\ 2\pi r\cos 2\pi u \\ p \end{vmatrix} = \begin{vmatrix} r(\cos 2\pi u - 2\pi v\sin 2\pi u) \\ r(\sin 2\pi u + 2\pi v\cos 2\pi u) \\ (u + v)p \end{vmatrix}$$

$$(6.40)$$

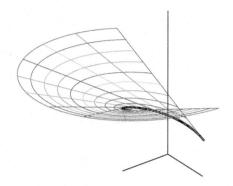

Figure 6.11 The surface of tangents to a helix

Fig. 6.11 shows the surface patch for $u \in [0,\ 0.5]$, $v \in [0, 0.5]$, $r = 10$, and $p = 30$. The script HelixTangents.m that produced the figure is stored on the MATLAB site of the book. We encounter similar surfaces in the drawings of screw propellers.

6.7.4 A Doubly-Ruled Surface, the Hyperboloid of One Sheet

The *hyperboloid of one sheet* is generated by rotating a hyperbola around the axis of symmetry that does not intersect the curve. The parametric equation of the resulting surface is (see Fig. 6.12)

$$\mathbf{P} = \begin{vmatrix} \cos 2\pi u \\ \sin 2\pi u \\ 0 \end{vmatrix} + v \begin{vmatrix} -\sin 2\pi u \\ \cos 2\pi u \\ 1 \end{vmatrix} = \begin{vmatrix} \cos 2\pi u - v \sin 2\pi u \\ \sin 2\pi u + v \cos 2\pi u \\ v \end{vmatrix},\ u \in [0,\ 1] \tag{6.41}$$

As the first row of Eq. (6.41) represents the coordinate x, the second row the coordinate y, and the third the coordinate z, squaring these rows, adding side by side the first and the second results and substituting in the sum the third one we obtain the Monge form

$$x^2 + y^2 = 1 + z^2 \tag{6.42}$$

Now, let us modify Eq. (6.41) by changing the signs of the x- and y-components of the direction vector

$$\mathbf{P} = \begin{vmatrix} \cos 2\pi u + v \sin 2\pi u \\ \sin 2\pi u - v \cos 2\pi u \\ v \end{vmatrix},\ u \in [0,\ 1] \tag{6.43}$$

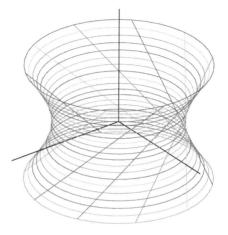

Figure 6.12 A hyperboloid of one sheet, first set of rulings

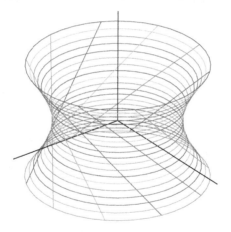

Figure 6.13 A hyperboloid of one sheet, second set of rulings

The Monge form of this equation is again Eq. (6.42). It follows that Eqs (6.41) and (6.13) represent the same hyperboloid of one sheet. The set of generators is, however, different and we see this in Fig. 6.13. The hyperboloid of one sheet is a *doubly-ruled* surface.

6.8 GEODESIC CURVATURE

In Section 6.5 we considered a curve **C** on a surface **R**. We called θ the angle between the normal to **C** at a point **P** and the normal to the surface

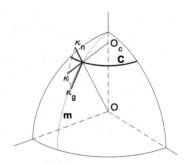

Figure 6.14 For the definition of geodesic curvature

at the same point. If the curvature of **C** at **P** is κ, the normal curvature of the surface at the same point is $\kappa \cos\theta$. Let us project the curve **C** on the plane tangent to the surface **R** at the point **P**. We call the curvature of the projected curve at **P** *geodesic* or *tangential curvature* and note it by κ_g. It can be shown that

$$\kappa_g = \kappa \sin\theta \tag{6.44}$$

As we did in Section 6.5 and in Example 6.3, we illustrate the above relationship on a sphere with radius R. In Fig. 6.14 the surface **R** is defined by its traces on the coordinate planes. The curve **C** is a parallel, a *small circle*. The angle between the principal normal to **C** and the normal to **R** is the latitude that we noted ϕ. The principal normal to **C** lies on the line that connects the point P to O_c, the centre of curvature of **C**. The normal to **R** lies on the line that connects P to O, the centre of the sphere. The latter point is the centre of curvature of all great circles of the sphere, in particular the meridian that passes through P. At the point **P** we show the curvatures κ, κ_n, κ_g as vectors. For the sake of readability the lengths of the vectors are exaggerated with respect to lengths on the sphere surface, but the ratios of the curvatures are preserved. The vector κ lies in the plane of the curve. The vector κ_n is perpendicular to the surface. We see it as perpendicular to the meridian **m** that passes through **P**. The vector κ_g lies in the plane tangent to **R** in P. This plane is perpendicular to $\overline{OP}$. Another representation of the above relationships is shown in Fig. 6.15, a section of the sphere through the plane of the meridian **m**. The vector κ is the resultant of κ_n and κ_g

$$\kappa^2 = \kappa_n^2 + \kappa_g^2 \tag{6.45}$$

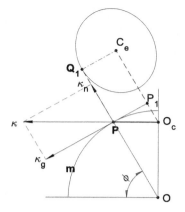

Figure 6.15 Meridian section of the sphere **R**

We calculated κ and κ_n in Example 6.3. Let us calculate now, by elementary means, the geodesic curvature of **C** at **P**. In Fig. 6.15 P_1 is the projection of O_c on the plane tangent in P. We project in the same way the whole circle **C**, rotate it by 90° and, for the sake of readability, translate it along the projection lines. Finally, the projection of **C** on the plane tangent in P is the ellipse with the centre C_e; its semi-axes are

$$a = R \cos \phi \sin \phi$$
$$b = R \cos \phi$$

As shown in Chapter 5, Example 5.1, the radius of curvature of the ellipse at Q_1 is b^2/a. Substituting the relevant values yields

$$\kappa_g = \frac{R \cos \phi \sin \phi}{R^2 \cos^2 \phi} = \frac{\sin \phi}{R \cos \phi} = \kappa \sin \phi$$

as expected. This verifies

$$\kappa_n^2 + \kappa_g^2 = \kappa^2 \cos^2 \phi + \kappa^2 \sin^2 \phi = \kappa^2 \tag{6.46}$$

6.9 DEVELOPABLE SURFACES

A developable surface is a surface that can be unbent on a plane without stretching, shrinking, or tearing. A surface is developable if it is a ruled surface and the tangent plane in a point of a generator remains the same when the point moves along the generator. It can be proved that a surface that fulfills these conditions has zero Gaussian curvature everywhere. This important result is a particular case of a famous theorem due to Gauss and known as **Theorema egregium**, translated as *The remarkable theorem* (see Gauss, 1828). The original statement in Latin is

'itaque art. praec. sponte perducit ad egregium
THEOREMA. Si superficies curva in quamcunque aliam superficias explicatur,
mensura curvaturas in singulis punctis invarianta manet.'

A commonly-accepted translation is

Thus the formula of the preceding article leads itself to the remarkable theorem.
If a curved surface is developed upon any other surface whatever, the measure
of curvature in each point remains unchanged.

In particular, a developable surface and a plane have the same Gaussian curvature, i.e. zero. The distance between two points of the surface is preserved on the developed surface. So are angles between curves and areas. Therefore, mathematicians say that the relationship between the given surface and its development is an *isometric mapping*. Obviously, the qualifier 'isometric' has here another meaning than in the term 'isometric projection' encountered in Chapter 1. The question whether a surface is developable or not is of paramount importance in ship construction. It is easier to produce a shell plate that is developable than a plate that is not developable. Usually, the ship shell is composed of plates that can be classified into three categories: flat plates, single-curvature plates, and double-curvature plates. Flat plates appear in the sides, possibly also in the bottom of parallel middlebodies. Such plates need no bending. *Single-curvature* plates is another name for plates that are developable; in most cases they are parts of a cylindrical surface, for example in the bilge region. Sometimes single-curvature are parts of a conical surface and can be found towards the ship extremities. *Double-curvature* plates are not developable; usually they are parts of the ship extremities. The percentage of each category of plates depends on the hull lines. For a certain tanker with 336 plates Lourenço (2010) indicates that 66% of them are planar, 23% single-curvature, and 11% double-curvature plates.

While all developable surfaces are ruled surfaces, not all ruled surfaces are developable. For example, the hyperboloid of one sheet and the hyperbolic paraboloid are ruled surfaces, in fact doubly-ruled surfaces, but are not developable surfaces. The only developable surfaces are the cylinders,

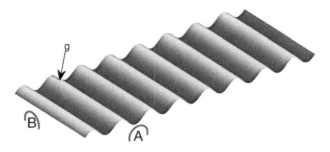

Figure 6.16 A corrugated panel

the cones and the surfaces of tangents to a space curve. The most important example of non-developable surface is, perhaps, the sphere. Because of this it is impossible to produce a geographical map that preserves distances, angles, and areas. Geographical maps are projections that preserve only part of the above properties. For example, *conformal maps* preserve angles. This is the case of the *Mercator projection* presented in 1569 by the Flemish Gerhard Mercator (1512–1594). On this map curves that make the same angle with the Northern direction appear as straight lines. In navigation these curves are known as *rhumb lines* or *loxodromes*. The shortest path between two points on the Earth surface does not lie on a rhumb line, but, as we already know, on a great circle. The Mercator projection is probably the most popular Earth map, but it distorts severely some measures. We invite the reader to take a look at a Mercator map. All parallels have the same length although it is obvious that their lengths decrease from that of the Equator to zero at the poles. The two poles are sent to infinity. Greenland appears larger than Australia although the converse is true. More on maps can be read, for example, in McCleary (1997).

We mentioned above why the property of being developable is important in shipbuilding. Not a few researchers studied and devised methods for designing ship hull surfaces that are developable. A manual method due to Kilgore is mentioned by Konesky (2005), computer-aided methods are treated in Clements (1981), Branco and Guedes Soares (2005), Konesky (2005). For a short review of the related literature and more comments see also Bertram (2004). The production of plates that are not developable require some plastic deformation as described, for example, by Letcher (1993), or Shin and Ryu (2000).

Example 6.5 (A corrugated panel). In Fig. 6.16 we consider a corrugated panel whose ruled surface belongs to the family of cylinders; it can be described by Eq. (6.38) with a permutation of coordinates. Such panels made of steel, composite materials, or cardboard are used in various fields, for instance for improvised roofs. The Gaussian curvature is zero because one of the principal curvatures, namely that of the rulings, is zero. If the material is flexible it is easy to bend the panel around one of its rulings, for example the one marked 'g' in the figure, because the resulting surface maintains the curvature. The way of bending so is shown schematically at A. It is not the same if one tries to bend the panel as shown schematically at B because then the curvature of the ruling will take a value different from zero. The resulting surface would have a Gaussian curvature different from zero and the initial panel cannot be bent over it without stretching and shrinking.

6.10 GEODESICS AND PLATE DEVELOPMENT

Given a surface **S** and two points on it, the shortest path on **S** that connects them is along a **geodesic** of **S**. However, the definition of a geodesic as the line of shortest distance on a surface causes some difficulties. For example, the geodesics of a sphere are its great circles. Given two points on the sphere, P_1 and P_2, the two points divide the great circle passing through them into two arcs. One of the arcs is the shortest distance between P_1 and P_2, while the other is not. Obviously, an exception occurs when the two points are *antipodal*, that is situated on the same diameter. Then, the lengths of the two arcs are equal. To avoid such difficulties it is usual to define the geodesics as lines of zero geodesic curvature. It follows from Eq. (6.45) that when the geodesic curvature of a curve is zero, the curvature of the curve is equal to the normal curvature of the surface in the same point. Then the principal normal of the curve coincides with the normal of the surface in the given point. Thus, the normals of a geodesic coincide everywhere with the normals of the surface. For a simple example let us return to the sphere. The geodesic curvature of the parallel **C** is null only for $\phi = 0$. Then **C** coincides with the Equator. At all points the principal normal of the Equator coincides with the normal to the sphere.

If a surface is developable its geodesics are mapped onto straight lines on the developed surface. This property can be used in developing the plates of a ship hull; detailed procedures are described, for example, in Mitamura (1961), Clements (1984), Zhao and Wang (2008), and Gotman (no year indicated).

Example 6.6 (A geodesic on a cone). Fig. 6.17 shows a right, circular cone and two points, A, B, on its surface. What is the shortest distance

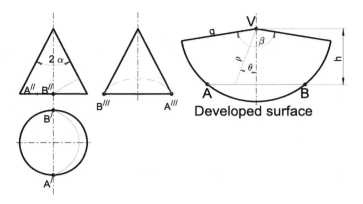

Figure 6.17 A geodesic on a cone

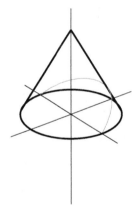

Figure 6.18 A geodesic on a cone, axonometric view

between these points? At the right of the figure we see the developed cone surface. The angle of the circular sector, β, is defined by $\beta = 360 \sin \alpha$ degrees, where α is the semi-angle at the cone vertex (see Section 1.18.3). Being diametrically opposed, each one of the points A, B appears on the development at an angle $\theta = \beta/4$ with reference to the axis of symmetry passing through the vertex V. The shortest path on the developed surface is, obviously, the straight-line segment joining the two points. Bending back the developed surface, the shortest path appears as the curve shown as a dashed line, red in the electronic version of the book, in three orthographic projections in the left-hand side of the figure and in axonometric projection in Fig. 6.18. We strongly recommend the reader to make an enlarged copy of the developed surface and to cut it out, leaving for gluing a short strip on one side. Bend the figure into a cone and convince yourself that the shown curve is, indeed, the geodesic that joins the given points. What about the symmetric curve?

Imagine now that you have a solid cone, similar to that in Fig. 6.17, and you stretch an elastic thread between the points *A* and *B*. The thread will take the shape of the geodesic shown in Figs 6.17 and 6.18. We discover thus a way of characterizing a geodesic without mathematics:

Given two points on a surface, the geodesic passing through them is the curve assumed by an elastic thread stretched on the surface between the two points.

We consider this fact as an application of the notion of geodesic and not as a definition because it would be very awkward to derive from it the mathematical properties of geodesics. Let us look again at Figs 6.17 and 6.18. The two points lie on a horizontal section of the cone, a circle. However, the shortest path on the surface, from one point to the other, is not along the circle as one could wrongly guess from the sketch of the cone. To follow the shortest path one has to 'climb' up the cone till mid–way and from there return down. In the case of the cone it is easy to explain why. On the developed surface it is obvious that the circular arc between the two points is longer than the straight line that represents the developed geodesic. The circle that is the base of the cone is not a geodesic. First, the projection of this circle on a tangent plane is an ellipse. The curvature of the ellipse is different from zero at all points. Second, the normals to the base circle do not coincide with the normals of the cone.

In real life we do not travel on the surface of a cone, but on that of the Earth. The first approximation of the Earth surface is a sphere (a better approximation is an ellipsoid). Therefore, we may wish to explain in the same way how to travel on a sphere along a geodesic. We cannot because the sphere is not developable. We can, however, find an analogy by considering in the Northern hemisphere two distant locations on the same parallel. The shortest path between the two locations is not along the parallel, but passes over the Arctic.

6.11 ON THE NATURE OF SURFACE CURVATURE

In Chapter 5 we analyzed the curvature of plane curves as their deviation from a straight line. The straight line is another object of the plane and this means that to measure the curvature of a one-dimensional curve we had to consider it embedded in a two-dimensional plane. In this chapter we discussed the curvatures of two-dimensional surfaces, among them that of the Earth sphere. Historically, the curvature of the Earth was discovered by ancient Greek mathematicians and astronomers by observing it as embedded in the three-dimensional space. For example, Aristotle (384–322 BC) wrote that one sees in Cyprus and Egypt stars that cannot be seen in Greece, a fact that can be explained only if the Earth is a sphere, and not a big one. In our terms he suggested thus that the curvature of the Earth is important. By referring to stars, objects that are not on the Earth surface, Aristotle considered the Earth as embedded in the three-dimensional space. Eratosthenes (276–194 BC) measured shadows of vertical rods and calculated the radius of the Earth with surprising precision. The rods did not belong to the Earth surface. School textbooks on geography contain more proofs, all referring to the Earth as embedded in the space. As planes flew higher and higher the man could see from above the curvature of the Earth and when astronauts landed on the Moon they took from there the whole picture of the Earth as a sphere.

It may be surprising, but it is possible to discover and measure the curvature of the sphere without the help of objects that lie outside the surface. Specifically, all the information is contained in the surface and it is sufficient to measure distances and angles on it. If we return to Eq. (6.35) we see that the Gaussian curvature, K, depends on the coefficients of the first and second fundamental forms. However, Gauss himself proved that K can be calculated using only the coefficients of the first fundamental form and their first and second derivatives. This fact seemed so surprising that Gauss described the resulting theorem as *egregium*, i.e. *remarkable*. The coefficients of the first fundamental form can be calculated from measurements on the two-dimensional surface without any reference to its embedment in the three-dimensional space. Therefore, the Gaussian curvature is an **intrinsic** property of a surface. For the genesis of this theorem see Merker (2014). To give an interpretation to this property we can imagine a two-dimensional observer, such as the one created by Abbot (1992) in his famous *Flatland*. Such a being cannot 'feel' the third dimension, but still can detect the curvature of the Earth by observations in two dimensions. We are going to show this with the help of *thought experiments*. The term originally coined in German is *Gedankenexperimente*.

In a first experiment we suppose that an observer travels on the Euclidean plane any distance a, turns 90° to the right, continues a distance a, turns 90° to the right, travels a distance a and stops. The final position is a distance a afar from the starting point. Next, suppose that the observer is travelling on the Earth sphere. Starting from a given point the observer travels on a parallel 90 degrees of longitude to the West, turns 90° to the right, travels a distance b to the North until reaching the Northern Pole, turns 90° to the right and travels a distance b in the Southern direction. The observer returns to the starting point! The difference between the two results is due to the difference between the curvature of the plane and that of the sphere.

In a second Gedankenexperiment the observer fixes a rod on the Earth surface and tightens to it a rope of length L. The observer pulls the rope and keeping it tightened on the Earth surface turns 360 degrees around marking a circle on the surface. Next, the observer measures the length of the circle. The expected length is $2\pi L$, the measured one is shorter. Obviously, the larger the length L, the greater is the difference between the expected and the measured length. Imagine a meridian section containing the rope and explain why the above difference is due to the curvature of the surface.

In a third experiment we consider a *spherical triangle*. This figure on the sphere surface is composed of three arcs of great circles. For an immediately obvious example let us suppose that the coordinates of the vertices are

0° E	0° N
90° E	0° N
0° E	90° N

Each angle in this triangle is right; the sum of the angles is three right angles. We know from elementary plane geometry that for any triangle the sum of its angles equals two right angles. We do not always remember that this statement is true only if the fifth postulate of Euclid is true. The geometry on the sphere is not Euclidean.

We do not continue our experiments although this is possible, but we like to return to the last result. A generalized statement is that for a triangle on a given surface the sum of its angles equals

- two right angles if the Gaussian curvature of the surface is zero;
- more than two right angles if the Gaussian curvature is positive. Example: the sphere;
- less than two right angles if the Gaussian curvature is negative. Example: the hyperboloid of one sheet.

6.12 SUMMARY

A surface can be represented mathematically in three ways, as described below.
1. — By implicit equations of the form $f(x, y, z) = 0$.
2. — By explicit equations like $z = f(x, y)$.
3. — By parametric equations of the form

$$\mathbf{X}(u, v) = \begin{vmatrix} f_1(u, v) \\ f_2(u, v) \\ f_3(u, v) \end{vmatrix}$$

The lines of constant-u and the lines of constant-v form a net of coordinates on the surface. A relationship between the parameters u and v defines a curve on the surface, for example $\mathbf{C} = X(u(t), v(t))$. Given a surface $\mathbf{X}$, let us consider the partial derivatives

$$\mathbf{X}_u = \partial \mathbf{X}/\partial u$$
$$\mathbf{X}_v = \partial \mathbf{X}/\partial v$$

When these derivatives are calculated at a point P, the first one represents the tangent vector at P to the constant-v curve that passes through P, and the second the tangent vector at P to the constant-u curve that passes through the same point. The unit normal to the surface is given by

$$\mathbf{N} = \frac{X_u \times X_v}{||X_u \times X_v||}$$

The square of an element of length on the surface is

$$ds^2 = Edu^2 + 2Fdudv + Gdv^2 \tag{6.47}$$

where

$$E = \mathbf{X}_u \cdot \mathbf{X}_u, \quad F = \mathbf{X}_u \cdot \mathbf{X}_v, \quad G = \mathbf{X}_v \cdot \mathbf{X}_v$$

The expression in Eq. (6.47) is known as the *first fundamental form* and is noted I. If we call α the angle between the constant-u and the constant-v lines, we have

$$\cos \alpha = \frac{F}{\sqrt{EG}}$$

The lines of coordinates are orthogonal if $F = 0$. For a more general theorem let us consider the lines defined by $\mathbf{X}(u_1(t), v_1(t))$ and $\mathbf{X}(u_2(t), v_2(t))$. The angle between the two curves is given by

$$\cos\alpha = \frac{Edu_1\,du_2 + F(du_1\,dv_2 + du_2\,dv_1) + Gdv_1\,dv_2}{\sqrt{Edu_1^2 + 2Fdu_1\,dv_1 + Gdv_1^2}\sqrt{Edu_2^2 + 2Fdu_2\,dv_2 + Gdv_2^2}}$$

The condition for orthogonality is

$$Edu_1\,du_2 + F(du_1\,dv_2 + du_2\,dv_1) + Gdv_1\,dv_2 = 0$$

The area of an elemental parallelogram on the surface is

$$dA = \sqrt{EG - F^2}\,dudv$$

With the notations

$$e = \mathbf{R}_u \cdot \mathbf{M}_u, \;\; 2f = \mathbf{R}_u \cdot \mathbf{M}_v + \mathbf{R}_v \cdot \mathbf{M}_u, \; g = \mathbf{R}_v \cdot \mathbf{M}_v$$

the *second fundamental form is defined by*

$$II = edu^2 + 2fdudv + gdv^2$$

Let $\mathbf{C}$ be a curve on a surface $\mathbf{R}$, $\mathbf{N}$ the principal normal of the curve in a point P, $\mathbf{M}$ the normal to $\mathbf{R}$ in the same point, and θ the angle between $\mathbf{N}$ and $\mathbf{M}$. The curvature of $\mathbf{C}$ in P is given by

$$\kappa\cos\theta = \frac{\cos\theta}{\rho} = \frac{edu^2 + 2fdudv + gdv^2}{Edu^2 + 2Fdudv + Gdv^2}$$

On a given surface, in a non–singular point P, there is only one normal to the surface, but infinitely many tangents. The normal and each tangent define a plane normal to the surface that cuts the surface along a certain curve. The curvature of that curve in the given point is the *normal curvature* of the surface in the direction of the chosen tangent. Euler proved that there are two directions, perpendicular one to the other, such that the normal curvature for one of the directions is minimal, for the other maximal. Let us note the former κ_{min}, the latter κ_{max}. The *Gaussian curvature* of the surface is defined by

$$K = \kappa_{min}\kappa_{max}$$

and the *mean curvature* by

$$H = (\kappa_{min} + \kappa_{max})/2$$

Using the coefficients of the first and second fundamental forms these curvatures are given by

$$K = \frac{eg - f^2}{EG - F^2}, \quad H = \frac{1}{2} \frac{eG - 2fF + gE}{EG - F^2}$$

A **ruled surface** can be defined by one of the following equivalent statements:

1. A surface such that through each point of it passes a straight line that is fully contained in the surface.
2. A surface generated by the motion of a straight line.
3. The set of a family of straight lines depending on a parameter that spans a set of real numbers.

The straight lines mentioned in the definitions are the *generators* or *rulings* of the surface. The general equation of a ruled surface is

$$\mathbf{P}(u, v) = \mathbf{C}(u) + v\mathbf{r}(u)$$

We distinguish several categories of ruled surface, as listed below.

Cylindrical — the rulings are parallel to a given direction.

Conical — the rulings pass through a given point.

Surfaces of tangents to a space curve with the general equation

$$\mathbf{P}(u, v) = \mathbf{C}(u) + v\mathbf{C}_u(u)$$

Let us consider a curve on a surface and the plane tangent to the surface in a given point of the curve. The curvature of the projection of the curve on the tangent plane is the *geodesic curvature*. A *geodesic* is a curve whose geodesic curvature is zero. At each point of the curve its principal normal coincides with the normal to the surface. Given two points on a surface the shortest path, on the surface, between them is along a geodesic. If the surface is developable, its geodesics map onto straight lines on the developed surface. Several methods for developing plates of the ship shell are based on the use of geodetics.

Two surfaces that have the same Gaussian curvature can be bent one over the other without stretching, shrinking, or tearing. A developable surface is a surface that can be unbent on a plane without stretching, shrinking, or tearing. A surface is developable if it is a ruled surface and the tangent plane in a point of a generator remains the same when the point moves along the generator. A developable surface has zero Gaussian curvature everywhere, like the plane. The distance between two points of the surface is preserved on the developed surface. So are angles between curves and areas. Therefore, mathematicians say that the relationship between a given surface and its development is an *isometric mapping*. The only developable surfaces are the cones, the cylinders, and the surfaces of tangents to a space curve. Not all ruled surfaces are also developable. For example, the hyperbolic paraboloid is a doubly ruled surface, yet it is not developable. Perhaps the most important example of non-developable surface is the sphere. The consequence is that is impossible to produce a map of an Earth region that preserves distances, angles and areas at the same time. Maps are projections that preserve only part of the mentioned properties. For example, the Mercator projection preserves angles, but distorts areas.

For picturesque and intuitive interpretations of the material explained in this chapter we recommend the reader to refer to Hilbert and Cohn-Vossen (1944).

6.13 EXERCISES

Exercise 6.1 (Cylinder — Tangents and normal). This exercise refers to the cylinder represented by Eqs (6.1). Let $r = 1$, $h = 2$. You are asked to

1. — write the equation of the unit tangents to the constant-u lines;
2. — write the equation of the unit tangents to the constant-v lines;
3. — write the equation of the unit normals to the constant-v lines;
4. — draw a line of constant v and show on it the unit tangents in the points corresponding to $u = 0.125$ and 0.25;
5. — on the same figure show the unit normal corresponding to $u = 0.125$.

Exercise 6.2 (A cone surface). We refer to the cone surface shown in Fig. 6.19. Your tasks are

1. to write the parametric equations of the surface;
2. to find the equation of the vector tangent to a v-constant line;
3. to find the equation of the vector tangent to a u-constant line;
4. find the equation of the unit normal to the surface.

Exercise 6.3 (A cone surface). As for Exercise 6.2 but for the cone shown in Fig. 6.20.

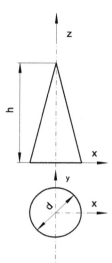

Figure 6.19 A cone surface

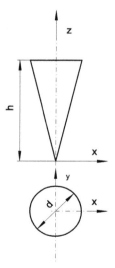

Figure 6.20 A cone surface

Exercise 6.4 (A cone surface). As for Exercise 6.2 but for the cone shown in Fig. 6.21.

Exercise 6.5 (Geographic coordinates — Orthogonality). Prove that the net of parallels and meridians is orthogonal.

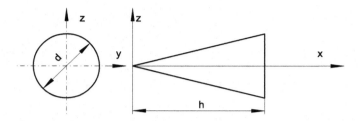

Figure 6.21 A cone surface

Exercise 6.6 (Cylinder area). Consider a right circular cylinder with radius r and height h. Find the area of the surface using Eq. (6.16).

Exercise 6.7 (Ruled surface — Cylinder). The exercise refers to the cylindrical surface described by Eq. (6.38). Your tasks are listed below.
1. Prove that the normal unit vector does not depend on the parameter v. This means that the tangent plane is constant along the generatrix.
2. Prove that the Gaussian curvature equals everywhere zero.
3. Calculate the mean curvature and show that it equals, indeed, the mean of the curvatures of the directrix and the generatrix.

Exercise 6.8 (Ruled surface — An elliptic cone). This exercise refers to the elliptic cone shown in Fig. 6.10. Your tasks are listed below.
1. Prove that the normal unit vector does not depend on the parameter v. This means that the tangent plane is constant along the generatrix.
2. Derive the Monge equation $z = f(x, y)$ of the surface.

Exercise 6.9 (Bending a cone on a cylinder). Cones and cylinders have the same Gaussian curvature; that is zero. Therefore, a patch of a cone surface can be bent on a cylinder surface without stretching, shrinking or tearing. As an example refer to the developed surface in Fig. 6.17. Make a sketch showing how would this circular segment look if bent on a cylinder having the same radius as the initial cone. Include the geodesic between the two given points.

APPENDIX 6.A   A FEW MULTISURF TOOLS FOR WORKING WITH SURFACES

The programs used in hull design include extensive tools for generating and viewing surfaces. As a help for fairing the ship lines such programs also provide displays of surface curvatures. We are going to exemplify a

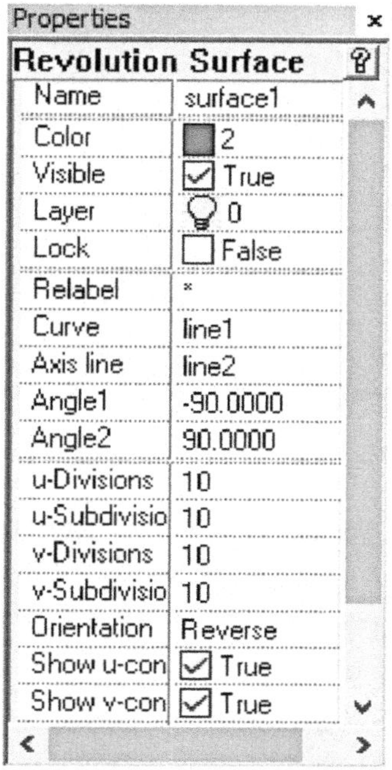

Properties	✕
Revolution Surface	❓
Name	surface1
Color	■ 2
Visible	☑ True
Layer	💡 0
Lock	☐ False
Relabel	✻
Curve	line1
Axis line	line2
Angle1	-90.0000
Angle2	90.0000
u-Divisions	10
u-Subdivisio	10
v-Divisions	10
v-Subdivisio	10
Orientation	Reverse
Show u-con	☑ True
Show v-con	☑ True

Figure 6.22 The properties manager of the surface of revolution

few tools of MultiSurf. In this software the relationship to the theory is particularly visible. As we have done in the previous chapters, we do not use yet the advanced curves and surfaces that will be treated in later chapters of the book. Our example is a circular cylinder generated as a surface of revolution. We start by defining four points

$$
pt1 = \begin{vmatrix} 0 \\ 0 \\ 0 \end{vmatrix}, \; pt2 = \begin{vmatrix} 0 \\ 0 \\ 1 \end{vmatrix}, \; pt3 = \begin{vmatrix} 10 \\ 0 \\ 0 \end{vmatrix}, \; pt4 = \begin{vmatrix} 10 \\ 0 \\ 1 \end{vmatrix}
$$

Next, we choose pt3 and pt1 to define line1, and pt4 and pt2 to define line2. We generate the cylinder by rotating the straight–line segment line1 around the axis line2 with

Insert → Surface → Revolution Surface

and set the properties of the surface as displayed in Fig. 6.22. After selecting the surface by clicking on it, the view displays the origin of the parameters

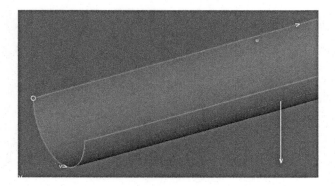

Figure 6.23 Notations on the surface

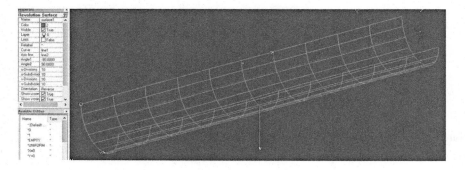

Figure 6.24 Wireframe view of a circular cylinder

marked by 'O', the directions in which the parameters u and v increase, and the normal to the surface at the point $u = 0.5$, $v = 0.5$. See Fig. 6.23. The constant-u and -v lines can be visualized in the wireframe display shown in Fig. 6.24. This view can be obtained by either clicking on the wireframe icon or by opening the View menu, choosing in it Display, and in the submenu Wireframe. To experiment a little with the parametrization of the surface let us place a point on it. In the terminology of MultiSurf a point bound to a surface is called *magnet*. A magnet is defined by selecting the supporting surface and following the sequence

Insert → Point → Magnet

By default the point is placed at the position corresponding to $u = 0.5$, $v = 0.5$. This position can be changed by dragging the point with the mouse

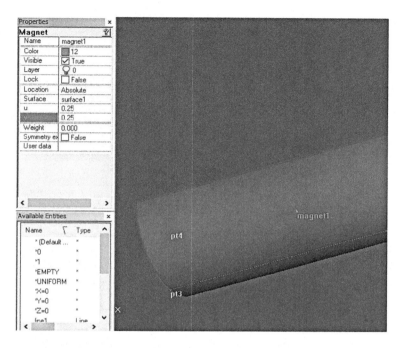

Figure 6.25 A magnet point on the cylindrical surface

Figure 6.26 Curvature options

or by setting other values in the `Properties` manager. Fig. 6.25 corresponds to $u = 0.25$, $v = 0.25$.

To examine the curvatures of the surface one has to select it and go through

$$\text{View} \rightarrow \text{Display} \rightarrow \text{Surface Curvature}$$

The dialogue window that opens is shown in Fig. 6.26. The default settings produce Fig. 6.27. The constant-v lines are parallel to the cylinder

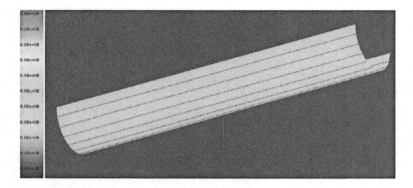

Figure 6.27 Normal curvature in the direction of constant *u* lines

Figure 6.28 The Gaussian curvature of a circular cylinder

generatrix and the displayed colour corresponds to zero curvature. Changing the angle to 90 produces a display of half-circles parallel to the directrix of the cylindrical surface. The displayed colour of the normal curvature in their direction corresponds, indeed, to 1. If we check the Mean curvature, the colour corresponds practically to 0.5, the mean of 1 and 0. Let us return to the shaded view and the Surface Curvature dialogue window. We choose now Gaussian; the result is shown in Fig. 6.28 and the displayed colour corresponds to the zero value. As we already knew from Chapter 1, and as we learned more details in this chapter, this surface is developable.

PART 3

Computer Methods

CHAPTER 7

Cubic Splines

Contents

7.1 INTRODUCTION

Many ship sections cannot be described by a single simple equation. As explained in Section 2.1, this fact was known in another form a few hundred years ago and shipbuilders devised empirical procedures for drawing transverse ship sections as a succession of arcs with tangent continuity at the joining points. This is the basic idea of the mathematical objects that are called today *splines*. The term comes obviously from the drawing instrument specific to Naval Architecture and it was Schoenberg (Issac Jacob, born in Romania 1903, died in USA 1990) who borrowed it for the functions developed by him. The general idea is to divide a curve arc into several segments and fit to each segment its own function, while imposing continuity conditions at the joining points. The preferred functions are polynomials; they have several useful features such as

- there is an efficient algorithm, Horner's scheme, for calculating polynomial values at given points;
- it is easy to find the coefficients of a polynomial that passes through given points;
- given a polynomial it is easy to differentiate or integrate it.

On the other hand, from a certain degree up polynomials can oscillate wildly, a phenomenon known as *polynomial inflexibility*. To give an example with a Naval-Architectural flavour, let us draw a station such as can be encountered in the aft of many ships. We choose Station 1 of the ship

Geometry for Naval Architects
https://doi.org/10.1016/B978-0-08-100328-2.00018-3
Copyright © 2019 Elsevier Ltd.
All rights reserved.
305

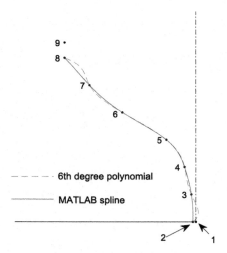

Figure 7.1 Fitting curves to an aft station

model C1189 tested in the Rome basin (INSEAN, 1965). The coordinates, in millimeters, of nine points along the station are

```
Point   Draught   Half-breadth
  1       0.0         0.0
  2       0.0         6.7
  3      60.5         9.7
  4     121.1        24.2
  5     181.6        63.0
  6     242.2       157.4
  7     302.7       227.7
  8     363.3       280.9
  9     397.2       280.9
```

These points are shown as black circles in Fig. 7.1. The arcs between points 1 and 2, and between points 8 and 9 are straight-line segments; for the moment we do not treat them. We fit a polynomial to the points 2 to 8. To do this in MATLAB we build first the array of half breadths, B10, and the array of draughts, T10. Next, we call the function polyfit with three input arguments, T10, B10, and the degree of the requested polynomial. There are seven points, they define a 6th-degree polynomial. The code is

```
c   = polyfit(T10, B10, 6);
```

The output argument, c, is the array of polynomial coefficients in the descending order of the exponents. The MATLAB function `plot` draws straight-line segments between given points. To plot a smoothly-looking curve we must interpolate enough points between the given ones. We call for this the function `polyval` with two input arguments, the array of coefficients and an array of interpolating points

```
Tmx = max(T10);
Ti  = 0: Tmx/50: Tmx;
B1i = polyval(c, Ti);
plot(B1i, Ti, 'r--')
```

The result is shown in Fig. 7.1 as a dash-dash line, red in the electronic version of the book. The arcs between the points 2 and 3, and between the points 7 and 8 cannot be considered as faired. This polynomial solution is not satisfactory.

7.2 CUBIC SPLINES

In the preceding section we have learned that it may not be good to use high-degree polynomials. There is also a lower limit of the degree, 3. Second-degree polynomials do not have a second derivative; they cannot represent a curve with an inflexion point such as is the one close to point 6 in Fig. 7.1. Having chosen the 3d degree polynomial the procedure for building a **cubic spline** is based on the following considerations.

- Given $n + 1$ points with the coordinates x_i, y_i, fit between each pair of points a cubic polynomial, in total n polynomials. Let us call $s_i(x_i)$ the polynomial between points i and $i + 1$. Each polynomial is defined by four coefficients. To find all coefficients we need $4n$ conditions.
- Impose the condition that the spline passes through the given points

$$S_i(x_i) = y_i, \ S_i(x_{i+1}) = y_{i+1}$$

There are two such conditions for each polynomial, in total $2n$ conditions.

- At the joining point, x_i, of two polynomials impose the condition that the two adjacent arcs have the same tangent (C^1 continuity)

$$\frac{dS_i(x_i)}{dx} = \frac{dS_{i+1}(x_i)}{dx}$$

As there are $n - 1$ joining points, we obtain $n - 1$ conditions.

- At the joining point, x_i, of two polynomials impose the condition that the two adjacent arcs have the same curvature. If the preceding condition is fulfilled, Eq. (5.8) shows that it remains to impose

$$\frac{d^2 S_i(x_i)}{dx^2} = \frac{d^2 S_{i+1}(x_i)}{dx^2}$$

This means C^2 continuity. As there are $n-1$ joining points, we obtain another $n-1$ conditions.

- At this point we obtained $4n-2$ conditions out of the required $4n$. It remains to impose two additional conditions. A frequent choice is to let the curvature at the end points be zero. This is what happens when drawing with a *spline* and the instrument is forced to pass through the end points, while it is left free beyond those points. Therefore, with this condition the result is called *natural spline*. Alternatively, it is possible to specify the slopes at the end points and then we talk about a *clamped spline*.

7.3 THE MATLAB SPLINE

In MATLAB the cubic spline is implemented by the function `spline`. For a set of points defined by the arrays of coordinates x and y, the function is called with three arguments, in this order: the array x, the array y, and the array x_i of x-coordinates at which we want to interpolate y-values. Thus, for the points exemplified in Section 7.2 the code is

```
B1s = spline(T10, B10, Ti);
plot(B1s, Ti, 'b-')
```

The resulting curve is shown in Fig. 7.1 as a solid line, blue in the electronic edition of this book.

The equation of the spline is

$$S_i = c(i, 1)(x - x_i)^3 + c(i, 2)(x - x_i)^2 + c(i, 3)(x - x_i) + c(i, 4) \qquad (7.1)$$

The array of coefficients can be retrieved with the commands

```
pp = spline(T10, B10)
[ breaks, c, 11, kk, dd ] = unmkpp(pp)
```

where `breaks` are the x-coordinates of the given points, and c is the array of coefficients

```
c =
   1.0e+02 *
  0.000000013720311  -0.000018182488479   0.000550595238095  -0.066601562500000
  0.000000013720311  -0.000015690322581  -0.001500297619048  -0.096875000000000
 -0.000000641444844  -0.000013198156682  -0.003249404761905  -0.242187500000000
  0.000001133589971  -0.000129710599078  -0.011902083333333  -0.629687500000000
 -0.000000292185799   0.000076195391705  -0.015142261904762  -1.574218750000000
 -0.000000292185799   0.000023122580645  -0.009128869047619  -2.276562500000000
```

The output argument 11 is the number of cubic polynomials; it is equal to the number of given points minus one. We can use the output of unmkpp to plot the spline

```
for k = 1:11
  xi = breaks(k) : 0.05: breaks(k+1);
  zi = c(k, 1)*(xi - breaks(k)).^3 + ...
       c(k, 2)*(xi - breaks(k)).^2 + ...
       c(k, 3)*(xi - breaks(k)).^1 + ...
       c(k, 4)*(xi - breaks(k)).^0
  plot(zi, xi, 'g-')
end
```

The result is identical to that shown in Fig. 7.1. The information yielded by the function unmkpp can be used for performing other operations, for instance the calculation of the area under the curve, or the area of the section represented by the spline (see Biran, 2011, pp. 157–65).

Example 7.1 (The sine function — Plot and area under the curve). The following MATLAB code plots first 10 points of the sine curve. As seen in Fig. 7.2, the maximum value, 1, is not given. Next, the MATLAB spline function is used to interpolate more points between those given. Interesting, the spline yields the maximum value. Printing it with four decimal digits we obtain, indeed, 1. However, using the MATLAB command format long reveals a value slightly under 1.

```
x = 0: pi/9: pi;
y = sin(x);
xi = 0: pi/200: pi;
yi = spline(x, y, xi);
plot(x, y, 'ro', xi, yi, 'k-')
xlabel('\phi, rad')
ylabel('sin \phi')
disp('y maximum')
max(yi)
```

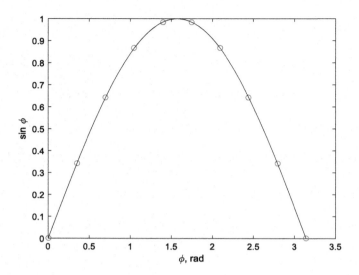

Figure 7.2 Using `spline` to plot the sine function

```
disp('Area under the curve')
trapz(xi, yi)
```

Finally, the code exemplifies how to use the spline to calculate the area under the curve by the *trapezoidal rule*. The exact value is given by

$$\int_0^\pi \sin x \, dx = 2$$

This is what we see in the display with four decimal digits. Again, using the MATLAB command `format long` reveals a value slightly under 2. This approximation can be improved by taking more points in `xi`. At a first glance this procedure may not look like the best one, but in cases such as that exemplified in Exercise 7.6 it is very convenient.

7.4 WORKING WITH PARAMETRIC SPLINES

Fig. 7.3 shows a number of points along a typical wall-sided transverse ship section. The bilge is a quarter of circle. The points are produced by the following MATLAB script.

```
%WALLSIDED Produces and plots points along a wall-sided,
% round-bilge transverse section.
```

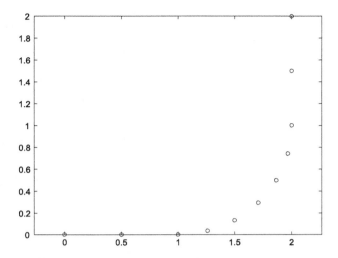

Figure 7.3 Points on a wall-sided transverse section

```
% define bilge
r     = 1;              % bilge radius
alpha = 270: 15: 360;   % axis of angles, deg
yb    = 1 + r*cosd(alpha);
zb    = 1 + r*sind(alpha);
y  = [ 0 0.5 yb 2 2 ];
z  = [ 0 0 zb 1.5 2 ];
plot(y, z, 'ko'), axis equal
```

If we try to apply the MATLAB spline function to the above points we'll elicit an error message that contains the sentence

```
The data sites should be distinct.
```

We get this message because two points have the same y-coordinate.

The first solution that may cross our mind is to divide the section into three segments, use a spline for the curved arc and straight lines for the other parts. We remark, however, that there is a continuity of tangents at the points where the round bilge joins the straight-line segments of the station. Therefore, in this case there is a much better solution and it consists in fitting one spline to the x-coordinates, and another spline to the y-coordinates. We speak then about *parametric splines*. To do this we define a vector of parameter values in the interval [0 1] and fit splines that represent each coordinate as a function of the chosen parameter. The following script implements this solution.

```
%DOUBLESPLINE1  Demonstrates the efficiency of defining separate
%    splines for each coordinate in part, by drawing a section
%    with flat bottom, circular bilge, and wall side.

% define bilge
r     = 1;              % bilge radius
alpha = 270: 15: 360;   % axis of angles, deg
yb    = 1 + r*cosd(alpha);
zb    = 1 + r*sind(alpha);
% define curve
t  = 0: 1/10: 1;        % curve parameter
ti = 0: 1/100: 1;       % interpolation parameter
y  = [ 0 0.5 yb 2 2 ]; z  = [ 0 0.0 zb 1.5 2 ];
Y  = spline(t, y, ti); Z  = spline(t, z, ti);
subplot(1, 3, 1)        % y(t)
    plot(ti, Y, 'k-')
    title('a'), xlabel('ti'), ylabel('Y')
subplot(1, 3, 2)        % z(t)
    plot(ti, Z, 'k-')
    title('b'), xlabel('ti'), ylabel('Z')
subplot(1, 3, 3)        % z(y)
    plot(Y, Z, 'k-', 'LineWidth', 1.5)
    axis equal
    title('c'), xlabel('Y'), ylabel('Z')
```

The results are displayed in Fig. 7.4.

7.5 SPACE CURVES

In Chapter 4 we learned how to plot a parametric curve in 3D space. In this section we show how to do it using parametric splines. Our example is the deck line of a real vessel. We omit the aft end because it has a discontinuity. The *deck-at-side* line is sometimes called *sheer line*. The coordinates at Stations 2–10 and at the forward extremity are

x	y	z
3000	2700	3590
4500	2790	3565

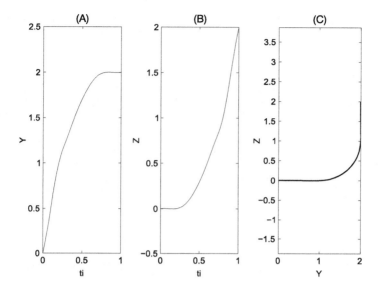

Figure 7.4 A parametric spline

6000	2880	3560
7500	2925	3560
9000	2940	3580
10500	2835	3660
12000	2670	3725
13500	2340	3820
15000	1770	3940
17580	0	4140

The following code interpolates more points and plots the axonometric view of the deck line and its projections on the body, sheer, and waterlines plans. The result is shown in Fig. 7.5.

```
xi = 3000: 10: 17580;
yi = spline(x, y, xi); zi = spline(x, z, xi);
plot3(xi, yi, zi, 'k-', 'LineWidth', 1.5), grid
axis equal, box
hold on
plot3(xi, yi, zeros(size(xi)), 'r-')
plot3(xi, 3000*ones(size(xi)), zi, 'r-')
plot3(18000*ones(size(yi)), yi, zi, 'r-')
xlabel('x'), ylabel('y'), zlabel('z')
hold off
```

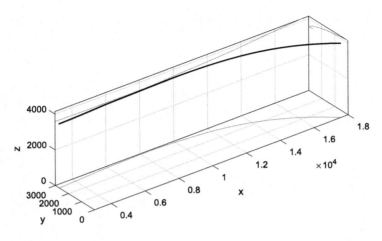

Figure 7.5 A space curve

7.6 CHORD-LENGTH PARAMETRIZATION

In the preceding section we defined a parameter that spans uniformly the interval [0 1]. In other words, the parameter values are equally spaced in the interval [0 1]. The corresponding points on the curve may not be equally spaced along the curve. To better relate the parameter values to the location of points along the curve we must know the lengths of arcs along the curve. We may say that we need the *curvilinear abscissa* of the corresponding curve points. In general it may be difficult or even impossible to obtain them. The literature of specialty cites an article in which Epstein (1976) proposes as an approximate approach the **parametrization by chord length**. Below we are going to apply this idea to the case studied in Section 7.4. First, we write a function that creates the array of parameter values.

```
function t = ChordLparam(points)
%CHORDLPARAM Values for chord-length parametrization
%    Given a sequence of points this function generates a sequence
%    of parameter values distributed in the interval [ 0 1 ]
%    proportionally to the distances between the input points.
%    Input: array of point coordinates in the format
%             [ x
%               y ]
%    Output: row vector of t values in the interval [ 0 1 ]
```

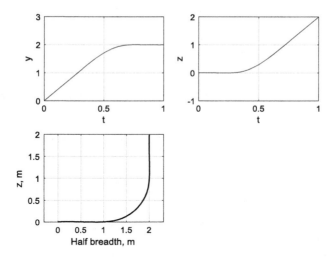

Figure 7.6 Points on a wall-sided transverse section obtained by chord-length parametrization

```
dt      = diff(points, 1, 2)      % calculates vector of differences
[ m n ] = size(dt);
dl      = zeros(1, n);            % allocates memory
for k = 1:n
    dl(k) = norm(dt(:, k))        % calculates chord lengths
end
t0 = cumsum([ 0 dl ])             % sums chord lengths
t  = t0/t0(end)                   % normalize
end                               % end of function
```

The above function is called in the following script and the resulting $x(t)$, $y(t)$ and $y(x)$ curves are shown in Fig. 7.6. The bottom and the side display slight deviations from straight lines. The deviations can be reduced by inserting more points on the respective segments. Other solutions are shown in Section 7.7 and in the Appendix.

```
%DOUBLESPLINE3 Version with chord-length parametrization

% define bilge
r     = 1;                    % bilge radius
alpha = 270: 15: 360;
yb    = 1 + r*cosd(alpha); zb = 1 + r*sind(alpha);
```

```
% define curve
y  = [ 0 0.5 yb 2 2 ]; z  = [ 0 0 zb 1.5 2 ];
P  = [ y; z ];
% define curve parameter
t = ChordLparam(P);
ti = 0: 1/100: 1;        % interpolation parameter
Y  = spline(t, y, ti); Z  = spline(t, z, ti);
plot(Y, Z, 'LineWidth', 1.5), grid, axis equal
xlabel('Half breadth'), ylabel('z')
```

7.7 CENTRIPETAL PARAMETRIZATION

MATLAB provides the function cscvn that parametrizes the spline by the square root of the chord length; it was claimed that it yields better results than parametrization by chord length. By some analogy with kinematics this method was called *centripetal parametrization*. The following code implements an example. We draw again the station shown in Fig. 7.1, but this time we can use a simple trick to treat the whole station as one curve: the points of discontinuity are entered twice in the input.

```
B = 775;          % actual moulded breadth, mm, on drawing 160 mm.
S = 775/160;      % standard scale is 1/S
% Enter arrays of measured draughts and half-breadths and scale
% Discontinuity points 2 and 8 are entered twice.
T1 =   S*[ 0.00 0.000 0.000 12.5 25.0 37.5 50.0 62.5 75.0 75.0 82.0 ];
B1 = -S*[ 0.00 1.375 1.375   2.0  5.0 13.0 32.5 47.0 58.0 58.0 58.0 ];
% plot BL and CL
plot([ 0 -B/2 ], [ 0 0 ], 'k-', [ 0 0 ], [ 0 0.6*B ], 'k-.')
axis equal, axis off
hold on
% 1 - plot given points
lT = length(T1);
pt = [ B1; T1 ];
for k = 1:lT
    point(pt(:, k), 2)
    if (k > 3)&(k<10)
        text((pt(1,k) - 25), pt(2, k), num2str(k-1))
    end
end
% Centripetal parameterization
points = [ B1; T1 ];
```

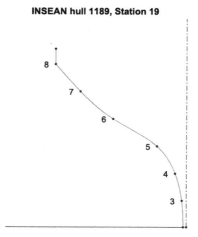

Figure 7.7 Station plotted with centripetal parametrization

```
[P, t] = fnplt(cscvn(points));
plot(P(1, :), P(2, :), 'b-'), grid, axis equal
title('INSEAN hull 1189, Station 19')
xlabel('y'), ylabel('z')
hold off
pause
subplot(1, 2, 1)
    plot(t, P(1, :)), grid, xlabel('t'), ylabel('y, mm')
subplot(1, 2, 2)
    plot(t, P(2, :)), grid, xlabel('t'), ylabel('z, mm')
```

The resulting plot of the station is shown in Fig. 7.7, the coordinates are displayed as functions of one parameter in Fig. 7.8. Researchers observed that splines built with centripetal parametrization do not behave well under certain scaling transformations. It so happens that such transformations are used in some methods of modifying ship lines.

7.8 SUMMARY

Many ship sections cannot be described by a single, simple equation. Therefore, it is necessary to divide such a section into several segments and fit to each segment its own function, while imposing continuity conditions at the joining points. The preferred functions are polynomials. High-degree polynomials may display large deviations from the desired curve, a phe-

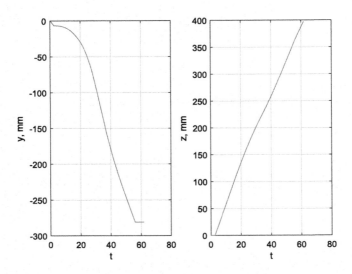

Figure 7.8 Coordinates obtained with centripetal parametrization

nomenon known as polynomial inflexibility. Therefore, the usual choice is that of cubic polynomials, the lowest degree that can represent points of inflection. MATLAB provides a function for cubic–spline interpolation, it is called `spline`. For example, given the coordinates y and z of a number of points along a ship station, the function yields the coordinates z_i of points interpolated at the coordinates y_i. This function fails, for example, if the station is wall sided. A solution is to treat the coordinates as functions of a parameter t. In this case we work with *parametric splines*. In the trivial solution the parameter is distributed uniformly in the interval [0, 1]. Better curve fitting can be obtained by distributing the parameter approximately proportional to the distance between the given points. This procedure is known as *chord-length parametrization*. It was claimed that it is still better to parametrize proportionally to the square root of the distance between the given points. This is *centripetal parametrization*; in MATLAB it is implemented by the function `cscvn`.

7.9 EXERCISES

Exercise 7.1 (Drawing a ship section, 1). The following points belong to Station 19 of the previously cited INSEAN model 1189. The dimensions are in mm.

Point	Draught	Half-breadth	Point	Draught	Half-breadth
1	0.0	0.0	7	242.2	55.7
2	12.1	24.2	8	302.7	79.9
3	24.2	29.1	9	363.3	111.4
4	60.5	33.9	10	423.8	145.3
5	121.1	38.8	11	484.4	188.9
6	181.6	43.6	12	523.1	218.0

Your tasks are:

1. Plot the points.
2. Fit the polynomial defined by the given points and plot it over the previous plot.
3. Add the corresponding MATLAB spline.

Exercise 7.2 (A forward station of a fishing vessel). The following points belong to station 20 of the INSEAN fishing-vessel model 647 (INSEAN, 1962). The dimensions are in metres.

z	0.192	0.192	0.262	0.421	0.525
y	0.000	0.015	0.015	0.086	0.141

1. Try to model this station using a cubic spline.
2. Calculate the half breadth at $z = 0.35$ m.

Exercise 7.3 (The design waterline of INSEAN model 1189). The following points approximate the design waterline of the INSEAN model 1189. The coordinates are measured in metres.

Station	Aft	0	5	10	15	20
x	−0.232	0.000	1.161	2.322	3.483	4.644
y	0.000	0.141	0.363	0.388	0.300	0.000

In this exercise we are going to check the influence of parametrization.

1. Use the spline $y = f(x)$ to plot the waterline and calculate the half breadths at stations 3, 7, 17. Do not forget the command `axis equal`.
2. Plot the waterline using parametric splines, $x(t)$, $y(t)$, with uniform parametrization. The resulting plot is inacceptable.
3. Plot the waterline, over the plot obtained at 1, using parametric splines with chord-length parametrization. Calculate the half breadths at stations 3, 7, 17. Identify the slight differences relative to the results obtained with the $y = f(x)$ spline.

Exercise 7.4 (The design waterline of the INSEAN fishing-vessel model 467). The following points approximate the design waterline of the INSEAN model 1189. The coordinates are measured in metres.

Station	1/2	1	5	10	15	20
x	0.0781	0.1563	0.7813	1.5625	2.3438	3.1250
y	0	0.0580	0.3330	0.4188	0.3784	0.0151

1. Use the spline $y = f(x)$ to plot the waterline and calculate the half breadths at stations 3, 7, 17. Do not forget the command `axis equal`.
2. Plot the waterline, over the plot obtained at 1, using parametric splines with chord–length parametrization. Use the MATLAB `spline` to calculate the half breadths at stations 3, 7, 17. Mind the slight differences relative to the answers to the first question.

Exercise 7.5 (Trawler sheer line). Below are the coordinates of six points belonging to the deck line of a trawler model (SNAME, 1966, Model 68). We omitted the aft extremity to simplify the exercise. Plot the deck line using the MATLAB function `spline`. If you can use a surface modelling program, such as MultiSurf, try to model the line in this software too and compare the results.

Station	0	5	10	15	20	Stem
x	0.000	0.619	1.238	1.860	2.470	2.510
y	0.130	0.249	0.260	0.200	0.020	0.000
z	0.320	0.308	0.308	0.340	0.420	0.430

Exercise 7.6 (Curve of statical stability). In this chapter we have dealt with cubic splines as a tool for drawing ship lines. Cubic splines, however, are a general tool for interpolating points for plotting. An important application in Naval Architecture is in the drawing of the curve of *statical stability*. For this concept see, for example, Biran and López–Pulido (2014), Chapter 5.

1) Use the MATLAB function `spline` to plot the curve of the righting arms given below; they belong to a real vessel. **Hint:** intervals of 2.5 degrees are usually suitable for a smooth appearance.

Heel angle ϕ, degrees	Righting, arm $\overline{GZ}$, m	Heel angle ϕ, degrees	Righting, arm $\overline{GZ}$, m
0.000	0.000	50.000	1.102
5.000	0.212	60.000	1.083
10.000	0.403	70.000	0.911
20.000	0.697	80.000	0.671
30.000	0.904	90.000	0.389
40.000	1.037		

2) Calculate the area in m·rad between 0 and 30 degrees, and the area between 0 and 40 degrees. These data are required by Rahola's criterion of stability and the *IMO code of intact stability* (see, for example, Biran and López–Pulido, 2014, Chapter 8).

Exercise 7.7 (Curve of resistance, INSEAN model 1189). The data shown below are derived from the basin tests of the INSEAN model 1189 (INSEAN, 1965). F_n is the Froude number and R_D the non-dimensional resistance defined by

$$F_n = \frac{V}{\sqrt{gL}}, \quad R_D = \frac{R_t}{\Delta}$$

where V is the model speed, g the gravity acceleration, R_t the total resistance, and Δ the displacement of the model. Use the MATLAB spline to plot R_D against F_n.

F_n	0.305	0.327	0.347	0.365	0.383	0.390	0.400
R_D	10.129	11.857	14.276	18.011	22.504	24.883	28.499

Exercise 7.8 (Curve of resistance, Trawler model). The data below result from the basin tests of a trawler model (SNAME, 1966, Model 68). Plot the curve of non-dimensional resistance versus Froude number using the MATLAB spline for interpolation.

F_n	R/Δ	F_n	R/Δ
0.10	1.17	0.24	7.07
0.12	1.67	0.26	8.53
0.14	2.26	0.28	10.17
0.16	2.96	0.30	12.35
0.18	3.78	0.32	15.04
0.20	4.74	0.34	18.35
0.22	5.83	0.36	28.59

APPENDIX 7.A MULTISURF — CUBIC SPLINE, POLYCURVE

In this appendix we show how to draw in MultiSurf the station exemplified in Sections 7.1 and 7.3. After inserting the nine given points we proceed as shown below.

1. Select points 1 and 2 and then Insert → Curve → Line. The resulting straight-line segment gets the name line1.
2. Select points 2 to 8 and then Insert → Curve → C-spline Curve. The resulting curve is named curve1. This curve is shown in Fig. 7.9 where we can see the origin and the positive direction of the parameter t. The Properties manager shows that the curve is a cubic spline.
3. Select points 8 and 9 and then Insert → Curve → Line. The resulting straight-line segment gets the name line2.

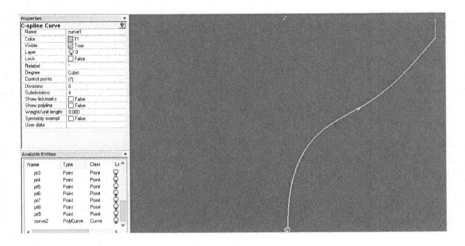

Figure 7.9 A cubic spline in MultiSurf

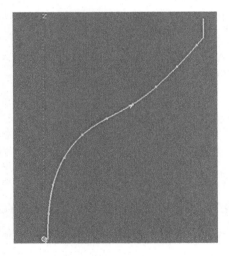

Figure 7.10 Station defined in MultiSurf as a *PolyCurve*

4. Select line1, curve1, line2, in this order. Next Insert → Curve → PolyCurve. The result has the name curve2.

The whole station is shown in Fig. 7.10; it is composed of three entities, but in the end it is defined as one entity. To show that curve2 can be treated as a single entity we insert a point that is constrained to stay on this curve. In MultiSurf terminology a point constrained on a curve is called a *bead*; to

Figure 7.11 Bead properties

create it select `curve2` and use the commands

$$Insert \rightarrow Point \rightarrow Bead$$

Fig. 7.11 shows the properties of the inserted bead. The parameter t is defined for the whole station. By default it equals 0.5. Confirm the choice and the point will be displayed. Drag the point with the mouse or set various values of the parameter in the `Properties` manager and see that $t = 0$ brings the bead over point 1, while $t = 1$ moves the bead over point 9. In the figure we can see the points corresponding to $t = 0, 0.1, 0.2, ..., 1$.

We have thus shown how to model a problematic planar section. With the same tools we can define also space curves. To show this we return to the deck line exemplified in Section 7.5. We input the points using the coordinate system of MultiSurf and carry on the sequence

$$Insert \rightarrow Curve \rightarrow C\text{-spline Curve}$$

The axonometric view of the deck line is shown in Fig. 7.12. In the `Properties` manager we see that the curve is a cubic spline. We can check other projections, for example that on the waterlines plan shown in Fig. 7.13. MultiSurf allows the user to rotate the axonometric view. Doing this slowly it is possible to find a position in which the deck line looks nearly like a straight line. If it were exactly a straight line this would have proved that the line is planar and the projection on the screen is the *edge*

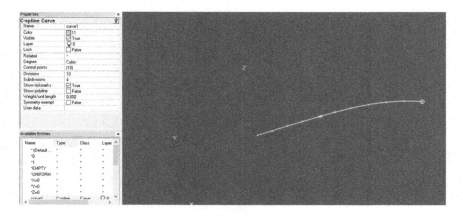

Figure 7.12 Part of a deck line, axonometric view

Figure 7.13 Part of a deck line projected on the waterline plan

view of the curve plane. Naval Architects say that a well-looking sheer line should be planar.

We have exemplified here an essential feature that distinguishes *surface modelling* from *solid modelling*. While in the latter the user starts with a planar figure and only later extends it to the third dimension, in the former it is possible to start directly in three dimensions.

CHAPTER 8

Geometrical Transformations

Contents

8.1 INTRODUCTION

In Chapter 2 we treated two transformations of the coordinate axes in the plane: the translation and the rotation. In this chapter we deal with transformation of points in the plane and the 3D space. See Table 8.1. In computer graphics the final aim is the transformation of whole figures, and in Naval Architecture we often need to transform the whole hull model. Transformations are used in computer graphics to change the view. In Naval Architecture the scaling transformation is probably the simplest way to derive new ship lines from the given lines of a parent ship. If we consider a figure as a set of points, and apply the same transformation to all its points, we obtain the transformation of the figure. The computational procedures

Geometry for Naval Architects
https://doi.org/10.1016/B978-0-08-100328-2.00019-5

Copyright © 2019 Elsevier Ltd.
All rights reserved.

Table 8.1 Terminology of geometric transformations

English	French	German	Italian
transformation	transformation	Transformation	trasformazione
translation	translation	Verschiebung	traslazione
rotation	rotation	Rotation, Drehung	rotazione
reflection	symétrie réflexion	Reflexion, Spiegelung	riflessione, simmetria
scaling	changement d'échelle	Skalierung	trasformazione di scala
homothety, homothecy	homothétie	Homothetie	omotetia
shearing	cisaillement	Scherung	taglio
homogeneous coordinates	coordonnés homogènes	homogene Koordinaten	coordinate omogenee
vanishing point	point de fuite	Fluchtpunkt	punto di fuga

are based on elementary linear algebra. A point is defined as a vector of coordinates, a transformation is performed with the aid of a square matrix. In a first approach translations are an exception as they are done by the addition of the point vector and a vector of displacements. The point can be defined either as a row or a column vector. We choose the latter possibility for reasons that we will explain immediately. Thus, in the plane a point $\mathbf{P}_i$ is defined by

$$\mathbf{P}_i = \left| \begin{array}{c} x_i \\ y_i \end{array} \right|$$

and in 3D space by

$$\mathbf{P}_i = \left| \begin{array}{c} x_i \\ y_i \\ z_i \end{array} \right|$$

Except for translations, and with the above mentioned choice of point representation, transformations are carried on by left multiplication by a transformation matrix $\mathbf{T}$. The general formulation in the plane is

$$\mathbf{Q}_i = T\mathbf{P}_i = \left| \begin{array}{cc} t_{11} & t_{12} \\ t_{21} & t_{22} \end{array} \right| \left| \begin{array}{c} x_i \\ y_i \end{array} \right| = \left| \begin{array}{c} t_{11}x_i + t_{12}y_i \\ t_{21}x_i + t_{22}y_i \end{array} \right|$$

The extension to 3D space is straightforward. As written above, the final aim is to transform figures. As a simple example let us consider a triangle with vertices

$$\mathbf{P}_1 = \begin{vmatrix} x_1 \\ y_1 \end{vmatrix}, \quad \mathbf{P}_2 = \begin{vmatrix} x_2 \\ y_2 \end{vmatrix}, \quad \mathbf{P}_3 = \begin{vmatrix} x_3 \\ y_3 \end{vmatrix}$$

With our choice of representation we can define the triangle by the horizontal concatenation of its vertices

$$\begin{vmatrix} \mathbf{P}_1 & \mathbf{P}_2 & \mathbf{P}_3 \end{vmatrix} = \begin{vmatrix} x_1 & x_2 & x_3 \\ y_1 & y_2 & y_3 \end{vmatrix}$$

The transformation of the triangle is performed by the multiplication

$$T \begin{vmatrix} \mathbf{P}_1 & \mathbf{P}_2 & \mathbf{P}_3 \end{vmatrix}$$

and the result is the 2-*by*-3 matrix of the transformed coordinates. Our representation suits well MATLAB and we are going to use extensively this environment for computing and displaying examples.

Often it is necessary to apply several transformations to the same point or figure, for example a rotation and a translation. The order in which such transformations are applied may influence the result. We'll explain this by examples. In continuation we'll show that it is possible to perform translations also by left multiplication by a square matrix. To do this we add a dimension. Specifically, we define points in the plane by a 3-*by*-1 vector, and points in the space by a 4-*by*-1 vector. Coordinates formulated in this way are called *homogeneous*. Correspondingly the transformation matrices are 3-*by*-3 in the plane, and 4-*by*-4 in the space. Working in homogeneous coordinates allows for a unified treatment of all transformations including projective ones.

Geometrical transformations may be applied to points defined by other points; we say that points defined so are *combinations* of given points. For example, given two points in space, $\mathbf{P}_1$ and $\mathbf{P}_2$, and two real numbers, μ and λ, the point $\mathbf{Q}$ is a *linear combination* of $\mathbf{P}_1$ and $\mathbf{P}_2$ if

$$\mathbf{Q} = \mu \mathbf{P}_1 + \lambda \mathbf{P}_2 = \mu \begin{vmatrix} x_1 \\ y_1 \\ z_1 \end{vmatrix} + \lambda \begin{vmatrix} x_2 \\ y_2 \\ z_3 \end{vmatrix}$$

If $\lambda + \mu = 1$ the point $\mathbf{Q}$ is an *affine combination* and, given a transformation matrix, T, it enjoys the property

$$T\mathbf{Q} = \mu T \mathbf{P}_1 + \lambda T \mathbf{P}_2$$

This means that we may either transform the affine combination of the given points, or make the affine combination of the transformations of the given points, and the result will be same. Moreover, in the case of two points their affine combination lies on the straight line that passes through the given points, and the affine combination of three non–collinear points lies in the plane defined by the given points. Given a set of mass points, their *centre of mass* is the affine combination of the mass points. The curves treated in Chapters 9 and 10 are affine combinations of given *control points*. More transformations are used in computer graphics, namely *projective* and *viewing transformations*. This chapter includes a short introduction to the mathematics of perspective transformations, a kind of projective transformation. The reader may find more details on these subjects in specialized books, such as Marsh (2000), or Eggerton and Hall (1999).

8.2 TRANSFORMATIONS IN THE PLANE

8.2.1 Translation

Let us suppose that we want to translate a point $\mathbf{P}$ a distance δx in parallel to the x-axis, and a distance δy in parallel to the y-axis. This transformation is performed by the addition

$$\mathbf{Q} = \mathbf{P} + \begin{vmatrix} \delta x \\ \delta y \end{vmatrix} \tag{8.1}$$

In the following MATLAB script we define an equilateral triangle with vertices $\mathbf{P}_1$, $\mathbf{P}_2$, $\mathbf{P}_3$, and translate it so that one of its vertices moves to the origin of the system of coordinates. We note the translated vertices by $\mathbf{Q}_1$, $\mathbf{Q}_2$, $\mathbf{Q}_3$.

```
%TRIANGLETRANSLATE Translates a triangle

% Define frame
plot([ -2 2 ], [ 0 0 ], 'k-', [ 0 0 ], [ -2 2 ], 'k-'), grid
axis equal
hold on
% Define and plot triangle
P1      = [ 1; 1 ]; P2 = [ 2; 1 ]; P3 = [ 1.5; (1 + sqrt(5))/2 ];
Triangle = [ P1 P2 P3 P1 ];
pline(Triangle, 1.5, 'k', '-')
text((P1(1) - 0.13), (P1(2) - 0.13), 'P_1');
```

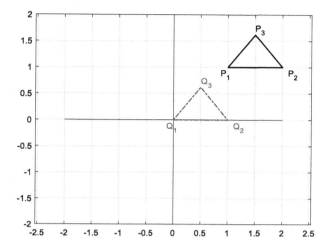

Figure 8.1 Translating a triangle

```
text((P2(1) + 0.1), (P2(2) - 0.13), 'P_2');
text(P3(1), (P3(2) + 0.1), 'P_3');
% Translation
d         = [ -1; -1 ];
Q1 = P1 + d; Q2 = P2 + d; Q3 = P3 + d;
TransTri = [ Q1 Q2 Q3 Q1 ];
pline(TransTri, 1.5, 'r', '-.')
text((Q1(1) - 0.13), (Q1(2) - 0.13), 'Q_1', 'Color', 'r');
text((Q2(1) + 0.1), (Q2(2) - 0.13), 'Q_2', 'Color', 'r');
text(Q3(1), (Q3(2) + 0.1), 'Q_3', 'Color', 'r');
hold off
```

The result is shown in Fig. 8.1.

8.2.2 Rotation Around the Origin

Let us suppose that given a point **P** we want to rotate it around the origin by an angle ϕ. In Fig. 8.2 we show the given and the rotated point and call the latter **Q**. The coordinates of the given and of the rotated point are

$$\mathbf{P} = \left| \begin{array}{c} x \\ y \end{array} \right| = \left| \begin{array}{c} \overline{\mathbf{OP}}\cos\alpha \\ \overline{\mathbf{OP}}\sin\alpha \end{array} \right|, \quad \mathbf{Q} = \left| \begin{array}{c} \overline{\mathbf{OP}}\cos(\alpha + \phi) \\ \overline{\mathbf{OP}}\sin(\alpha + \phi) \end{array} \right| \qquad (8.2)$$

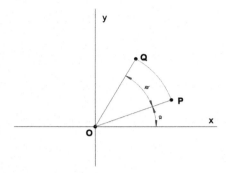

Figure 8.2 Rotation of a point around the origin

Expanding the second equation (8.2) yields

$$\mathbf{Q} = \overline{\mathbf{OP}} \begin{vmatrix} \cos\alpha\cos\phi - \sin\alpha\sin\phi \\ \sin\alpha\cos\phi + \cos\alpha\sin\phi \end{vmatrix}$$

$$= \begin{vmatrix} \cos\phi & -\sin\phi \\ \sin\phi & \cos\phi \end{vmatrix} \begin{vmatrix} x \\ y \end{vmatrix} \tag{8.3}$$

This defines the matrix of rotation around the origin as

$$T_R = \begin{vmatrix} \cos\phi & -\sin\phi \\ \sin\phi & \cos\phi \end{vmatrix} \tag{8.4}$$

A MATLAB implementation for an angle of rotation given in degrees is

```
function Tr1 = RotMatrix1(angle)
%Tr1 Matrix for rotation around origin
%    Input: angle, degrees
%    Left multiplies a 2-by-n array of point coordinates

Tr1 = [ cosd(angle) -sind(angle)
        sind(angle)  cosd(angle) ];
end
```

8.2.3 Rotation About an Arbitrary Point

Let us suppose that given a point **P** in a system of coordinates with origin **O** we want to rotate it about a point $\mathbf{O}_2$ different from the origin. We carry on this transformation in three steps as detailed below.

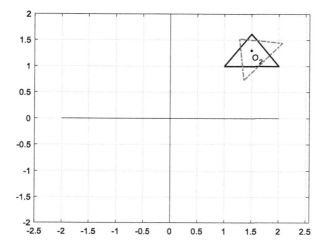

Figure 8.3 Rotation of a point around an arbitrary point

1. First, we translate the figure so that the point O_2 moves to the origin O.
2. Next, we rotate the translated figure around the origin.
3. Finally, we translate the whole figure so that the point O_2 returns to its original position.

The following script implements the above steps for the equilateral triangle used in Fig. 8.1. Fig. 8.3 shows the result.

```
%ROTABTANYPOINT - Rotates a triangle around an arbitrary point

% Define frame
plot([ -2 2 ], [ 0 0 ], 'k-', [ 0 0 ], [ -2 2 ], 'k-'), grid
axis equal
hold on
% Define and plot triangle
P1 = [ 1; 1 ]; P2 = [ 2; 1 ]; P3 = [ 1.5; (1 + sqrt(5))/2 ];
Triangle = [ P1 P2 P3 P1 ];
pline(Triangle, 1.5, 'k', '-')
% Define point around which the triangle will be rotated.
O2       = [ 1.5; 1.3 ];
point(O2, 0.02), text(O2(1), (O2(2) - 0.15), 'O_2')
% Translate point O2 to origin O
Tri1     = Triangle - [ O2 O2 O2 O2 ];
% Rotate around O2
Tr = RotMatrix1(45);
```

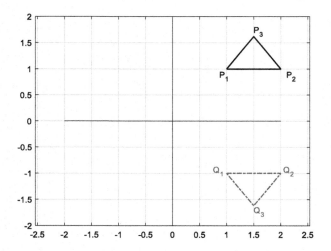

Figure 8.4 Reflecting a triangle in the x-axis

```
RotTri = Tr*Tri1;
% Translate back to initial position
Tri2    = RotTri + [ 02 02 02 02 ];
pline(Tri2, 1.5, 'r', '-.')
hold off
```

8.2.4 Reflection

When a point is *reflected in the x-axis* its x-coordinate remains unchanged while the y-coordinate changes sign. This transformation is carried out by left multiplication by the matrix

$$T_{rx} = \begin{vmatrix} 1 & 0 \\ 0 & -1 \end{vmatrix} \tag{8.5}$$

In Fig. 8.4 we see an application to the same equilateral triangle we used above. The triangle $\mathbf{Q_1Q_2Q_3}$ is a *mirror* image of the triangle $\mathbf{P_1P_2P_3}$. In Exercise 8.4 we ask the reader to write the matrix for reflection in the y-axis and apply it to the same equilateral triangle $\mathbf{P_1P_2P_3}$.

8.2.5 Isometries

The transformations of translation, rotation, and reflection preserve distances; therefore, they are characterized as *isometric transformations* or, simply, *isometries*. As defined in Section 3.3.1, a given figure and the figure obtained by its isometric transformation are *congruent*. Let us give a general proof for translations. Let

$$
\mathbf{P}_1 = \begin{vmatrix} x_1 \\ y_1 \end{vmatrix}, \quad \mathbf{P}_2 = \begin{vmatrix} x_2 \\ y_2 \end{vmatrix}, \quad \mathbf{d} = \begin{vmatrix} \delta x \\ \delta y \end{vmatrix}
$$

The translation of the two points by the vector **d** yields

$$
\mathbf{Q}_1 = \mathbf{P}_1 + \mathbf{d}, \quad \mathbf{Q}_2 = \mathbf{P}_2 + \mathbf{d}
$$

The distance between the translated points is

$$
d(Q_1 Q_2) = \sqrt{(x_2 + \delta x - x_1 - \delta x)^2 + (y_2 + \delta y - y_1 - \delta y)^2}
$$
$$
= \sqrt{(x_2 - x_1)^2 + (y_2 - y_1)^2}
$$

which is exactly the distance between the given points.

To show that the transformation of rotation preserves distances we can consider the same points $\mathbf{P}_1$, $\mathbf{P}_2$ as above. Using the matrix of rotation T_R detailed in Section 8.2.2, we obtain the rotated segment

$$
\overline{Q_1 Q_2} = T_R \begin{vmatrix} x_2 - x_1 \\ y_2 - y_1 \end{vmatrix}
$$
$$
= \begin{vmatrix} \cos\phi\,(x_2 - x_1) - \sin\phi\,(y_2 - y_1) \\ \sin\phi\,(x_2 - x_1) + \cos\phi\,(y_2 - y_1) \end{vmatrix}
$$

The length of this segment equals the length of the segment $\overline{P_1 P_2}$. We can check numerically that rotation preserves distances. For example, add the following lines to the script `RotAbtAnyPoint` detailed in Section 8.2.3

```
format compact
d1 = norm( P2 - P1);
t1       = [ 'Distance between P1 and P2 = ', num2str(d1) ];
sprintf(t1)
d2       = norm(Tri2(:, 2) - Tri2(:, 1));
t2       = [ 'Distance between Q1 and Q2 = ', num2str(d2) ];
sprintf(t2)
```

The display will be

```
ans =
Distance between P1 and P2 = 1
ans =
Distance between Q1 and Q2 = 1
```

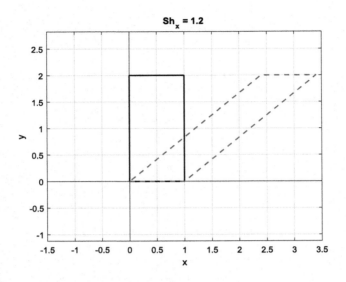

Figure 8.5 Shearing a rectangle in the *x*-direction

In Exercise 8.5 we ask the reader to prove that reflection in a coordinate axis preserves distances.

8.2.6 Shearing

In Fig. 8.5 we show a rectangle and its transformation by *shearing in the x-direction*. For a point with coordinates x, y this transformation is carried on as

$$
\begin{vmatrix} 1 & sh_x \\ 0 & 1 \end{vmatrix} \begin{vmatrix} x \\ y \end{vmatrix} = \begin{vmatrix} x + sh_x y \\ y \end{vmatrix}
\tag{8.6}
$$

8.2.7 Scaling About the Origin

The matrix T_s defined by Eq. (8.7) performs *scaling* about the origin.

$$
T_S = \begin{vmatrix} S_x & 0 \\ 0 & S_y \end{vmatrix}
\tag{8.7}
$$

When $S_x = S_y$ we say that the *scaling is uniform*. An example is shown in Fig. 8.6. The triangles $P_1 P_2 P_3$ and $Q_1 Q_2 Q_3$ are *similar*. Other terms used for this particular transformation are *homothety* or *homothecy*.

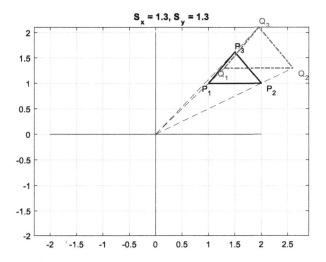

Figure 8.6 Scaling a triangle about the origin

8.2.8 Affine Transformations

The transformations described above are known collectively as **affine transformations**. Given a point with coordinates x_1, y_1, the general formulation of its affine transformation is

$$
\begin{vmatrix} x_2 \\ y_2 \end{vmatrix} = \begin{vmatrix} t_{11} & t_{12} \\ t_{21} & t_{22} \end{vmatrix} \begin{vmatrix} x_1 \\ y_1 \end{vmatrix} + \begin{vmatrix} \delta x \\ \delta y \end{vmatrix} \tag{8.8}
$$

Often it is necessary to perform several transformations; the order in which they are carried out may influence the results. For example, given the same equilateral triangle as in Fig. 8.1, let us first rotate it 90° counterclockwise around the origin, and next translate it a distance equal to -0.5 in the x-direction, and a distance equal to -0.8 in the y-direction. The result is shown in Fig. 8.7. If we change the order of transformations we obtain the result shown in Fig. 8.8. The difference is obvious.

Suppose that we want to combine two transformations that are carried on by multiplication. For example, given a point **P** we perform first the transformation

$$\mathbf{Q} = T_1 \mathbf{P}$$

and next the transformation

$$\mathbf{R} = T_2 \mathbf{Q}$$

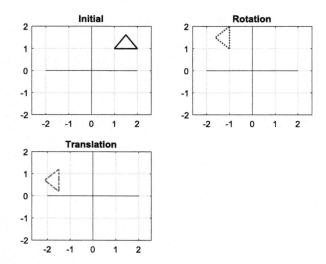

Figure 8.7 First rotation, next translation

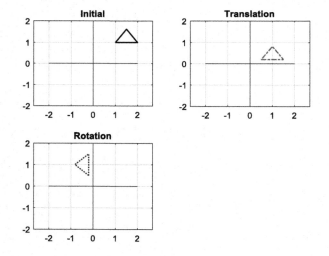

Figure 8.8 First translation, next rotation

Instead of proceeding so we may multiply the matrices

$$T = T_1 T_2$$

and carry on the combined transformation as

$$\mathbf{R} = T\mathbf{P}$$

As an example, given a point $\mathbf{P}$ with coordinates x_1, y_1, let us first rotate it counterclockwise around the origin by an angle θ, and next reflect it in the x-axis. In detail, we first calculate

$$\mathbf{Q} = \begin{vmatrix} \cos\theta & -\sin\theta \\ \sin\theta & \cos\theta \end{vmatrix} \begin{vmatrix} x_1 \\ y_1 \end{vmatrix} = \begin{vmatrix} x_1\cos\theta - y_1\sin\theta \\ x_1\sin\theta + y_1\cos\theta \end{vmatrix}$$

and next

$$\mathbf{R} = \begin{vmatrix} 1 & 0 \\ 0 & -1 \end{vmatrix} \begin{vmatrix} x_1\cos\theta - y_1\sin\theta \\ x_1\sin\theta + y_1\cos\theta \end{vmatrix} = \begin{vmatrix} x_1\cos\theta - y_1\sin\theta \\ -x_1\sin\theta - y_1\cos\theta \end{vmatrix}$$

Instead of proceeding so we first calculate the matrix that combines the transformations

$$T = \begin{vmatrix} 1 & 0 \\ 0 & -1 \end{vmatrix} \begin{vmatrix} \cos\theta & -\sin\theta \\ \sin\theta & \cos\theta \end{vmatrix} = \begin{vmatrix} \cos\theta & -\sin\theta \\ -\sin\theta & -\cos\theta \end{vmatrix}$$

and next we left multiply the coordinates of the given point $\mathbf{P}$ by the combined matrix T. The matrices appear in the reversed order of the transformations, e.g. the matrix of the first transformation is the first from the right. We leave to the reader to prove that the combined matrix yields the same result as the sequence of the simple transformations. Using the combined matrix we can reduce rounding errors and improve computational efficiency, especially when many points have to be transformed.

8.2.9 Homogeneous Coordinates

Until now we performed translations using the operation of addition, while all other transformations were carried out by multiplication. Proceeding so it is impossible to write a matrix that combines translation with another kind of transformation. This is not only a matter of mathematical elegance, but also of computational efficiency. A solution exists and it is a good one; it is based on the use of **homogeneous coordinates** defined as follows. Given a point defined in the usual *Euclidean coordinates* as

$$\mathbf{P} = \begin{vmatrix} x \\ y \end{vmatrix}$$

its representation in homogeneous coordinates is

$$\mathbf{P} = \begin{vmatrix} X \\ Y \\ w \end{vmatrix} \tag{8.9}$$

subject to the condition

$$x = X/w, \quad y = Y/w, \quad w \neq 0 \tag{8.10}$$

Then, for example, anyone of the following homogeneous coordinates

$$\begin{vmatrix} 2 \\ 3 \\ 1 \end{vmatrix}, \begin{vmatrix} 4 \\ 6 \\ 2 \end{vmatrix}, \begin{vmatrix} 6 \\ 9 \\ 3 \end{vmatrix}$$

represent the same point

$$\begin{vmatrix} 2 \\ 3 \end{vmatrix}$$

in *Euclidean coordinates*. The usual choice in computer graphics is $w = 1$.

In homogeneous coordinates it is immediately possible to perform a translation by multiplication. Given a point with Euclidean coordinates x, y, its translation by a distance δx in the x-direction, and δy in the y-direction is carried on in homogeneous coordinates as

$$\begin{vmatrix} 1 & 0 & \delta x \\ 0 & 1 & \delta y \\ 0 & 0 & 1 \end{vmatrix} \begin{vmatrix} x \\ y \\ 1 \end{vmatrix} = \begin{vmatrix} x + \delta x \\ y + \delta y \\ 1 \end{vmatrix} \tag{8.11}$$

Rotation around the origin is carried on in homogeneous coordinates as

$$\begin{vmatrix} \cos\phi & -\sin\phi & 0 \\ \sin\phi & \cos\phi & 0 \\ 0 & 0 & 1 \end{vmatrix} \begin{vmatrix} x \\ y \\ 1 \end{vmatrix} = \begin{vmatrix} x\cos\phi - y\sin\phi \\ x\sin\phi + y\cos\phi \\ 1 \end{vmatrix} \tag{8.12}$$

In Exercise 8.13 we ask the reader to write the matrix that performs scaling in homogeneous coordinates.

Historically, the homogeneous coordinates were introduced in mathematics more than a century before the birth of computer graphics and for quite other reasons than those described above. In homogeneous coordinates it is possible to unify the treatment of parallel and non-parallel lines and to deal with points at infinity. This subject belongs to projective geometry, here we give only a concise view of it. From Eqs (8.10) it is obvious that as the third coordinate, w, gets smaller, the Euclidean coordinates of the point grow larger. This happens also to the intersection of two straight lines when one of them rotates so that it tends to be parallel to the other. To show that parallelism is achieved when $w = 0$ we consider the equations

$$ax + by + c = 0$$
$$ax + by + d = 0, \quad c \neq d \tag{8.13}$$

They represent two parallel lines in the plane. The common slope is $-a/b$. If we try to find the intersection by Cramer's rule we discover immediately that the determinant of the system is zero. A solution exists only if $c = d$, but this contradicts our assumption that the lines are different. Let us see what happens if we use homogeneous coordinates. To do so let

$$x = \frac{X}{w}, \quad y = \frac{Y}{w} \tag{8.14}$$

Substituting into Eqs (8.13) yields

$$aX + bY + cw = 0$$
$$aX + bY + dw = 0 \tag{8.15}$$

Subtracting one of the above equations from the other we obtain

$$(c - d)w = 0$$

As we assumed that $c \neq d$, we conclude that $w = 0$. The solutions of Eqs (8.15) are of the form

$$\mathbf{P}_\infty = \begin{vmatrix} rb \\ -ra \\ 0 \end{vmatrix}$$

This is the **point at infinity** of the given lines. Substituting, for example, such a solution into the first equation (8.13) we get, indeed,

$$rab - rab + c \cdot 0 = 0$$

An immediate consequence of the above considerations is that all lines with a given direction $-a/b$ have the same point at infinity. The Euclidean plane completed with its points at infinity is called *projective plane*. The points at infinity are also called *ideal points*. When treating the central projection, in Section 1.4.2, we saw that parallel lines not parallel to the image plane intersect in *vanishing points*. The vanishing points are the visible projections of ideal points. We'll return to this subject in Section 8.4.

8.3 TRANSFORMATIONS IN 3D SPACE

Having explained why it is advantageous to work in homogeneous coordinates we continue to the 3D space using this representation. Thus, the translation of a point **P** by the vector

$$d = \begin{vmatrix} \delta x \\ \delta y \\ \delta z \\ 1 \end{vmatrix}$$

is carried on by the defined below matrix T_{3t} as follows

$$T_{3t}\mathbf{P} = \begin{vmatrix} 1 & 0 & 0 & \delta x \\ 0 & 1 & 0 & \delta y \\ 0 & 0 & 1 & \delta z \\ 0 & 0 & 0 & 1 \end{vmatrix} \begin{vmatrix} x \\ y \\ z \\ 1 \end{vmatrix} = \begin{vmatrix} x + \delta x \\ y + \delta y \\ z + \delta z \\ 1 \end{vmatrix} \tag{8.16}$$

Rotation of a point **P** around the x-axis is performed by the matrix T_{3rx} defined below

$$T_{3Rx}\mathbf{P} = \begin{vmatrix} 1 & 0 & 0 & 0 \\ 0 & \cos\phi & -\sin\phi & 0 \\ 0 & \sin\phi & \cos\phi & 0 \\ 0 & 0 & 0 & 1 \end{vmatrix} \begin{vmatrix} x \\ y \\ z \\ 1 \end{vmatrix} = \begin{vmatrix} x \\ y\cos\phi - z\sin\phi \\ y\sin\phi + z\cos\phi \\ 1 \end{vmatrix} \tag{8.17}$$

Rotation around the y-axis is carried on as

$$T_{3Ry}\mathbf{P} = \begin{vmatrix} \cos\phi & 0 & \sin\phi & 0 \\ 0 & 1 & 0 & 0 \\ -\sin\phi & 0 & \cos\phi & 0 \\ 0 & 0 & 0 & 1 \end{vmatrix} \begin{vmatrix} x \\ y \\ z \\ 1 \end{vmatrix} = \begin{vmatrix} x\cos\phi + z\sin\phi \\ y \\ -x\sin\phi + z\cos\phi \\ 1 \end{vmatrix} \tag{8.18}$$

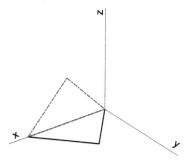

Figure 8.9 Rotation around the x-axis

and rotation around the z-axis as

$$
T_{3R_Y}\mathbf{P} = \begin{vmatrix} \cos\phi & -\sin\phi & 0 & 0 \\ \sin\phi & \cos\phi & 0 & 0 \\ 0 & 0 & 1 & 0 \\ 0 & 0 & 0 & 1 \end{vmatrix} \begin{vmatrix} x \\ y \\ z \\ 1 \end{vmatrix} = \begin{vmatrix} x\cos\phi - y\sin\phi \\ x\sin\phi + y\cos\phi \\ z \\ 1 \end{vmatrix} \quad (8.19)
$$

For reflection in the xy-plane we use

$$
\begin{vmatrix} 1 & 0 & 0 & 0 \\ 0 & 1 & 0 & 0 \\ 0 & 0 & -1 & 0 \\ 0 & 0 & 0 & 1 \end{vmatrix} \begin{vmatrix} x \\ y \\ z \\ 1 \end{vmatrix} = \begin{vmatrix} x \\ y \\ -z \\ 1 \end{vmatrix} \quad (8.20)
$$

Homogeneous scaling in 3D is obtained as

$$
\begin{vmatrix} S_x & 0 & 0 & 0 \\ 0 & S_y & 0 & 0 \\ 0 & 0 & S_z & 0 \\ 0 & 0 & 0 & 1 \end{vmatrix} \begin{vmatrix} x \\ y \\ z \\ 1 \end{vmatrix} = \begin{vmatrix} S_x x \\ S_y y \\ S_z z \\ 1 \end{vmatrix} \quad (8.21)
$$

Shearing in the x-direction is carried on as

$$
\begin{vmatrix} 1 & 0 & S_{hx} & 0 \\ 0 & 1 & 1 & 0 \\ 0 & 0 & 1 & 0 \\ 0 & 0 & 0 & 1 \end{vmatrix} \begin{vmatrix} x \\ y \\ z \\ 1 \end{vmatrix} = \begin{vmatrix} x + S_{hx}z \\ y \\ z \\ 1 \end{vmatrix} \quad (8.22)
$$

Example 8.1 (Rotating a triangle in 3D homogeneous coordinates). Fig. 8.9 illustrates the rotation of a plane triangle around the x-axis.

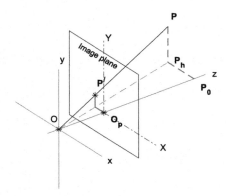

Figure 8.10 Central projection of point **P**

8.4 PERSPECTIVE PROJECTIONS

8.4.1 The Projection Matrix

In Chapter 1, Appendix 1.A, we have shown that orthographic projections can be obtained multiplying the array of coordinates by a projection matrix. In this section we explain how to use matrices for central projections. This section should be regarded as a continuation of Section 1.4. We use the term *perspective projection* as an alternative for the term *central projection*. To derive a transformation matrix refer to Fig. 8.10. The projection centre is the point **O**. We place in this point the origin of the *world* coordinates. The z-axis is horizontal. The line Oz is the *optical axis*. The object to be projected is the point **P**; its coordinates are x, y, z. The image plane is situated at the distance f from **O**. In the case of a camera the distance f is the *focal length*. The optical axis pierces the image plane in the point $\mathbf{O}_p$ called *principal point* or *image centre*. The central, or perspective projection of **P** is **P'**, the point in which the view line $\overline{OP}$ pierces the image plane. The coordinates of **P'** in the image plane are X, Y. $\mathbf{P}_h$ is the orthogonal projection of **P** on the plane xz and P_0 the orthogonal projection of $\mathbf{P}_h$ on the optical axis. From similar triangles we derive

$$\frac{f}{z} = \frac{X}{x} = \frac{Y}{y} \qquad (8.23)$$

From these equations we obtain the coordinates of **P'** in the image plane

$$X = \frac{xf}{z}, \; Y = \frac{yf}{z} \qquad (8.24)$$

Obviously $Z = f$. Arranged in matrix form Eqs (8.24) become

$$
\begin{vmatrix} X \\ Y \\ Z \\ 1 \end{vmatrix}
\begin{vmatrix} f/z & 0 & 0 & 0 \\ 0 & f/z & 0 & 0 \\ 0 & 0 & f/z & 0 \\ 0 & 0 & 1/z & 0 \end{vmatrix}
\begin{vmatrix} x \\ y \\ z \\ 1 \end{vmatrix}
=
\begin{vmatrix} fx/z \\ fy/z \\ f \\ 1 \end{vmatrix}
\tag{8.25}
$$

The third component is f as the projected point lies in the image plane. An example of function that calculates the perspective projection of a given point is shown below.

```
function Q = PerspPoint( P, f)
%PERSPPOINT Yields the central projection of point P.
%    Input arguments:
%         P, column vector of the homogeneous coordinates of the point P
%         f, distance between centre of projection and image plane,
%         for example the focal length of a camera.

% create projection matrix
PM = [ f    0   0   0
       0    f   0   0
       0    0   f   0
       0    0   1   0 ];
% calculate homogeneous coordinates of the projected point
P1   = PM*P;
% convert to Euclidean coordinates in the image plane
Q   = P1(1:2)/P(3);
end
```

The following MATLAB script calculates the perspective projection of a unit cube centred about the optical axis.

```
%PERSPCUBE1 Perspective view of unit cube centered about the optical axis.

% define frame
plot([ -0.05 0.05 ], [ 0 0 ], 'k-', [ 0 0 ], [ -0.05, 0.05 ], 'k-')
axis equal, axis off
hold on

% define cube vertices
P1 = [ -0.5; -0.5; 2; 1 ]; P2 = [  0.5; -0.5; 2; 1 ];
P3 = [  0.5;  0.5; 2; 1 ]; P4 = [ -0.5;  0.5; 2; 1 ];
d  = [ 0; 0; 1; 0 ];           % cube depth
```

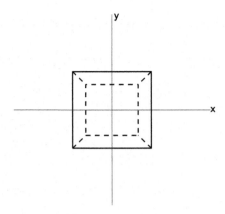

Figure 8.11 Central projection of a cube centred about the optical axis

```
P5 = P1 + d; P6 = P2 + d; P7 = P3 + d; P8 = P4 + d;
f  = 0.08;
Q1 = PerspPoint(P1, f); Q2 = PerspPoint(P2, f);
Q3 = PerspPoint(P3, f); Q4 = PerspPoint(P4, f);
Ff = [ Q1 Q2 Q3 Q4 Q1 ]    % front face projection in homogeneous coordinates
Ffe = Ff(1:2, 1:5)         % projection of front face in Euclidean coordinates
pline(Ffe, 1.5, 'k', '-')
Q5 = PerspPoint(P5, f); Q6 = PerspPoint(P6, f);
Q7 = PerspPoint(P7, f); Q8 = PerspPoint(P8, f);
Fb = [ Q5 Q6 Q7 Q8 Q5 ]    % back face projection in homogeneous coordinates
Fbe = Fb(1:2, 1:5)         % projection of back face in Euclidean coordinates
pline(Fbe, 1.5, 'k', '-')
pline([ Q2(1:2) Q6(1:2) ], 1.5,'k', '-')
pline([ Q3(1:2) Q7(1:2) ], 1.5,'k', '-')
pline([ Q4(1:2) Q8(1:2) ], 1.5,'k', '-')
pline([ Q1(1:2) Q5(1:2) ], 1.5,'k', '-')
hold off
```

The result is shown in Fig. 8.11. The vanishing point coincides with the origin of the coordinates x, y. To obtain a projection with two vanishing points we rotate the cube around the vertical axis of coordinates. To exemplify this add the following lines to the script PerspCube1.

```
% create matrix of rotation by -30 deg around the y-axis
TR = [ cosd(-30) 0 sind(-30)    0
            0     1    0        0
      -sind(-30) 0   cosd(-30)  0
            0     0    0         1 ];
```

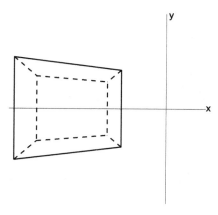

Figure 8.12 Central projection of a cube rotated around the vertical axis

```
% rotate
P1 = TR*P1; P2 = TR*P2; P3 = TR*P3;
P4 = TR*P4; P5 = TR*P5; P6 = TR*P6;
P7 = TR*P7; P8 = TR*P8;
```

8.4.2 Ideal and Vanishing Points

We consider a line passing through a point **P** and having the direction defined by a vector v. The parametric equation of this line, in the system of coordinates assumed in Fig. 8.10, is

$$\mathbf{L}_1 = \mathbf{P}_1 + vt = \begin{vmatrix} x_1 \\ y_1 \\ z_1 \\ 1 \end{vmatrix} + \begin{vmatrix} v_1 \\ v_2 \\ v_3 \\ 0 \end{vmatrix} t \qquad (8.26)$$

The equation of a line parallel to this and passing through a point $\mathbf{P}_2$ is

$$\mathbf{L}_2 = \mathbf{P}_2 + vt = \begin{vmatrix} x_2 \\ y_2 \\ z_2 \\ 1 \end{vmatrix} + \begin{vmatrix} v_1 \\ v_2 \\ v_3 \\ 0 \end{vmatrix} t \qquad (8.27)$$

As scaling the homogeneous coordinates of a point by the same factor does not change the corresponding Euclidean coordinates of that point we divide Eq. (8.26) by t and get

$$\mathbf{L}_1 = \begin{vmatrix} x_1/t \\ y_1/t \\ z_1/t \\ 1/t \end{vmatrix} + \begin{vmatrix} v_1 \\ v_2 \\ v_3 \\ 0 \end{vmatrix} \tag{8.28}$$

The point at infinity of the line is

$$\lim_{t\to\infty} \mathbf{L}_1 = \begin{vmatrix} v_1 \\ v_2 \\ v_3 \\ 0 \end{vmatrix} \tag{8.29}$$

In a similar way we obtain the same point at infinity for the line $\mathbf{L}_2$. We conclude that

1. the lines $\mathbf{L}_1$ and $\mathbf{L}_2$ intersect in the point defined by Eq. (8.29);
2. the coordinates of the point at infinity of a set of parallel line depend only on the direction of those lines. Actually, these coordinates represent the common direction.

As we made no restriction on the direction of the lines, conclusion 1 can be generalized to say that *all lines that have the same direction also have the same point at infinity*. We go now a step ahead and calculate the perspective projection of the point at infinity found above. As scaling all homogeneous coordinates by the same factor does not change the corresponding Euclidean coordinates, we use Eq. (8.25) multiplied by z and write

$$\mathbf{V} = \begin{vmatrix} f & 0 & 0 & 0 \\ 0 & f & 0 & 0 \\ 0 & 0 & f & 0 \\ 0 & 0 & 1 & 0 \end{vmatrix} \begin{vmatrix} v_1 \\ v_2 \\ v_3 \\ 0 \end{vmatrix} = \begin{vmatrix} fv_1 \\ fv_2 \\ fv_3 \\ v_3 \end{vmatrix} \tag{8.30}$$

Dividing all components by v_3 we get

$$\mathbf{V} = \begin{vmatrix} fv_1/v_3 \\ fv_2/v_3 \\ f \\ 1 \end{vmatrix} \tag{8.31}$$

Analyzing this result we reach algebraically conclusions that in Chapter 1 were presented geometrically. These conclusions are listed below.

Figure 8.13 A perspective view of a hexagon

VANISHING POINTS. If $v_3 \neq 0$ the ideal point is projected at a finite distance and is the vanishing point for all the lines with the direction defined by the vector v. The condition $v_3 \neq 0$ means that the lines are not parallel to the image plane.

LINES PARALLEL TO THE IMAGE PLANE. The condition $v_3 = 0$ means lines parallel to the image point. Eq. (8.31) shows that there is no visible vanishing point. Lines parallel to the image plane are projected as parallel lines.

LINES PERPENDICULAR TO THE IMAGE PLANE. For lines perpendicular to the image point $v_1 = v_2 = 0$. The vanishing point is the image centre, the point in which the optical axis pierces the image plane.

8.4.3 The Vanishing Line

In projective geometry it is taught that all the points at infinity of a plane lie on a straight line called the *line at infinity* of that plane. Correspondingly, in a perspective projection the vanishing points of a plane lie on the *vanishing line* of that plane. To give an example we show in Fig. 8.13 the perspective view of a hexagon that lies in a horizontal plane below the centre of projection. Starting from the vertex closest to the centre of projection, the first edge is parallel to the fourth, the second to the fifth, and the third to the sixth. The parallel edges are projected as lines that intersect in three vanishing points that are collinear. In perspective drawing the common line is called *horizon line*. This property is used in the drawing of perspective views. In a three-point perspective we have a vanishing point that does not lie on the horizon line. An example is shown in Fig. 1.14.

8.4.4 The Orthographic Projection as Limit of Perspective Projection

In Section 1.6 we have shown that the parallel projection is obtained when 'the projection centre is sent to infinity'. In Section 1.7 we went one step farther and introduced the orthogonal projection assuming 'that the projection rays are perpendicular to the image plane'. In this section we prove algebraically that the orthogonal projection can be derived, indeed, from the perspective projection by sending the projection centre (point **O** in Fig. 8.10) to infinity. We return to Eqs (8.24) remarking that we cannot use anymore the coordinates z because they go to infinity. Therefore, we move the origin of coordinates to the image plane and define the horizontal coordinate by

$$z = f + \zeta$$

Using L'Hôpital's rule we obtain from Eqs (8.24)

$$X = \lim_{f \to \infty} \frac{fx}{f + \zeta} = x, \quad Y = \lim_{f \to \infty} \frac{fy}{f + \zeta} = y \tag{8.32}$$

The image is projected at $Z = 0$. Finally, the orthogonal projection is obtained carrying on the transformation

$$
\begin{vmatrix} X \\ Y \\ Z \\ 1 \end{vmatrix} = \begin{vmatrix} 1 & 0 & 0 & 0 \\ 0 & 1 & 0 & 0 \\ 0 & 0 & 0 & 0 \\ 0 & 0 & 0 & 1 \end{vmatrix} \begin{vmatrix} x \\ y \\ \zeta \\ 1 \end{vmatrix} = \begin{vmatrix} x \\ y \\ 0 \\ 1 \end{vmatrix} \tag{8.33}
$$

8.5 AFFINE COMBINATIONS OF POINTS

8.5.1 Affine Combination of Two Points — Collinearity

Let

$$
P_1 = \begin{bmatrix} x_1 \\ y_1 \\ z_1 \end{bmatrix}, \quad P_2 = \begin{bmatrix} x_2 \\ y_2 \\ z_2 \end{bmatrix}
$$

be two given points. We associate to them a third point defined as

$$
P_3 = \lambda P_1 + \mu P_2 = \begin{vmatrix} \lambda x_1 + \mu x_2 \\ \lambda y_1 + \mu y_2 \\ \lambda z_1 + \mu z_2 \end{vmatrix} \tag{8.34}
$$

We say that P_2 is a **linear combination** of P_1 and P_2. The corresponding definition for a point in the x, y plane is evident. We are going to show that it is advantageous to impose the condition $\lambda + \mu = 1$. First, we consider two points in the plane and write the equation of the straight line defined by the points P_1, P_2 in the form

$$y = ax + b$$

Writing that this line passes through the two given points we obtain

$$y = \frac{\begin{vmatrix} y_1 & 1 \\ y_2 & 1 \end{vmatrix}}{\begin{vmatrix} x_1 & 1 \\ x_2 & 1 \end{vmatrix}} x + \frac{\begin{vmatrix} x_1 & y_1 \\ x_2 & y_2 \end{vmatrix}}{\begin{vmatrix} x_1 & 1 \\ x_2 & 1 \end{vmatrix}} \tag{8.35}$$

Let

$$P = \lambda P_1 + \mu P_2$$

be a linear combination of the given points. To lie on the straight line defined by the given points, P must satisfy Eq. (8.35), that is

$$\lambda y_1 + \mu y_2 = \frac{\begin{vmatrix} y_1 & 1 \\ y_2 & 1 \end{vmatrix} (\lambda x_1 + \mu x_2) + \begin{vmatrix} x_1 & y_1 \\ x_2 & y_2 \end{vmatrix}}{\begin{vmatrix} x_1 & 1 \\ x_2 & 1 \end{vmatrix}} \tag{8.36}$$

Multiplication by the denominator of the right-hand part yields

$$(\lambda y_1 + \mu y_2) \begin{vmatrix} x_1 & 1 \\ x_2 & 1 \end{vmatrix} = \begin{vmatrix} y_1 & 1 \\ y_2 & 1 \end{vmatrix} (\lambda x_1 + \mu x_2) + \begin{vmatrix} x_1 & y_1 \\ x_2 & y_2 \end{vmatrix} \tag{8.37}$$

Multiplying two determinants by the factors between parentheses we obtain

$$\begin{vmatrix} x_1 & \lambda y_1 + \mu y_2 \\ x_2 & \lambda y_1 + \mu y_2 \end{vmatrix} = \begin{vmatrix} y_1 & \lambda x_1 + \mu x_2 \\ y_2 & \lambda x_1 + \mu x_2 \end{vmatrix} + \begin{vmatrix} x_1 & y_1 \\ x_2 & y_2 \end{vmatrix}$$

Expanding the determinants and cancelling equal terms we remain with

$$(\lambda + \mu) x_1 y_2 - (\lambda + \mu) x_2 y_1 = x_1 y_2 - x_2 y_1 \tag{8.38}$$

Eq. (8.38) is fulfilled if $\lambda + \mu = 1$. We conclude that the affine combination of two points, $\mathbf{P}_1$, $\mathbf{P}_2$, lies on the straight line that passes through these points.

8.5.2 Alternative Proof of Collinearity

This time we consider two points in 3D space

$$P_1 = \begin{bmatrix} x_1 \\ y_1 \\ z_1 \end{bmatrix}, \quad P_2 = \begin{bmatrix} x_2 \\ y_2 \\ z_2 \end{bmatrix}$$

and their linear combination

$$Q = \lambda P_1 + \mu P_2 = \begin{bmatrix} \lambda x_1 + \mu x_2 \\ \lambda y_1 + \mu y_2 \\ \lambda z_1 + \mu z_2 \end{bmatrix}$$

We are going to show in another way that if $\lambda + \mu = 1$, then the point Q lies on the straight line defined by the points P_1 and P_2. Now, if the three points Q, P_1, P_2 are collinear, the area of the parallelogram defined by the vectors $\overrightarrow{QP_1}$ and $\overrightarrow{QP_2}$ is zero. These vectors are

$$\overrightarrow{QP_1} = P_1 - Q = \begin{bmatrix} (1-\lambda)x_1 - \mu x_2 \\ (1-\lambda)y_1 - \mu y_2 \\ (1-\lambda)z_1 - \mu z_2 \end{bmatrix}$$

and

$$\overrightarrow{QP_2} = P_2 - Q = \begin{bmatrix} -\lambda x_1 + (1-\mu)x_2 \\ -\lambda y_1 + (1-\mu)y_2 \\ -\lambda z_1 + (1-\mu)z_2 \end{bmatrix}$$

The area of the parallelogram defined by the above vectors is calculated as the cross product

$$\overrightarrow{QP_1} \times \overrightarrow{QP_2}$$
$$= \begin{bmatrix} \mathbf{i} & \mathbf{j} & \mathbf{k} \\ (1-\lambda)x_1 - \mu x_2 & (1-\lambda)y_1 - \mu y_2 & (1-\lambda)z_1 - \mu z_2 \\ -\lambda x_1 + (1-\mu)x_2 & -\lambda y_1 + (1-\mu)y_2 & -\lambda z_1 + (1-\mu)z_2 \end{bmatrix}$$

Assuming that $\lambda + \mu = 1$, the determinant of the cross product becomes

$$\overrightarrow{QP_1} \times \overrightarrow{QP_2} = \begin{bmatrix} \mathbf{i} & \mathbf{j} & \mathbf{k} \\ \mu x_1 - \mu x_2 & \mu y_1 - \mu y_2 & \mu z_1 - \mu z_2 \\ -\lambda x_1 + \lambda x_2 & -\lambda y_1 + \lambda y_2 & -\lambda z_1 + \lambda z_2 \end{bmatrix}$$

$$= -\lambda\mu \begin{bmatrix} \mathbf{i} & \mathbf{j} & \mathbf{k} \\ x_1 - x_2 & y_1 - y_2 & z_1 - z_2 \\ x_1 - x_2 & y_1 - y_2 & z_1 - z_2 \end{bmatrix} = 0$$

as two lines are identical.

Example 8.2 (An example of translation). We consider again two points, P_1 and P_2, either in plane or in 3D space, and their linear combination

$$Q = \lambda P_1 + \mu P_2$$

We assume now that the two points are subject to a translation by a distance d, where by d we mean a vector having the same dimensions as the given points. The translated points are

$$P_1^* = P_1 + d$$
$$P_2^* = P_2 + d \tag{8.39}$$

We consider the linear combination

$$Q^* = \lambda P_1^* + \mu P_2^*$$
$$= (\lambda P_1 + \mu P_2) + (\lambda + \mu)d = Q + (\lambda + \mu)d \tag{8.40}$$

We conclude that the point Q translates by the same vector as the points P_1 and P_2 if and only if

$$\lambda + \mu = 1$$

The effect of the translation under the above condition is exemplified in Fig. 8.14. In Fig. 8.15 we can see an example of what happens when the sum of weights is greater than one. Finally, in Fig. 8.16 we see the result of the transformation when the sum of the weights is less than one. In Exercise 8.18 we propose the reader to check what happens if one of the weights λ or μ is negative while the sum of the weights equals 1.

8.5.3 Affine Combination of Three Points — Coplanarity

Given three points, P_1, P_2, P_3, their affine combination is

$$P = \lambda P_1 + \mu P_2 + \nu P_3, \quad \lambda + \mu + \nu = 1 \tag{8.41}$$

The point **P** lies in the plane defined by the given points. To show this we consider the general equation of a plane

$$ax + by + cz + d = 0$$

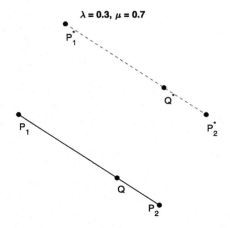

Figure 8.14 Affine combination of two points — Effect of translation

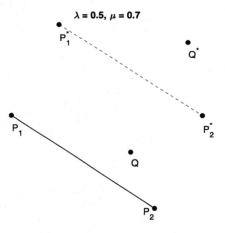

Figure 8.15 A linear, not affine combination of two points

The three given points must fulfill this equation, i.e.

$$ax_1 + by_1 + cz_1 = -d$$
$$ax_2 + by_2 + cz_2 = -d$$
$$ax_3 + by_3 + cz_3 = -d \tag{8.42}$$

Let us substitute the affine combination **P** in the general equation of the plane.

$$a(\lambda x_1 + \mu x_2 + \nu x_3) + b(\lambda y_1 + \mu y_2 + \nu x_3) + a(\lambda z_1 + \mu z_2 + \nu z_3) \tag{8.43}$$

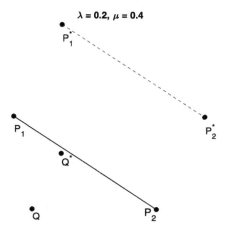

$\lambda = 0.2, \mu = 0.4$

Figure 8.16 Another linear, not affine combination of two points

$$= \lambda(ax_1 + by_1 + cz_1) + \mu(ax_2 + by_2 + cz_2) + v(ax_3 + by_3 + cz_3) \qquad (8.44)$$
$$= -d(\lambda + \mu + v) \qquad (8.45)$$

as each expression within parentheses equals $-d$. If $\lambda + \mu + v = 1$ the point **P** is coplanar with the three given points.

8.6 BARYCENTRES

We consider a set of mass points

$$\mathbf{P}_i = \begin{vmatrix} x_i \\ y_i \\ z_i \end{vmatrix}, \quad i = 1, 2, ..., n \qquad (8.46)$$

and let m_i be the mass of the point $\mathbf{P}_i$. Similar to the definitions of centres of areas in Section 3.4.1, the *centre of mass of the system is given by*

$$\mathbf{X}_{CG} = \frac{\sum_{i=1}^{n} m_i \mathbf{P}_i}{\sum_{i=1}^{n} m_i} \qquad (8.47)$$

If instead of masses, m_i, we talk in terms of weights, w_i, we refer to $\mathbf{X}_{CG}$ as the vector of coordinates of the *centre of gravity* of the system. The sum of the coefficients that multiply the coordinates of the mass points equals 1. This means that the centre of mass of a set of mass points is an affine combination of the mass points and enjoys the properties of affine combinations. Some examples are listed below.

1. Given a system of two mass points, $\mathbf{P}_1$, $\mathbf{P}_2$, the centre of mass of the system lies on the straight-line segment $\overline{P_1 P_2}$.

2. Given a system of three non–collinear mass points, $\mathbf{P}_1$, $\mathbf{P}_2$, $\mathbf{P}_3$, the centre of mass of the system lies in the plane defined by the given points.

3. Given a set of mass points $\mathbf{P}_i$, and a transformation T, the centre of mass of the set of transformed points $T\mathbf{P}_i$ is the same point as $T\mathbf{X}_{CG}$, where $\mathbf{X}_{CG}$ is defined as in Eq. (8.47).

In addition, affine combinations enjoy an associative property that is used in the calculation of centres of gravity. It is possible to group the mass points of a set, S, in disjoint subsets, S_i, calculate the centres of mass, $\mathbf{X}_i$, of the subsets, and obtain the centre of mass of the whole system, S, as the centre of mass of the system $\mathbf{X}_i$. In terms of set theory the subsets S_i should fulfill the conditions:

1. the *union* of the subsets should be the given set S;

2. each mass point should appear in only one subset.

We prove this associative property for a grouping in two subsets. Let $1 < k < n$. The centres of mass of the two subsets are

$$\mathbf{X}_1 = \frac{\sum_{i=1}^{k} m_i \mathbf{P}_i}{\sum_{i=1}^{k} m_i}, \quad \mathbf{X}_2 = \frac{\sum_{i=k+1}^{n} m_i \mathbf{P}_i}{\sum_{i=k+1}^{n} m_i} \tag{8.48}$$

The centre of mass of the system $\mathbf{X}_1$, $\mathbf{X}_2$ is

$$\frac{\mathbf{X}_1 \sum_{i=1}^{k} m_i + \mathbf{X}_2 \sum_{i=k+1}^{n} m_i}{\sum_{i=1}^{k} m_i + \sum_{i=k+1}^{n} m_i} = \frac{\sum_{i=1}^{n} m_i \mathbf{P}_i}{\sum_{i=1}^{n} m_i} \tag{8.49}$$

The associative property is used in the weight calculations of ship loading conditions. The centres of gravity of the lightship and of the deadweight are calculated separately and the centre of gravity of the loaded ship is obtained as the centre of gravity of these two components.

8.7 SUMMARY

We define a point by the column vector of its coordinates. This is a 2-by-1 array in the plane, 3-by-1 in space; let us call it $\mathbf{P}$. A geometric figure is defined by concatenating points. For example, a triangle with vertices $\mathbf{P}_1$,

$\mathbf{P}_2$, $\mathbf{P}_3$ is defined by the array $|\mathbf{P}_1 \, \mathbf{P}_2 \, \mathbf{P}_3|$. An *affine transformation* of $\mathbf{P}$ is performed by

$$T\mathbf{P} + T_t$$

where T is a *transformation matrix* of dimensions 2-by-2 in the plane, 3-by-3 in space, and T_t is the vector of *translation*. Transformations like *translation*, *rotation*, and *reflection* do not change distances or angles. Therefore, they are called *isometric transformations*, or, shortly, *isometries*. Do not confound with the term *isometric projection* defined in Chapter 1. Other transformations, not isometric, are *shearing* and *scaling*.

To perform all transformations by multiplication we use *homogeneous coordinates*. In homogeneous coordinates we augment the dimensions by one. Let us call the usual coordinates *Euclidean*. Given in the plane a point with Euclidean coordinates x, y, its homogeneous coordinates are

$$\begin{vmatrix} X \\ Y \\ w \end{vmatrix} = \begin{vmatrix} wx \\ wy \\ w \end{vmatrix}, \quad w \neq 0$$

A common assumption is $w = 1$. The extension to 3D space is

$$\begin{vmatrix} X \\ Y \\ Z \\ w \end{vmatrix} = \begin{vmatrix} wx \\ wy \\ wz \\ w \end{vmatrix}, \quad w \neq 0$$

In homogeneous coordinates a translation is performed as

$$\begin{vmatrix} 1 & 0 & 0 & \delta x \\ 0 & 1 & 0 & \delta y \\ 0 & 0 & 1 & \delta z \\ 0 & 0 & 0 & 1 \end{vmatrix} \begin{vmatrix} X \\ Y \\ Z \\ 1 \end{vmatrix} = \begin{vmatrix} X + \delta x \\ Y + \delta y \\ Z + \delta z \\ 1 \end{vmatrix}$$

When $w \to 0$ the point tends to infinity, and for $w = 0$ we have a *point at infinity*, or *ideal point*. All lines with the same directions have the same point at infinity. The Euclidean plane completed with its ideal points is called *projective plane*.

When several transformations must be performed on the same point or figure, the order in which they are carried on may influence the results. For example, the order of two consecutive rotations can be reversed, but the sequence rotation–translation yields another result than the sequence translation–rotation. Given a point or figure **P**, instead of carrying on the sequence

$$\mathbf{Q} = T_1 \mathbf{P}, \ \mathbf{R} = T_2 \mathbf{Q}$$

it is possible to perform the sequence

$$T = T_1 T_2, \ \mathbf{R} = T \mathbf{P}$$

The extension to a sequence of more transformations is straightforward. Working with a combined matrix T may reduce numerical errors and improve computational efficiency, especially when the number of points to be transformed is large.

The use of homogeneous coordinates unifies also the application of *projection matrices*, among them the matrices for *perspective projection*. By the latter term we mean here the central projection. In perspective projections parallel lines are projected as parallels only if they are parallel to the image plane. Otherwise parallel lines are projected as intersecting in points called *vanishing points*. These are the visible projections of points at infinity.

A point can be defined as the linear combination of other points, for example

$$\mathbf{P} = \lambda \mathbf{P}_1 + \mu \mathbf{P}_2$$

If $\lambda + \mu = 1$, then **P** is an *affine combination* of $\mathbf{P}_1$ and $\mathbf{P}_2$. The affine combination of two points lies on the line defined by the two points, and the affine combination of three points lies in the plane defined by the three points. Moreover, to apply a transformation to an affine combination we may either

- first calculate the affine combination and next transform it, or
- first transform the points and next calculate the affine combination.

Given a system of mass points, their *centre of mass* (alternatively *centre of gravity* or *barycentre*) is the affine combination of the mass points. Some *splines* used in computer graphics are affine combinations of given *control points*.

8.8 EXERCISES

Exercise 8.1 (Translation in the plane). Consider the triangle $\mathbf{P}_1 \mathbf{P}_2 \mathbf{P}_3$ exemplified in Section 8.2.1.

1. Translate the triangle so that the vertex $\mathbf{P}_2$ moves to the origin of coordinates.
2. Check that this translation does not change the distance between the points $\mathbf{P}_1$ and $\mathbf{P}_2$.

Exercise 8.2 (Translation in the plane). Consider the triangle $\mathbf{P}_1\mathbf{P}_2\mathbf{P}_3$ exemplified in Section 8.2.1.
1. Translate the triangle so that the vertex $\mathbf{P}_3$ moves to the origin of coordinates.
2. Check that this translation does not change the distance between the points $\mathbf{P}_2$ and $\mathbf{P}_3$.

Exercise 8.3 (The product of two rotations). Starting from the definition of the matrix of rotation T_R in Section 8.2.2 show that

$$T_R(\theta_1)\,T_R(\theta_2) = T_R(\theta_2)\,T_R(\theta_1) = T_R(\theta_1 + \theta_2)$$

Exercise 8.4 (Reflection in the y-axis). **1.** Write the matrix of reflection in the y-axis.
2. Use the above matrix to reflect in the y-axis the triangle exemplified in Fig. 8.4.

Exercise 8.5 (Reflection as an isometric transformation). Prove that the reflection in the x-axis preserves distances.

Exercise 8.6 (Reflection in the origin). The matrix for reflection in the origin, in the plane, is

$$\begin{vmatrix} -1 & 0 \\ 0 & -1 \end{vmatrix}$$

Use this matrix to transform the triangle $\mathbf{P}_1\mathbf{P}_2\mathbf{P}_3$ exemplified in Section 8.2.1.

Exercise 8.7 (Scaling with equal factors). Prove that the triangles $P_1P_2P_3$ and $Q_1Q_2Q_3$ in Section 8.2.7 are similar.

Hint: show that the ratios of the side lengths of the scaled triangle to the side lengths of the initial triangle are equal to the scale factor.

Exercise 8.8 (Scaling a triangle). Scale the triangle $P_1P_2P_3$ in Section 8.2.7 using the scales $S_x = 1.5$, $S_y = 1.8$.

Exercise 8.9 (Rotating and translating a figure). Let the vertices of a plane figure be

$$\mathbf{P}_1 = \begin{vmatrix} 1.5 \\ 1.5 \end{vmatrix}, \ \mathbf{P}_2 = \begin{vmatrix} 3.0 \\ 1.5 \end{vmatrix}, \ \mathbf{P}_3 = \begin{vmatrix} 3.0 \\ 3.0 \end{vmatrix}, \ \mathbf{P}_4 = \begin{vmatrix} 1.5 \\ 3.0 \end{vmatrix}$$

1. Plot the figure $\mathbf{P}_1\mathbf{P}_2\mathbf{P}_3\mathbf{P}_4$. What kind of figure is it?
2. Translate the figure by $\delta x = 1$, $\delta y = 2$ and afterwards rotate it counterclockwise around the origin by 45°. Add the transformed figure to the plot produced for the preceding question. Call the transformed figure $\mathbf{Q}_1\mathbf{Q}_2\mathbf{Q}_3\mathbf{Q}_4$.
3. Rotate the figure $\mathbf{P}_1\mathbf{P}_2\mathbf{P}_3\mathbf{P}_4$ counterclockwise around the origin by the angle 45° and afterwards translate it by $\delta x = 1$, $\delta y = 2$. Add the resulting figure to the existing plot. Call the transformed figure $\mathbf{R}_1\mathbf{R}_2\mathbf{R}_3\mathbf{R}_4$. Is the order of transformations important?
4. Use homogeneous coordinates and calculate the transformation matrices that perform the combined transformations specified in questions 2 and 3. Identify the differences.
5. Carry on the transformations using this time the combined matrices calculated in Question 4.

Exercise 8.10 (Shearing in the y-direction). **1.** Write the matrix that performs the transformation shown in Fig. 8.17.
2. Write a MATLAB script that produces Fig. 8.17.

Exercise 8.11 (Shearing in the x-direction). Consider the initial rectangle shown in Fig. 8.17. Transform it by shearing with $Sh_x = -1.2$.

Exercise 8.12 (Shearing is not an isometry). Consider shearing in the x-direction. The distances between which points are preserved?

Exercise 8.13 (Scaling in homogeneous coordinates). Write the matrix for scaling in homogeneous coordinates in the plane.

Exercise 8.14 (Transformations of a ship waterplane). Fig. 8.18 shows a schematic ship waterplane. Your tasks are:
1. plot the figure;
2. rotate the figure counterclockwise 45 degrees around the origin of coordinates;
3. translate the initial figure 3 units in the positive z-direction;
4. reflect the initial figure in the x-axis. Is the result plausible?

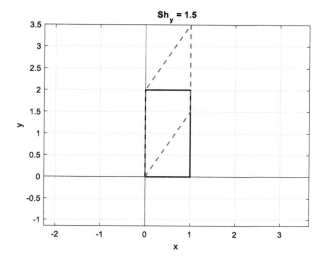

Figure 8.17 Shearing in the y-direction

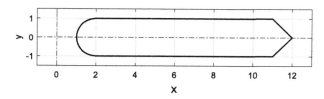

Figure 8.18 A schematic ship waterplane

Exercise 8.15 (Perspective view of unit cube). This exercise refers to the unit cube defined in the script PerspCube1, Section 8.4.1. Rotate this cube 45° around its vertical axis and apply the perspective transformation.

Exercise 8.16 (Ideal points in the projective plane). **1.** What is the ideal point of the lines parallel to the x-axis?
2. What is the ideal point of the lines parallel to the y-axis?

Exercise 8.17 (Identification of vanishing points). Make an enlarged copy of Fig. 8.12 and identify on it the vanishing points.

Exercise 8.18 (Affine combinations of two points). Given the two points

$$P_1 = \begin{vmatrix} 1 \\ 5 \end{vmatrix}, \quad P_2 = \begin{vmatrix} 3 \\ 10 \end{vmatrix}$$

Point	Mass	x	y	z
P_1	1.0	1.0	1.0	0.0
P_2	1.3	2.3	3.0	0.5
P_3	2.1	2.0	4.5	1.7
Q_1	3.0	0.4	0.7	3.1
Q_2	3.5	0.4	0.8	3.7
Q_3	4.0	0.8	1.2	3.9

Figure 8.19 A system of mass points

plot in MATLAB the segment $\overline{P_1 P_2}$ and the points

$$P_3 = -0.5P_1 + 1.5P_2$$
$$P_4 = 1.2P_1 - 0.2P_2$$

Exercise 8.19 (Centres of mass). Given the system of mass points defined in Fig. 8.19 you are asked to

1. calculate the centre of mass, P, of the system P_1, P_2, P_3;
2. calculate the centre of mass, Q, of the system Q_1, Q_2, Q_3;
3. calculate the centre of mass of the system P, Q;
4. calculate the centre of mass of the system P_1, P_2, P_3, Q_1, Q_2, Q_3;
5. compare the results.

CHAPTER 9

Bézier Curves

Contents

9.1 INTRODUCTION

English	French	German	Italian
binomial coefficient	coefficient binomial	Binomialkoeffizient	coefficiente binomiale
n choose i	i parmi n	n über i	n sopra i
convex hull	enveloppe convexe	konvexe Hülle	inviluppo convesso
geometric continuity	continuité géométrique	geometrische Stetigkeit	continuità geometrica

The cubic splines treated in Chapter 7 pass through all the points that define them. Therefore, these curves belong to the category of *interpolating splines*. In this chapter, and in the following, we are going to deal with splines that do not pass through all the points that define them. Given n *control points*, these curves pass only through the first and the last of them. We can join in a given order the control points by straight–line segments and obtain thus a *control polygon*. The curves are smooth approximations of the control polygons. Therefore, the curves discussed in this chapter and

Geometry for Naval Architects
https://doi.org/10.1016/B978-0-08-100328-2.00020-1

Copyright © 2019 Elsevier Ltd.
All rights reserved.

in the next one belong to the category of *approximating splines*. The shape of these splines can be easily modified by moving one or more control points. The application that interests us is in the drawing of *freeform* curves, that is smooth curves defined by their shape and not by specifying *a priori* a particular equation or a kind of classic curve, such as circle, ellipse, parabola, a.s.o. This is certainly the case in naval architecture, but also in the car and aeronautical industries. The development of the curves we are going to deal with started with De Casteljau and Bézier.

Paul de Faget de Casteljau was born in Eastern France in 1930. He studied mathematics and physics and was engaged by Citroën, in Paris, in 1958. De Casteljau's task was to develop methods for modelling car bodies and implement them on machines controlled at that time by analogue computers. De Casteljau's idea was to define and modify curves with the help of *control points*. He called his invention *courbes à pôles*. For a long time the company considered the work as a secret and De Casteljau published nothing about his achievements. Pierre Etienne Bézier (1910–1999) was in that period a member of the managing staff at Renault. He had some knowledge of what was going on at Citroën, but did not know how things were done. He developed his own curves, similar to those of De Casteljau. Bézier was allowed to publish his results and thus the new mathematical objects became known as **Bézier curves**. Both De Casteljau and Bézier were met with incredulity, even hostility. Townsend (2014) cites reactions like 'If your system were that good, the Americans would have invented it first!' A rather sarcastic account of this story is due to De Casteljau himself (De Casteljau, 1999). He received due recognition (see Bieri and Prautsch, 1999) and his and Bézier's work are considered as fundamental in present-day computer graphics. An intuitive introduction to their methods is given by *Bézier* in the first chapter of Farin (2002).

Further developments are the *rational Bézier* curves also treated in this chapter, and the *B-splines, rational B-splines*, and *NURBS* briefly discussed in the next chapter. We describe to some extent the Bézier curves because they are the simplest, easiest to understand of all the mentioned curves, and already present the main features of the more advances splines. The understanding of these features and of the way of manipulating Bézier curves is a good introduction to the ship-hull design programs available on the market. We do not go into more details because this would increase too much the size of this book. The interested reader can refer to one of the classic books wholly dedicated to the subject, for example Rogers and Adams

(1990), Hoschek and Lasser (1993), Mortenson (1997), Farin (1999), Marsh (2000), Farin (2002), Kim et al. (2002).

In this chapter sometimes we note the control points by $\mathbf{P}_1, \mathbf{P}_2, \ldots$, other times by $\mathbf{P}_0, \mathbf{P}_1, \ldots$. The former notation suits MATLAB arrays that do not allow the index 0. The latter notation slightly simplifies equations.

9.2 THE FIRST-DEGREE BÉZIER CURVES

Two points, P_1, P_2, define the first-degree Bézier curve

$$\mathbf{P} = (1 - t)\mathbf{P_1} + t\mathbf{P_2} \tag{9.1}$$

In the plane Eq. (9.1) stands for

$$x = (1 - t)x_1 + tx_2$$
$$y = (1 - t)y_1 + ty_2$$

The extension to the three-dimensional space is obvious. As $(1 - t) + t = 1$, we conclude that the point P is the affine combination of the points P_1, P_2 and it enjoys the properties of affine combinations. For example, P lies on the line passing through P_1 and P_2, and for $t \in [0, 1]$, P lies between P_1 and P_2. A function that calculates a point on the first-degree Bézier curve is shown below.

```
function  P  = Bezier1(P1, P2, t)
%BEZIER1 Finds the coordinates of point P for given t
%    Input arguments:
%        P1, P2, coordinates of the first point and second points
%        given as column vectors;
%        t, parameter. For points between P1 and P_2 t is in the
%        interval [ 0, 1 ];
%    Output argument: coordinates of point P given as column vector

P = (1 - t)*P1 + t*P2;
end
```

9.3 THE SECOND-DEGREE BÉZIER CURVES

Three points, P_1, P_2, P_3, define the second-degree Bézier curve

$$\mathbf{P} = (1 - t)^2\mathbf{P_1} + 2(1 - t)t\mathbf{P_2} + t^2\mathbf{P_3} \tag{9.2}$$

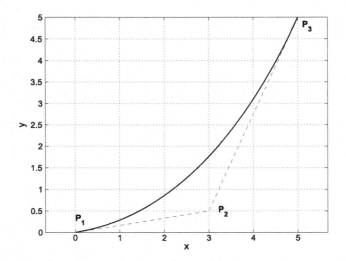

Figure 9.1 A second-degree Bézier curve

The coefficients that multiply the coordinates of the three points, $\mathbf{P}_1$, $\mathbf{P}_2$, $\mathbf{P}_3$, are the terms of a binomial expansion and fulfill the condition

$$((1 - t) + t)^2 = 1^2 = 1$$

Therefore, the point $\mathbf{P}$ is the affine combination of the points $\mathbf{P}_1$, $\mathbf{P}_2$, $\mathbf{P}_3$. As proven in Section 8.5.3, the point $\mathbf{P}$ lies in the plane defined by the three points $\mathbf{P}_1$, $\mathbf{P}_2$, $\mathbf{P}_3$. The points, $\mathbf{P}_1$, $\mathbf{P}_2$, $\mathbf{P}_3$ are called **control points** of the curve, and the line $\mathbf{P_1P_2P_3}$ is the **control polygon** of the curve. Fig. 9.1 shows a second-degree Bézier curve and its control polygon. Curves like this are arcs of parabola. Instead of giving a general, formal proof, in Exercise 9.2 we invite the reader to confirm experimentally our statement.

9.4 THE THIRD-DEGREE BÉZIER CURVES

Four points, $\mathbf{P}_1$, $\mathbf{P}_2$, $\mathbf{P}_3$, $\mathbf{P}_4$, define the third-degree Bézier curve

$$\mathbf{P} = (1 - t)^3\mathbf{P_1} + 3(1 - t)^2 t\mathbf{P_2} + 3(1 - t)t^2\mathbf{P_3} + t^3\mathbf{P_4} \qquad (9.3)$$

The coefficients that multiply the coordinates of the four points, $\mathbf{P}_1$, $\mathbf{P}_2$, $\mathbf{P}_3$, $\mathbf{P}_4$, are the terms of a binomial expansion and fulfill the condition

$$((1 - t) + t)^3 = 1^3 = 1$$

Therefore, the point **P** is the affine combination of the points $\mathbf{P}_1$, $\mathbf{P}_2$, $\mathbf{P}_3$, $\mathbf{P}_4$. Below is a function that calculates points on a cubic Bézier curve.

```
function B3 = Bezier3(P)
%BEZIER3 calculates 51 points on the third-degree Bezier curve
%    Input argument, P, array of coordinates of the four control
%    points, 2x4 in plane, 3x4 in space.
%    Output argument, B3, array of the coordinates of points on the
%    Bezier curve

%  Check input
[ m, n ] = size(P);
if n ~= 4
    error('Data not for four points')
end
t   = 0: 0.02: 1;                  % parameter
C1 = (1 - t).^3;                   % Bernstein polynomials
C2 = 3*t.*(1 -t).^2;
C3 = 3*t.^2.*(1 -t);
C4 = t.^3;
C   = [ C1; C2; C3; C4 ];
B3  = P*C;
```

9.5 THE GENERAL DEFINITION OF BÉZIER CURVES

Given $n+1$ control points, $\mathbf{P}_0, \mathbf{P}_1, \ldots, \mathbf{P}_i, \ldots, \mathbf{P}_n$, they define a Bézier curve of degree n. The general equation of this curve is

$$\mathbf{P} = \sum_{i=0}^{n} B_i^n(t)\mathbf{P}_i \tag{9.4}$$

where B_i^n is a *Bernstein polynomial* defined by

$$B_i^n(t) = \left(\begin{array}{c} n \\ i \end{array} \right) t_i(1 - t)^{n-i} \tag{9.5}$$

The first factor is the *binomial coefficient* calculated as

$$\left(\begin{array}{c} n \\ i \end{array} \right) = \left\{ \begin{array}{ll} \frac{n!}{i!(n-i)!} & \text{if } 0 \leq i \leq 1 \\ 0 & \text{otherwise} \end{array} \right. \tag{9.6}$$

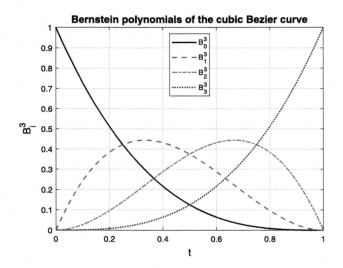

Figure 9.2 The Bernstein polynomials of the cubic Bézier curve

The common way of calling this expression is 'n choose i'. It is shown in combinatorics that given a set of n objects, Eq. (9.6) yields the number of subsets of i objects that can be chosen from it. Fig. 9.2 shows the curves of the four Bernstein polynomials of the cubic Bézier curve. In this figure it is possible to appreciate the influence of each control point on the whole curve. The Bernstein polynomials enjoy the following recursion property

$$B_i^n(t) = (1-t)B_i^{n-1}(t) + tB_{i-1}^{n-1}(t) \tag{9.7}$$

To prove this relationship we treat the right-hand side of Eq. (9.7) as shown below.

$$(1-t)\begin{pmatrix} n-1 \\ i \end{pmatrix}(1-t)^{n-1-i}t^i + \tag{9.8}$$

$$t\begin{pmatrix} n-1-i+1 \\ i-1 \end{pmatrix}(1-t)^{(n-1-i+1)}t(i-1)$$

$$= (1-t)^{(n-i)}t^i\left\{\begin{pmatrix} n-1 \\ i \end{pmatrix}+\begin{pmatrix} n-1 \\ i-1 \end{pmatrix}\right\} \tag{9.9}$$

The sum between the curled parentheses is

$$\begin{pmatrix} n-1 \\ i \end{pmatrix} + \begin{pmatrix} n-1 \\ i-1 \end{pmatrix} = \frac{(n-1)!}{(n-1-i)!i!} + \frac{(n-1)!}{(n-1-i+1)(i-1)!}$$

$$= (n-1)! \left\{ \frac{1}{(n-1-i)!i!} + \frac{i}{(n-i)!i!} \right\}$$

$$= \frac{n!}{(n-i)!i!} = \begin{pmatrix} n \\ i \end{pmatrix} \tag{9.10}$$

This proves that Eq. (9.10) yields the same result as Eq. (9.5).

9.6 INTERACTIVE MANIPULATION OF BÉZIER CURVES

Bézier curves can be modified interactively by dragging control points. This facility is provided by all CAD and drawing programs available today. Demos can be found on the web, some of them programmed in GeoGebra. A MATLAB script that allows the user to modify a cubic Bézier curve is shown below.

```
%DEMOBEZ Interactive demo that plots a cubic Bezier curve
%     and allows the user to change the curve by manipulating
%     the control points.
%     USAGE. Point with the mouse the control point that you want
%     to change and press down the left button. Point to the
%     desired new position and press again the left button.
%     To end the changing mode point far from the control polygon
%     and press the button.

t  = [ 0: 0.02: 1 ]'; % parameter
C1 = (1 - t).^3;        % calculate Bernstein polynomials
C2 = 3*t.*(1 - t).^2;
C3 = 3*t.^2.*(1 - t);
C4 = t.^3;
C  = [ C1 C2 C3 C4 ];
P  = [ 0 1 3 5; 0 3 5 5.5 ]; % initial control polygon
PlotBez(C, P)                % invoke nested function
d2 = 1;
while min(d2) < 2
        x0 = P(1, :);   % x-coordinates of control points
        y0 = P(2, :);   % y-coordinates of control points
```

```
        [ x1, y1 ] = ginput(1);
        d2 = (x0 - x1).^2 + (y0 - y1).^2;
                        % distance to control points
        dm = min(d2); % find closest control point
        ii = find(d2 == dm);
        [ x2 y2 ] = ginput(1); % choose new position
        x0(ii)    = x2; y0(ii) = y2;
        P = [ x0; y0 ];
        hold off              % allow changed plot
        PlotBez(C, P)
end
msgbox('Iterations terminated', 'error')

function PlotBez(c, p)
            x0 = p(1, :);   % x-coordinates of control points
            y0 = p(2, :);   % y-coordinates of control points
            % calculate coefficients of cubic polynomials
            xB = c*x0'; yB = c*y0';
            % Plot control polygon
            plot(x0, y0, 'r--', x0, y0, 'r*');
            grid, axis equal, axis([ -1 7 -1 7 ])
            title('Cubic Bezier curve')
            xlabel('x'), ylabel('y')
            % Plot Bezier curve
            hold on
            plot(xB, yB, 'k-');
end
```

The function `DemoBez` generates Fig. 9.3.

9.7 DE CASTELJAU'S ALGORITHM

An interesting and famous procedure for generating a Bézier curve is *de Casteljau's algorithm*. We are going to exemplify it for a third-degree curve. The steps are shown schematically in Fig. 9.4 and they are:

1. We are given four control points, P_1, P_2, P_3, P_4. We choose a value for the parameter t in the interval [0 1].
2. The points P_1, P_2 define a first-degree Bézier curve. For the chosen t value we obtain the point P_{12}. The points P_2, P_3 define a second first-degree Bézier curve. For the chosen t value we obtain the point P_{23}. The points P_3, P_4 define a third first-degree Bézier curve. For the chosen t value we obtain the point P_{34}.

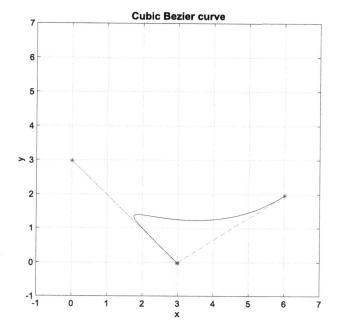

Figure 9.3 The screen produced by the DemoBez script

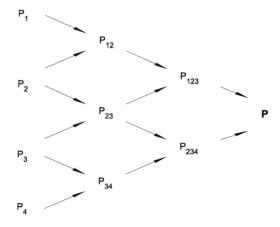

Figure 9.4 The scheme of de Casteljau's algorithm

3. The points P_{12}, P_{23} define a first-degree Bézier curve. For the chosen t value we obtain the point P_{123}. The points P_{23}, P_{34} define a second first-degree Bézier curve. For the chosen t value we obtain the point P_{234}.

4. The points P_{123}, P_{234} define a first-degree Bézier curve. For the chosen t value we obtain the point P. This is the final result.

The following MATLAB script implements the algorithm in steps corresponding to $t = 0, 0.1, \ldots, 1$.

```
%CASTELJAU Animation of De Casteljau's algorithm for a third-degree
%    Bezier curve. To run the curve in steps of t equal to 0.1
%    press ENTER. This script calls the functions pline and point.

P1 = [ 1; 2 ]; P2 = [ 3; 4 ];
P3 = [ 5; 3.5 ];   P4 = [ 7; 1.7 ];
pline([ P1 P2 P3 P4 ], 0.8, 'r', '-')
axis equal, axis off
ht = title('The algorithm of De Casteljau');
set(ht, 'FontSize', 14)
hold on
point(P1, 0.03), text(0.9, 2.15, 'P_1')
point(P2, 0.03), text(2.9, 4.15, 'P_2')
point(P3, 0.03), text(4.9, 3.65, 'P_3')
point(P4, 0.03), text(6.9, 1.85, 'P_4')
% calculate and plot Bezier curve
P = Bezier3([ P1 P2 P3 P4 ]);
pline(P, 0.8, 'k', '-')
hp1 = plot([ ], [ ]); hp2 = plot([ ], [ ]);
for t = 0.1: 0.1: 0.9;              % begin loop over t
    % first interpolation;
    P12 = Bezier1(P1, P2, t); P23 = Bezier1(P2, P3, t);
    P34 = Bezier1(P3, P4, t);
    delete(hp1)
    hp1 = plot([P12(1) P23(1) P34(1)], [P12(2) P23(2) P34(2)],...
               'b-', 'LineWidth', 1.5);
    %    second interpolation
    P123 = Bezier1(P12, P23, t); P234 = Bezier1(P23, P34, t);
    delete(hp2)
    hp2  = plot([ P123(1) P234(1)], [ P123(2) P234(2)], 'g-',...
               'LineWidth', 1.5);
    % third interpolation
    P    = Bezier1(P123, P234, t);
    point(P, 0.03)
    text(P(1), (P(2) - 0.2), [ 't = ' num2str(t) ])
    pause
end                     % end loop over t
hold off
```

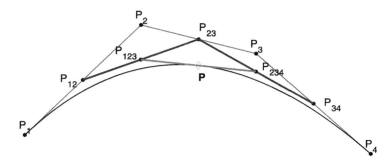

Figure 9.5 De Casteljau's algorithm, $t = 0.5$

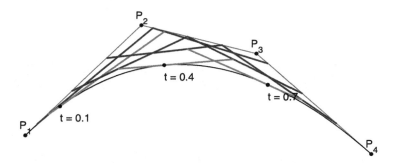

Figure 9.6 De Casteljau's algorithm, general picture

Fig. 9.5 shows the situation corresponding to $t = 0.5$. We have chosen this value because for it the reader can easily check the position of the points. Fig. 9.6 shows several steps of the algorithm.

9.8 SOME PROPERTIES OF BÉZIER CURVES

9.8.1 The First and the Last Point of the Curve

Given a control polygon, $\mathbf{P}_1, ... \mathbf{P}_n$, the corresponding Bézier curve passes only through the first and last control points and not through the others. This is easily seen if we substitute $t = 0$ and $t = 1$ in the equation of the curve.

9.8.2 End Tangents

Given a control polygon, $\mathbf{P}_1$, $\mathbf{P}_2$, ...$\mathbf{P}_{n-1}$, $\mathbf{P}_n$, the corresponding Bézier curve is tangent to the first segment, $\overline{\mathbf{P}_1\mathbf{P}_2}$, and to the last one, $\overline{\mathbf{P}_{n-1}\mathbf{P}_n}$. Let us prove this statement for the third degree Bézier curve. The slopes of the tangents are given by

$$
\frac{d\mathbf{P}}{dt} = -3(1-t)^2\mathbf{P}_1 + 3[-2(1-t)t + (1-t)^2]\mathbf{P}_2
$$
$$
+ 3[-t^2 + 2t(1-t)]\mathbf{P}_3 + 3t^2\mathbf{P}_4 \tag{9.11}
$$

For $t = 0$ we get

$$
\frac{d\mathbf{P}}{dt} = 3(\mathbf{P}_2 - \mathbf{P}_1)
$$

which means that the tangent to the curve in the point $\mathbf{P}_1$ has the direction of the segment $\overline{\mathbf{P}_1\mathbf{P}_2}$. Similarly, for $t = 1$ we get

$$
\frac{d\mathbf{P}}{dt} = 3(\mathbf{P}_4 - \mathbf{P}_3)
$$

which means that the tangent to the curve in the point $\mathbf{P}_4$ has the direction of the segment $\overline{\mathbf{P}_3\mathbf{P}_4}$. This property is used in Section 9.9 in which we learn how to join two Bézier curves.

9.8.3 Convex Hull

Given a set of points, $\mathbf{P}_1, \mathbf{P}_2, \ldots \mathbf{P}_n$, their **convex hull** is the set of all affine combinations $\sum_{i=1}^{n} \alpha_i\mathbf{P}_i$. In pictorial terms the convex hull of a coplanar set of points is described as the area within the polygon defined by a rubber band tightened around the points of the set. For non–coplanar points in 3D space the notion can be referred to as the volume inclosed by an elastic balloon tightened around a set of points. Two examples of convex hulls are shown in Fig. 9.7. Any Bézier curve is fully contained within the convex hull of its control polygon. To understand this property we return to the de Casteljau's algorithm. Without loss of generality let us consider a 3d degree Bézier curve with control points $\mathbf{P}_1, \ldots \mathbf{P}_4$. For a given value of the parameter t the points $\mathbf{P}_{12}$, $\mathbf{P}_{23}$, P_{34} lie on the control polygon. Next, the points $\mathbf{P}_{123}$, $\mathbf{P}_{234}$ lie on segments that are, by construction, inside the control polygon. Finally, the Bézier-curve point $\mathbf{P}$ lies on the segment $\overline{\mathbf{P}_{123}\mathbf{P}_{234}}$ that is, by construction, inside the control polygon. The convex-hull property ensures that a Bézier curve does not exhibit large oscillations like those that can result from a polynomial interpolation, or even a cubic spline.

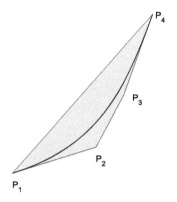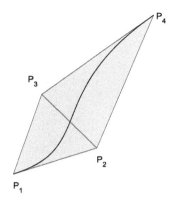

Figure 9.7 The convex-hull property

9.8.4 Variance Diminishing Property

A straight line intersects a Bézier curve at most the same number of times that it intersects the corresponding control polygon. We invite the reader to check this property on a few examples.

9.8.5 Invariance Under Affine Transformations

The Bézier curves are invariant under affine transformations. By this statement we mean that we obtain the same result if we build the curve and apply to it the transformation, or if we apply the transformation to the control polygon and afterward build the curve. Applications of this propriety are illustrated in Example 9.1 and Exercises 9.6 to 9.11. To illustrate this fact we consider a cubic Bézier curve viewed as in Figs 9.8. The centre of projection (eye or camera lens) is at height 0. The plane of the curve is at height -1 in the first figure, and -5 in the second. The projections of the control polygons, the curves defined by the projected polygons, and the projections of the given curves are shown in Figs 9.9 and 9.10. The differences are visible, as well as the influence of the 'focal distance' f. The Bézier curves described until now, known also as *integral Bézier curves*, are not invariant under perspective transformations.

Example 9.1 (Invariance under translation). We consider a control polygon $\mathbf{P}_0, \mathbf{P}_1, \dots, \mathbf{P}_i, \dots, \mathbf{P}_2$, and the corresponding Bézier curve

$$\mathbf{P} = \sum_{i=0}^{n} B_i^n(t)\mathbf{P}_i$$

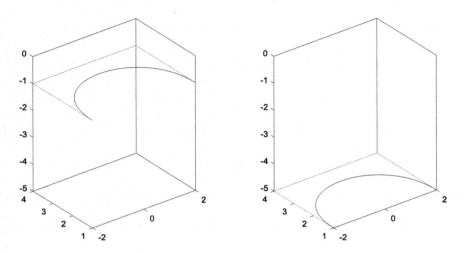

Figure 9.8 Figures to be projected

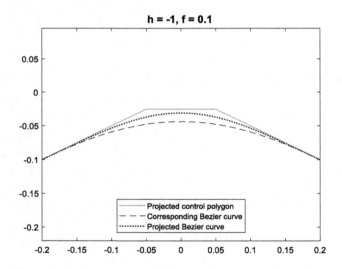

Figure 9.9 Perspective projection of Bézier curve

If we translate the control points by a vector $\mathbf{d}$ and build afterward the Bézier curve we obtain

$$\mathbf{Q} = \sum_{i=0}^{n} B_i^n(t)(\mathbf{P}_i + \mathbf{d}) = \sum_{i=0}^{n} B_i^n(t)\mathbf{P}_i + \sum_{i=0}^{n} B_i^n(t)d = \sum_{i=0}^{n} B_i^n(t)\mathbf{P}_i + \mathbf{d}$$

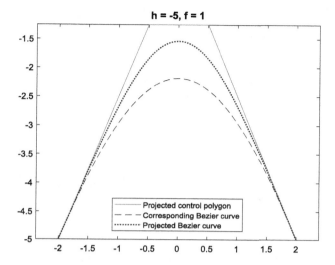

Figure 9.10 Perspective projection of Bézier curve

as the sum of the Bernstein polynomials equals 1. If we build first the Bézier curve and translate it by the vector d we have

$$\sum_{i=0}^{n} B_i^n(t)\mathbf{P}_i + \mathbf{d} = \mathbf{Q}$$

This proves the invariance of Bézier curves under translation.

9.9 JOINING TWO BÉZIER CURVES

Let us suppose that we want to join two Bézier curves. Without loss of generality we assume second-degree curves. For continuity the last point of the first curve must coincide with the first point of the second curve. As Bézier curves pass through the first and last points of their control polygons, we conclude that the last point of the first polygon must coincide with the first point of the second polygon. Then, let the first control polygon be $\mathbf{P}_1\mathbf{P}_2\mathbf{P}_3$, and the second polygon $\mathbf{P}_3\mathbf{P}_4\mathbf{P}_5$. Thus we ensure what is called C_0-continuity. To achieve C_1-continuity the two curves should have the same first derivatives at point $\mathbf{P}_3$. However, for the aspect of the composed curve this condition may be too strong and it may be sufficient to ask only that the tangents to the two curves, at the joining point, have the same direction. As Bézier curves are tangent to the first and last segments of

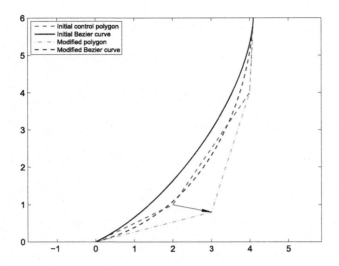

Figure 9.11 Moving a control point of a Bézier curve

their control polygons, this means that the segments $\mathbf{P}_2\mathbf{P}_3$ and $\mathbf{P}_3\mathbf{P}_4$ must belong to the same straight line. In other words, the points $\mathbf{P}_2$, $\mathbf{P}_3$, $\mathbf{P}_4$ must be collinear. Then, instead of C_1-continuity we speak of G_1-continuity. This is a case of *geometric continuity*. An application to third-degree curves is shown in Exercise 9.12, Fig. 9.14.

9.10 MOVING A CONTROL POINT

An important drawback of Bézier curves is that moving a control point produces a global change. The curves displayed in Fig. 9.2 explain well this fact. An example of the global influence is shown in Fig. 9.11. This behaviour limits the possibilities of manipulating the curve and justifies the development of more complex curves, first of all those called *B-splines*.

9.11 RATIONAL BÉZIER CURVES

The second-degree Bézier curves are arcs of parabola. In Exercise 9.2 we ask the reader to prove this statement experimentally. It is not possible to describe other conic sections by the Bézier curves described above. This drawback can be avoided by recurring to *rational Bézier curves* defined as

$$\mathbf{P} = \frac{\sum_{i=0}^{n} w_i B_i^n(t)\mathbf{P}_i}{\sum_{i=0}^{n} w_i B_i^n(t)} \tag{9.12}$$

where w_i is a *weight*, and B_i^n a *Bernstein polynomial*. The Bézier curves described before are a special case of the rational curves in which all the weights are equal to 1. To distinguish between the curves, the non-rational curves are known also as *integral Bézier curves*. Using second-degree rational Bézier curves it is possible to draw any conic section. To exemplify this in MATLAB we start by describing the following function.

```
function B = RatBez2(Pi, w)
%RATIONALBEZIER2 calculates 51 points on a 2nd-degree rational
% Bezier curve.
%    Input arguments:
%                Pi, array of coordinates of three control points
%                w,  array 1x3 of weights
%    Output argument, P, array 2x51 of the coordinates of points
%    on the Bezier curve

t  = 0: 0.02: 1;           % parameter
C1 = w(1)*(1 - t).^2       % weighted Bernstein polynomials
C2 = w(2)*2*t.*(1 -t)
C3 = w(3)*t.^2
C  = [ C1; C2; C3 ];
Den = C1 + C2 + C3
P   = Pi*C;
B = P./Den;
```

Using the control points

$$\mathbf{P}_1 = \begin{bmatrix} 1 \\ 0 \end{bmatrix}, \mathbf{P}_2 = \begin{bmatrix} 1 \\ 1 \end{bmatrix}, \mathbf{P}_3 = \begin{bmatrix} 0 \\ 0 \end{bmatrix} \tag{9.13}$$

and the equation

$$\mathbf{P} = \frac{(1-t)^2\mathbf{P}_1 + 2t(1-t)\mathbf{P}_2 + 2t^2\mathbf{P}_3}{(1-t)^2 + 2t(1-t) + 2t^3} \tag{9.14}$$

a quarter of circle can be drawn with the following MATLAB script.

```
%RATBEZCIRC Plots a quarter of a circle as a second-degree
%    rational Bezier curve.

% control points
P1 = [ 1; 0 ]; P2 = [ 1; 1 ]; P3 = [ 0; 1 ];
```

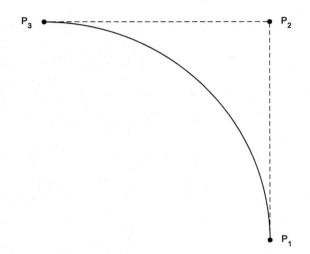

Figure 9.12 A quarter of circle drawn as a rational Bézier curve

```
% plot control polygon
pline([ P1 P2 P3 ], 1, 'r', '--')
axis equal, axis off
ht = xlabel('x'); set(ht, 'FontSize', 14)
ht = xlabel('y'); set(ht, 'FontSize', 14)
hold on
point(P1, 0.01), ht = text((P1(1) + 0.05), P1(2), 'P_1')
set(ht, 'FontSize', 14)
point(P2, 0.01), ht = text((P2(1) + 0.05), P2(2), 'P_2')
set(ht, 'FontSize', 14)
point(P3, 0.01), ht = text((P3(1) - 0.1), P3(2), 'P_3')
set(ht, 'FontSize', 14)
w   = [ 1 1 2 ];              % weights
P   = RatBez2([ P1 P2 P3 ], w);
pline(P, 1.5, 'k', '-')
hold off
```

See Fig. 9.12.

The rational Bézier curves inherit the properties of the Bézier curves listed in Section 9.8, but, in addition, they enjoy also the invariance under projective transformations.

Example 9.2 (Ellipse drawn as a rational Bézier curve). In a second example of conic section, in Fig. 9.13 we draw three quarters of ellipse by means of rational Bézier curves with the weights $w_1 = 3$, $w_2 = 2$, $w_3 = 3$. The

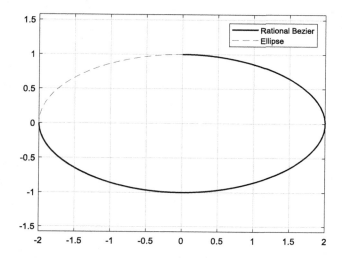

Figure 9.13 Ellipse drawn as a rational Bézier curve

fourth quarter is drawn by using directly the parametric equation of the ellipse. Obviously, the four arcs belong to the same ellipse. The MATLAB script that plots the figure is shown below.

```
%RATBEZELLIPSE Plots an ellipse as a second-degree
%    rational Bezier curve.

%%%%%%%%%%%% PLOT BEZIER CURVE %%%%%%%%%
P1  = [ -2    -2    0
          0    -1   -1 ];        % control points
w   = [ 3 2 3 ];                 % weights
Pi1 = RatBez2(P1, w);
axis equal
P2  = [  0     2    2
         -1    -1    0 ];        % control points
Pi2 = RatBez(P2, w);
P3  = [  2     2    0
          0     1    1 ];        % control points
Pi3 = RatBez2(P3, w);
pline([ Pi1 Pi2 Pi3 ], 1.5, 'k', '-'), grid
axis equal, hold on
%%%%%%%%%%%%%% PLOT ELLIPSE %%%%%%%%%%%%%%%
te = 0.25: 0.01: 0.5;
a  = 2; b  = 1;
```

```
xe = a*cos(2*pi*te); ye = b*sin(2*pi*te);
plot(xe, ye, 'r--')
legend('Rational Bezier', 'Ellipse')
hold off
```

9.12 SUMMARY

The *Bézier curves* are parametric curves defined by *control points*. Given $n+1$ control points, $\mathbf{P}_0, \mathbf{P}_1, \ldots, \mathbf{P}_n$, and the *Bernstein polynomials*

$$B_i^n(t) = \binom{n}{i} t^i (1-t)^{n-i}, \; i = 0, \ldots, n$$

the Bézier curve of degree n is defined by

$$\mathbf{P} = \sum_{i=0}^{n} B_i^n(t) \mathbf{P}_i$$

The first factor in the equation of a Bernstein polynomial is the *binomial coefficient*

$$\binom{n}{i} = \begin{cases} \frac{n!}{i!(n-i)!} & \text{if } 0 \le i \le 1 \\ 0 & \text{otherwise} \end{cases}$$

In particular we have the

First degree Bézier curve. $\mathbf{P} = (1-t)\mathbf{P}_0 + t\mathbf{P}_1$;

Second degree Bézier curve. $\mathbf{P} = (1-t)^2\mathbf{P}_0 + 2(1-t)t\mathbf{P}_1 + t^2\mathbf{P}_2$;

Cubic Bézier curve. $\mathbf{P} = (1-t)^3\mathbf{P}_0 + 3(1-t)^2 t\mathbf{P}_1 + 3(1-t)t^2\mathbf{P}_2 + t^3\mathbf{P}_3$.

In the case of the first-degree curve the point $\mathbf{P}$ lies on the straight line defined by the control points, and for $t \in [0\ 1]$ between $\mathbf{P}_0$ and $\mathbf{P}_1$. In the case of the second-degree curve the point $\mathbf{P}$ lies in the plane defined by the three control points. The Bézier curves enjoy the properties listed below.

End points — A Bézier curve passes only through the first and the last control points.

End tangents — A Bézier curve is tangent to the first and the last segments of the control polygon, that is to the segments $\overline{P_0 P_1}$ and $\overline{P_{n-1} P_n}$.

Convex hull — Any Bézier curve is fully contained within the convex hull of its control polygon.

Variation diminishing property — A straight line crosses a Bézier curve at most the same number of times that it crosses its control polygon.

Affine invariance Given a control polygon and an affine transformation T, we obtain the same result

- if we build first the corresponding Bézier curve and apply the transformation T to the curve,

- or, if we apply first the transformation to the control points and afterward we build the Bézier curve corresponding to the transformed control points.

The second-degree Bézier curve, known also as *quadratic Bézier curve*, is an arc of parabola. The Bézier curves cannot be used to describe arcs of other conic sections. This drawback can be avoided by using *rational Bézier curves* defined as

$$\mathbf{P} = \frac{\sum_{i=0}^{n} w_i B_i^n(t) \mathbf{P}_i}{\sum_{i=0}^{n} w_i B_i^n(t)}$$

where w_i is a *weight*. For example, a quarter of a circle can be displayed by the curve

$$\mathbf{P} = \frac{(1-t)^2 \mathbf{P}_1 + 2t(1-t)\mathbf{P}^2 + 2t^2 \mathbf{P}_3}{(1-t)^2 + 2t(1-t) + 2t^3}$$

The rational Bézier curves inherit the properties of the non-rational Bézier curves that are called also *integral Bézier curves*, but, in addition, enjoy also the property of invariance under perspective transformations.

9.13 EXERCISES

Exercise 9.1 (Approximating a waterline). Let the coordinates of a waterline be those listed below. Using the MATLAB `spline` function plot the waterline as a parametric cubic spline parametrized by chord length. Approximate the waterline by two cubic Bézier curves.

x	y	x	y
−103.425	0.000	0.000	14.250
−98.500	3.081	9.850	14.250
−88.650	7.189	19.700	14.250
−78.800	10.077	29.550	14.250
−68.950	12.324	39.400	14.250
−59.100	13.608	49.250	14.122
−49.250	14.121	59.100	13.351
−39.400	14.250	68.950	11.811
−29.550	14.250	78.800	9.089
−19.700	14.250	88.650	5.006
−9.850	14.250	98.500	0.000

Exercise 9.2 (The parabola as quadratic Bézier curve). As proven in Section 1.16.4, the canonic equation of the parabola is

$$y^2 = 4px$$

Your tasks are listed below.

1. Plot the parabola with parameter $p = 2$ in the interval $x = [0\ x_0]$, where $x_0 = 3$.
2. There are two points on the parabola for which $x = x_0$. Let us call them $\mathbf{P}_0$ and $\mathbf{P}_2$. Calculate the slopes of the tangents in these points.
3. Find the point in which the above-mentioned tangents intersect the x-axis. Let us call this point $\mathbf{P}_1$.
4. Calculate and plot the quadratic Bézier curve defined by the control polygon $\mathbf{P}_0\mathbf{P}_1\mathbf{P}_2$. This curve should coincide with the parabola plotted at 1.
5. Explain the construction of the control polygon.
6. Play with different values of p and x_0.

Exercise 9.3 (De Casteljau's algorithm). In this exercise we refer to a cubic Bézier curve. Prove algebraically that the curve is generated, indeed, by De Casteljau's algorithm. To do this begin by writing the equation of the point $\mathbf{P}$ as belonging to the first-degree Bézier curve defined by the points $\mathbf{P}_{123}$ and $\mathbf{P}_{234}$. Next, substitute the expression of the point $\mathbf{P}_{123}$ as defined by the points $\mathbf{P}_{12}$ and $\mathbf{P}_{23}$. Continue in this way till you introduce the control points $\mathbf{P}_1, \ldots, \mathbf{P}_4$.

Exercise 9.4 (Bernstein polynomials). Prove that for $n = 3$ Eq. (9.5) yields, indeed, the coefficients that multiply the coordinates of the control points in Eq. (9.3).

Exercise 9.5 (Invariance of Bézier curves under translation). Given the control points

$$\mathbf{P}_0 = \begin{vmatrix} 1 \\ 0.5 \end{vmatrix}, \mathbf{P}_1 = \begin{vmatrix} 2.75 \\ 1.5 \end{vmatrix}, \mathbf{P}_2 = \begin{vmatrix} 4 \\ 3 \end{vmatrix}, \mathbf{P}_3 = \begin{vmatrix} 2 \\ 4 \end{vmatrix}$$

your tasks are listed below.
1. Draw the control polygon and the corresponding Bézier curve.
2. Translate the curve 1.8 units in the direction x and -1.5 units in the direction y. Plot the translated curve on the same figure as the initial curve. Let's call the transformed curve $\mathbf{P}$.
3. Translate the control polygon as above and calculate the corresponding Bézier curve. Let's call this curve $\mathbf{Q}$.
4. Check that $\mathbf{Q}$ is identical to $\mathbf{P}$.

Exercise 9.6 (Invariance of Bézier curves under reflection in the x-axis). Given the control points

$$\mathbf{P}_0 = \begin{vmatrix} 1 \\ 0.5 \end{vmatrix}, \mathbf{P}_1 = \begin{vmatrix} 2.75 \\ 1.5 \end{vmatrix}, \mathbf{P}_2 = \begin{vmatrix} 3.5 \\ 3 \end{vmatrix}, \mathbf{P}_3 = \begin{vmatrix} 2.5 \\ 4 \end{vmatrix}$$

your tasks are listed below.
1. Draw the control polygon and the corresponding Bézier curve.
2. Reflect the curve in the x-axis and plot it on the same figure as the initial curve. Let's call this curve $\mathbf{P}$.
3. Reflect the control polygon in the x-axis and calculate the corresponding Bézier curve. Let's call this curve $\mathbf{Q}$.
4. Check that $\mathbf{Q}$ is identical to $\mathbf{P}$.

Exercise 9.7 (Invariance of Bézier curves under reflection in the y-axis). As in Exercise 9.6, but transform the curve by reflection in the y-axis.

Exercise 9.8 (Invariance of Bézier curve under reflection in the origin). As in Exercise 9.6, but transform the curve by reflection in the origin of coordinates.

Exercise 9.9 (Invariance of Bézier curves under rotation). Given the control points

$$\mathbf{P}_0 = \begin{vmatrix} 1 \\ 0.5 \end{vmatrix}, \mathbf{P}_1 = \begin{vmatrix} 2.75 \\ 1.5 \end{vmatrix}, \mathbf{P}_2 = \begin{vmatrix} 3.5 \\ 3 \end{vmatrix}, \mathbf{P}_3 = \begin{vmatrix} 2.5 \\ 4 \end{vmatrix}$$

your tasks are listed below.

1. Draw the control polygon and the corresponding Bézier curve.
2. Rotate the curve counterclockwise by 30° and plot it on the same figure as the initial curve. Let's call this curve **P**.
3. Rotate the control polygon counterclockwise by 30° and calculate the corresponding Bézier curve. Let's call this curve **Q**.
4. Check that **Q** is identical to **P**.

Exercise 9.10 (Invariance of Bézier curves under shearing). Given the control points

$$\mathbf{P}_0 = \begin{vmatrix} 1 \\ 0.5 \end{vmatrix}, \mathbf{P}_1 = \begin{vmatrix} 2.75 \\ 1.5 \end{vmatrix}, \mathbf{P}_2 = \begin{vmatrix} 3.5 \\ 3 \end{vmatrix}. \mathbf{P}_3 = \begin{vmatrix} 2.5 \\ 4 \end{vmatrix}$$

your tasks are listed below.
1. Draw the control polygon and the corresponding Bézier curve.
2. Transform the curve by shearing in the x-direction by the factor $S_x = 2$. Let's call this curve **P**.
3. Transform the control polygon by shearing as above and calculate the corresponding Bézier curve. Let's call this curve **Q**.
4. Check that **Q** is identical to **P**.

Exercise 9.11 (Invariance of Bézier curves under scaling). Given the control points

$$\mathbf{P}_0 = \begin{vmatrix} 1 \\ 0.5 \end{vmatrix}, \mathbf{P}_1 = \begin{vmatrix} 2.75 \\ 1.5 \end{vmatrix}, \mathbf{P}_2 = \begin{vmatrix} 3.5 \\ 3 \end{vmatrix}. \mathbf{P}_3 = \begin{vmatrix} 2.5 \\ 4 \end{vmatrix}$$

your tasks are listed below.
1. Draw the control polygon and the corresponding Bézier curve.
2. Scale the curve about the origin by the factors $S_x = 2$, $S_y = 2$. Let's call this curve **P**.
3. Scale the control polygon as above and calculate the corresponding Bézier curve. Let's call this curve **Q**.
4. Check that **Q** is identical to **P**.

Exercise 9.12 (Section composed of two Bézier curves). The curve in Fig. 9.14 is composed of two Bézier curves:
B_1 with control points $\mathbf{P}_1, \mathbf{P}_2, \mathbf{P}_3, \mathbf{P}_4$;
B_2 with control points $\mathbf{P}_4, \mathbf{P}_5, \mathbf{P}_6, \mathbf{P}_7$.

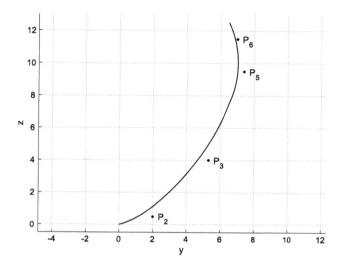

Figure 9.14 A section composed of two Bézier curves

Not all control points appear in the figure. Your tasks are:
1. to find the degree of the curves;
2. to mark the points $\mathbf{P}_1$, $\mathbf{P}_4$, $\mathbf{P}_7$;
3. to draw the control polygon.

Exercise 9.13 (Rational Bézier curves as affine combinations). Prove that the points of a rational Bézier curve are affine combinations of the control points.

CHAPTER 10

B-Splines and NURBS

Contents

10.1 INTRODUCTION

The Bézier curves studied in the preceding chapter have two disadvantages:
1. moving one control point affects the whole curve;
2. the number of control points and the degree of the curve are related. If the number of points is K, the degree of the curve is $K - 1$.

In this chapter we introduce more complex curves that do not have such disadvantages. These are the *B-splines*, a term coined by Schoenberg in 1946; its meaning is *Basic splines*. Like the Bézier curves, the B-splines are defined by control points, but their formulation and calculation is more complex. Also like for Bézier curves, there is an extension to *rational B-splines*. There are several variants that provide various degrees of flexibility, the most general one being the *non-uniform B-splines*, shortly **NURBS**. Unlike the Bézier curves, the B-splines do not pass through the first and last control points unless they are forced to do so. There are several ways of defining a B-spline that begins in the first control point and ends in the last one, the best one being the *clamped spline*.

Piegl (1991) starts his paper with a short note on the history of B-splines and NURBS, more details can be found in Rogers (2001). In a preface to Piegl and Tiller (1997) Rogers writes that a paper presented by him in 1977 'was arguably the first to examine the use of B-spline curves for ship design'.

Geometry for Naval Architects
https://doi.org/10.1016/B978-0-08-100328-2.00021-3

Copyright © 2019 Elsevier Ltd.
All rights reserved.

The development of the equations that yield points on a B-spline is tedious and the calculations may be rather difficult although there are efficient algorithms that do the job. We suppose that the reader interested in B-splines, especially NURBS, will use ready software available on the market. Therefore, we do not describe the development of the equations, but refer the reader to the relevant literature. To explain some of the important features of the B-splines, and enable the reader to experiment with them in MATLAB, we choose only the simplest variants that can be easily calculated in matrix form. We do not include proofs of the main properties of the described curves, but invite the reader to experiment with them in a few exercises.

10.2 B-SPLINES

A B-spline is composed of one or more segments calculated as polynomials of the degree $K - 1$, where K is called the *order* of the spline. For a Bézier curve the number of control points is equal to the order of the curve, that is the degree plus one. This is the minimum of control points for a B-spline, but there may be more. The segments of the spline join at points $t = t_0, \ldots, t_i, \ldots, t_n$ called knots. In general, at the joining points the segments enjoy $k - 2$ continuity. The set of the t-values is called *knot vector*. Given $n + 1$ control points, $\mathbf{P}_0, \ldots, \mathbf{P}_1, \ldots, \mathbf{P}_n$, the B-spline is defined by

$$\mathbf{P}(u) = \sum_{i=0}^{n} \mathbf{P}_i N_{i,K}(u) \tag{10.1}$$

where u is the parameter of the spline. The *basis functions*, $N_{i,k}$, are polynomials defined recursively by the *Cox-De Boor* relations

$$N_{i,1}(u) = \begin{cases} 1 & \text{if } t_i \leq u \leq t_{i+1} \\ 0 & \text{otherwise} \end{cases} \tag{10.2}$$

$$N_{i,k}(u) = \frac{(u - t_i)N_{i,k-1}(u)}{t_{i+k-1} - t_i} + \frac{(t_{i+k} - u)N_{i+1,k-1}(u)}{t_{i+k} - t_{i+1}} \tag{10.3}$$

where k assumes integer values ranging from 2 to K. When reading Eqs (10.1) to (10.3) please pay attention to the distinction between lower case k and upper case K. When the knots t_i are equally spaced the spline is called *uniform*, otherwise we say that the spline is non-uniform. To help the reader who may want to read more in the technical literature, we follow the traditional notation in which the indexes of the first control point and first knot are 0. When programming in MATLAB some changes may be necessary as this environment does not accept the index 0. The basis functions are non–zero only in certain t-intervals and this explains why a control point influences only a part of the curve. Examples of calculating basis functions can be found in Mortenson (1997); Piegl and Tiller (1997), and Rogers (2001). We skip this subject and go directly to practical procedures that use the results of such calculations.

10.3 QUADRATIC B-SPLINES

In Section 5.1 of his book Mortenson (1997) develops matrix equations for calculating B-splines of degree 2 and 3. We show below the equation that calculates the i-th segment of a second-degree, or *quadratic* B-spline. A second-degree, or *quadratic* B-spline requires at least three control points.

$$\mathbf{B}_i(u) = \frac{1}{2} \begin{bmatrix} u^2 & u & 1 \end{bmatrix} \begin{vmatrix} 1 & -2 & 1 \\ -2 & 2 & 0 \\ 1 & 1 & 0 \end{vmatrix} \begin{vmatrix} \mathbf{P}_{i-1} \\ \mathbf{P}_i \\ \mathbf{P}_{i+1} \end{vmatrix} \tag{10.4}$$

Thus, the i-th segment of the B-spline is defined by the three control points $\mathbf{P}_{i-1}$, $\mathbf{P}_i$, $\mathbf{P}_{i+1}$. To calculate the next segment we move one point further and apply Eq. (10.4) to the points $\mathbf{P}_i$, $\mathbf{P}_{i+1}$, $\mathbf{P}_{i+2}$, and so on. Within each segment the parameter u runs from 0 to 1. A MATLAB function that implements this procedure is

```
function BsplineQuadC(P, c, l)
%BSPLINEQUADC calculates and plots quadratic uniform B-spline
%    Input:
%              P, array of control points;
%              c, line colour given as character string, e.g. 'r';
%              l, line style given as character string, e.g. '--'.

% analyze input data
[ m,n ] = size(P);
if n < 3
    errordlg('Less than three control points', 'Input error')
else
    t1 = (0: 0.02: 1)';     t2 = t1.^2; t0 = ones(size(t1));
    M2 = [ 1   1 0
```

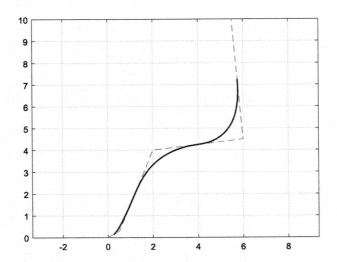

Figure 10.1 A quadratic B-spline

```
        -2   2 0
         1 -2 1 ]/2;
    for k = 1:(n-2)
        B = [ t0 t1 t2 ]*M2*[ P(:, k) P(:, (k+1)) P(:, (k+2)) ]';
        pline(B', 1.5, c, 1)
        axis equal, grid
    end % of loop over curve segments
end     % of conditional structure
end     % of function
```

In a first example we call the function BsplineQuad with the following control polygon as argument

```
P1 =
    0    0.5000    2.0000    6.0000    5.5000
    0    0.3000    4.0000    4.5000   10.0000
```

The result is shown in Fig. 10.1. In Exercise 10.1 we ask the reader to prove that the curve begins at the midpoint of the first two control points, and ends at the midpoint of the last two control points, a property that can be checked visually in the figure. One frequent way of forcing the curve to pass through the first and the last control points is to repeat these points. To give an example we use the augmented control polygon

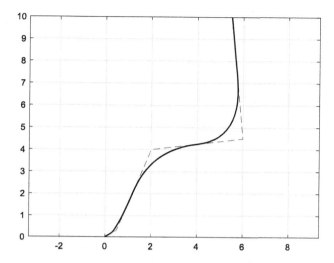

Figure 10.2 A quadratic B-spline with repeated end points

P2 =

0	0	0.5000	2.0000	6.0000	5.5000	5.5000
0	0	0.3000	4.0000	4.5000	10.0000	10.0000

Fig. 10.2 shows the result; it is a non-uniform B-spline. This approach has one drawback: the curve begins and ends with straight-line segments that can be sometimes unacceptable. Also, a repeated point reduces the degree of continuity. It is possible to avoid the straight-line segments by using *phantom points* as explained in Section 10.6. The preferred solution, however, is the **clamped** spline based on a knot vector in which the first and the last knots have a multiplicity equal to the order of the curve. Mortenson (1997) calls such a spline *open and non-uniform*. Piegl (1991) calls this spline *non-uniform and non-periodic*. Piegl and Tiller (1997) deals with the clamped spline throughout their book and only in the 12th chapter discuss shortly unclamped splines. Rogers (2001) gives examples of both types. In MultiSurf the B-splines and the NURBS are clamped splines. Letcher (2009) describes only B-splines of the clamped type. In our opinion the use of clamped splines is justified for at least one practical reason. The user may start with a control polygon supposed to be an approximation of the desired curve. It is impossible to do this without knowing the first and the last point of the curve. For example, a transverse section of a ship, like a station, starts at the keel and ends at the side of the deck. Analogously, a horizontal section, such as a waterline, starts and ends in the centreplane.

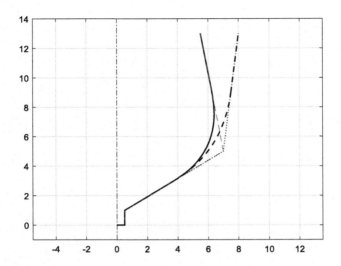

Figure 10.3 Changing a control point of a quadratic B-spline

10.4 MOVING A CONTROL POINT

Fig. 10.3 shows as a solid line what could have been a transverse section of a ship more than two centuries ago. The section is drawn as a single B-spline using the control polygon

```
P = [ 0 0 0.5 0.5 0.5 0.5 7  5.5   5.5
      0 0 0   0   1   1   5 13    13 ];
```

Let us move the last point by changing the input to

```
P1 = [ 0 0 0.5 0.5 0.5 0.5 7 8  8
       0 0 0   0   1   1   5 13   13 ];
```

The changed curve is shown as a dash–dash line. The change is *local*, not *global* as would have happened with a Bézier curve. Why? In Exercise 10.3 we ask a question related to this subject.

10.5 A CUBIC B-SPLINE

As proven, for example, by Mortenson (1997), the *i*-th segment of a third-degree, or *cubic* B-spline is given by

$$\mathbf{B}_i(u) = \frac{1}{6} [\, u^3 \quad u^2 \quad u \quad 1 \,] \begin{vmatrix} -1 & 3 & -3 & 1 \\ 3 & -6 & 3 & 0 \\ -3 & 0 & 3 & 0 \\ 1 & 4 & 1 & 0 \end{vmatrix} \begin{vmatrix} \mathbf{P}_{i-1} \\ \mathbf{P}_i \\ \mathbf{P}_{i+1} \\ \mathbf{P}_{i+2} \end{vmatrix} \tag{10.5}$$

This equation can be implemented by the following function.

```
function  BsplineCubC(P, c, 1)
%BSPLINECUBC calculates and plots cubic B-spline.
%    Input:
%         P - array of control points;
%         c - colour given as character string, e.g. 'k';
%         1 = line style given as character string, e.g. '--'.

% analyze input data
[ m,n ] = size(P);
if n < 4
    errordlg('Less than four control points', 'Input error')
else
    t1 = (0: 0.02: 1)'; t2 = t1.^2; t3 = t1.*t2;
    t  = [ t0 t1 t2 t3 ];
    t0 = ones(size(t1));
    M3 = [ 1   4   1 0
          -3   0   3 0
           3  -6   3 0
          -1   3  -3 1 ]/6;
    for k = 1:(n-3)
        B = t*M3*[ P(:, k) P(:, (k+1)) P(:, (k+2)) P(:, (k+3)) ]';
        pline(B', 1.5, c, 1)
        axis equal, grid
    end % of loop over curve segments
end        % of conditional structure
end        % of function
```

Before ending this section we mention the comments of Piegl and Tiller on such as those equations developed in Mortenson's book.

'For periodic B-splines there was no need to store a knot vector, the Cox-de Boor algorithm was not required, ... A simple matrix formulation of the ... basis function, together with a special control point indexing scheme were sufficient ... Although more efficient, this special treatment of periodic B-splines has all but disappeared from practice, probably due to the fact that the time required for software development and maintenance has become much more expensive than CPU time.'

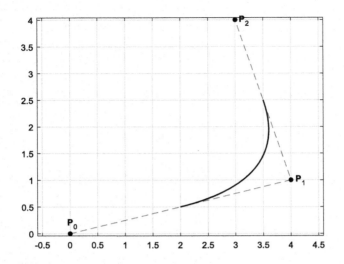

Figure 10.4 Quadratic B-spline with three control points

10.6 PHANTOM POINTS

We consider the control points

$$\mathbf{P}_0 = \begin{vmatrix} 0 \\ 0 \end{vmatrix}, \ \mathbf{P}_1 = \begin{vmatrix} 4 \\ 1 \end{vmatrix}, \ \mathbf{P}_2 = \begin{vmatrix} 3 \\ 4 \end{vmatrix}$$

Using these control points as the input argument of the function `BsplineQuadC` described in Section 10.3 we obtain the curve shown in Fig. 10.4. As already mentioned, in Exercise 10.1 we ask the reader to prove that the curve begins at $(\mathbf{P}_0 + \mathbf{P}_1)/2$, and ends at $(\mathbf{P}_1 + \mathbf{P}_2)/2$. This fact can be checked visually in the figure. We can use this property to move the beginning of the curve to $\mathbf{P}_0$ and the end to $\mathbf{P}_2$, without using multiple points. To do so, instead of the point $\mathbf{P}_0$ we use a **phantom point $\mathbf{P}_{-1}$**, and instead of $\mathbf{P}_2$ a *phantom point* $\mathbf{P}_3$ calculated from

$$\frac{\mathbf{P}_{-1} + \mathbf{P}_1}{2} = \mathbf{P}_0, \ \frac{\mathbf{P}_1 + \mathbf{P}_3}{2} = \mathbf{P}_2$$

that is

$$\mathbf{P}_{-1} = \begin{vmatrix} -4 \\ 0 \end{vmatrix}, \ \mathbf{P}_3 = \begin{vmatrix} 2 \\ 7 \end{vmatrix}$$

Now we call the function `BsplineQuadC` with the control polygon $[\mathbf{P}_{-1} \ \mathbf{P}_1 \ \mathbf{P}_3]$ as input argument and obtain the spline shown in Fig. 10.5. In

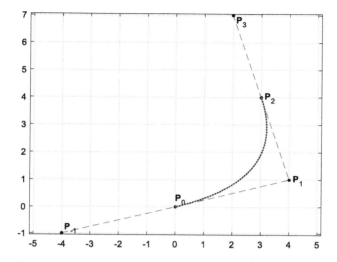

Figure 10.5 Quadratic B-spline identical to a quadratic Bézier curve

the same figure we plot the second degree Bézier curve defined by the initially given control polygon [$\mathbf{P}_0$ $\mathbf{P}_1$ $\mathbf{P}_2$]; it is shown as a dot-dot line, green in the electronic version of the book. The two curves coincide. In this case the quadratic B-spline reduces to a second-degree Bézier curve. We also observe that in this case the number of control points is equal to the degree of the curve plus one, that is the order of curve. The generalization to higher degrees is straightforward. Rogers (2001) uses the term *pseudovertex* instead of *phantom point* and remarks that 'Generally, these pseudovertices are neither displayed nor can a user manipulate them'. From here the qualifier 'phantom'.

10.7 SOME PROPERTIES OF THE B-SPLINES

In Chapter 8 we have seen that the Bernstein polynomials of a Bézier curve sum up to 1 for all the values of the parameter. Similarly, the basis functions $N_{i,K}(u)$ appearing in Eq. (10.1) sum up to 1 for all values of the parameter u. This property is known as *partition of unity*. It may be rather difficult to give a general proof starting from the definitions given in Section 10.2. It is easy, however, to prove this property for the simple cases dealt with in Sections 10.3 and 10.5. In Exercise 10.7 we ask the reader to do this. The proof is suggested by the following properties of the matrices M_2 and M_3 that appear in the above-mentioned sections:

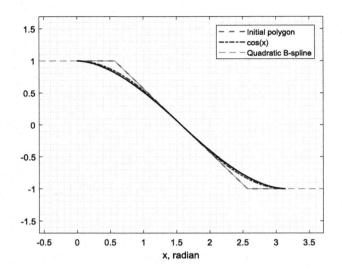

Figure 10.6 Cosine function approximated by a B-spline

- all rows except the last sum up to 0;
- the sum of the elements in the last row is 1.

It follows that the points of a B-spline are affine combinations of the control points and, therefore, they enjoy the properties of affine combinations. Among others, the B-splines are invariant under affine transformations. In a picturesque analogy, if we consider that, for a given $u = u_C$ value, the points P_i have the masses $N_{i,K}(u_C)$, then $P(u_C)$, is their barycentre.

Example 10.1 (Cosine approximated by a B-spline). In this example we approximate the cosine function by a non-uniform B-spline. The MATLAB script that solves this problem is shown below, and the result appears in Fig. 10.6.

```
%COSBSPLINE B-spline that approximates the cosine function

P = [ 0 (pi/2-1) (pi/2 + 1) pi
      1    1        -1      -1 ];
x  = 0: pi/50: pi; y  = cos(x);
plot(P(1, :), P(2, :), 'r--', x, y, 'b-.', 'LineWidth', 1.5)
xlabel('x, radian')
grid on, grid minor
axis equal
hold on
```

```
P1 = P(:, 1); P2 = P(:, 2); P3 = P(:, 3); P4 = P(:, 4);
% create phantom points to interpolate the ends
Ps = 2*P1 - P2; Pe = 2*P4 - P3;
P  = [ Ps P2 P3 Pe ];
BsplineQuad(P)
legend('Initial polygon', 'cos(x)', 'Quadratic B-spline')
hold off
```

10.8 NURBS

A *non-uniform rational B-spline*, shortly **NURBS**, is defined by

$$\mathbf{P}(u) = \frac{\sum_{i=1}^{n+1} \mathbf{P}_i w_i N_{i,K}(u)}{\sum_{i=1}^{n+1} w_i N_{i,K}(u)} \tag{10.6}$$

where $\mathbf{P}_i$ are the control points, w_i *weights*, and $N_{i,k}(u)$ the basis functions defined in Section 10.2. In this chapter we deal with plane NURBS. The extension to 3D curves is straightforward. To calculate points on a non-uniform rational B-spline we convert the coordinates of the control points to homogeneous coordinates, using the weights as the additional coordinate

$$\mathbf{Ph}_i = \begin{vmatrix} w_i \mathbf{P}_i(1) \\ w_i \mathbf{P}_i(2) \\ w_i \end{vmatrix}$$

We calculate the corresponding B-spline, **B**, and project the curve on a two-dimensional plane. The Euclidean coordinates of the curve in this plane are

$$\mathbf{B}_i = \begin{vmatrix} \mathbf{Ph}_i(1)/\mathbf{Ph}_i(3) \\ \mathbf{Ph}_i(2)/\mathbf{Ph}_i(3) \end{vmatrix}$$

A function that implements this procedure is shown below.

```
function B = NURBSfun(P, w)

%NURBSFUN  Calculates and plots a quadratic, plane NURBS
%    NURBSfun(P, w) plots the quadratic NURBS defined
%    by the control polygon P and the weights vector w.

% analyze input data
[ m, n ] = size(P); [ r, s ] = size(w);
```

```
if n < 3
    errordlg('Less than three control points', 'Input error')
elseif s ~= n
    errordlg('Incorrect number of weight values', 'Input error')
else
    % plot control polygon
    pline(P, 1, 'r', '--')
    axis equal, grid
    ht = xlabel('x'); set(ht, 'FontSize', 14)
    ht = ylabel('y'); set(ht, 'FontSize', 14)
    tt = [ 'w = ' ];
    for k = 1:(n-1)
        tt = [ tt num2str(w(k)) ', ' ];
    end
    tt = [ tt num2str(w(n)) ];
    title(tt)
    hold on
    % convert control points to homogeneous coordinates
    for k = 1:n
        P(:, k) = w(k)*P(:, k);
    end % of loop over weight values
    Ph  = [ P; w ];
    % calculate rational spline
    t1 = (0: 0.02: 1)'; t2 = t1.^2; t0 = ones(size(t1));
    M2 = [ 1   1 0
          -2   2 0
           1  -2 1 ]/2;
    for k = 1:(n-2)
        B1 = [ t0 t1 t2 ]*M2*[ Ph(:, k) Ph(:, (k+1)) Ph(:, (k+2)) ]';
        % convert to Euclidean coordinates
        B2 = [ B1(:, 1)./B1(:, 3) B1(:, 2)./B1(:, 3) ];
        pline(B2', 1.5, 'k', '-')
    end % of loop over curve segments
    hold off
end     % of conditional structure
end     % of function
```

The non-uniform rational B-splines inherit the properties of non-rational B-splines and, in addition, are invariant under projective projections. In Chapter 9 we learned that second-degree integral Bézier curves are parabolas. Besides this case, integral Bézier curves cannot represent other

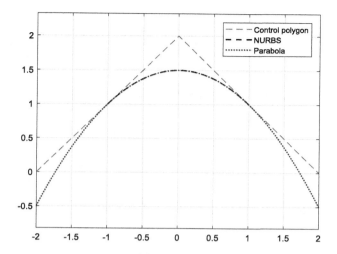

Figure 10.7 Parabola as non-uniform rational B-spline

conic sections, while rational Bézier curves can. B-splines too can represent parabolas, but no other conic sections, while NURBS can. As explained in Chapter 1, in analytic geometry all conic sections are described by second-degree equations. Therefore, it is natural to use quadratic NURBS to draw arcs of conic sections. To do this we need three control points and three weights. A current practice is to choose the value 1 for the first and third weight. Then, for the second weight we set

- $w(2) < 1$ for an ellipse;
- $w(2) = 1$ for a parabola;
- $w(2) > 1$ for a hyperbola.

In continuation we give two examples.

Example 10.2 (Parabola described by a non–uniform rational B-spline). In Fig. 10.7 we use a right-angled control polygon and the weight vector [1 1 1]. The corresponding rational B-spline is shown as a dash-dash line. We also plot in the same figure, as a dot-dot line (blue in the electronic version of the book) the parabola

$$y = -0.5x^2 + 1.5$$

It is easy to see that the two curves coincide. NURBS with all weights equal to one represent, indeed, parabolas. The MATLAB script that produced this figure is shown below.

```
%MYNURBSPARAB Draws a parabola as a NURBS defined
%    by a right-angled control polygon.

P1 = [ -2; 0 ]; P2 = [ 0; 2 ]; P3 = [ 2; 0 ];
w  = [ 1; 1; 1 ];
% plot control polygon
pline([ P1 P2 P3 ], 1, 'r', '--')
axis equal, grid
hold on
% convert control points to homogeneous coordinates
Ph1 = [ w(1)*P1; w(1) ]; Ph2 = [ w(2)*P2; w(2) ];
Ph3 = [ w(3)*P3; w(3) ]; Ph  = [ Ph1 Ph2 Ph3 ];
% calculate rational spline
    t1 = (0: 0.02: 1)'; t2 = t1.^2; t0 = ones(size(t1));
    M2 = [ 1   1 0
          -2   2 0
           1  -2 1 ]/2;
    B3 = [ t0 t1 t2 ]*M2*Ph';
    B2 = [ B3(:, 1)./B3(:, 3) B3(:, 2)./B3(:, 3) ];
    pline(B2', 1.5, 'k', '--')
x  = -2: 0.05: 2; y  = -0.5*x.^2 +1.5;
pline([x; y], 1.5, 'b', ':')
legend('Control polygon', 'NURBS', 'Parabola')
hold off
```

Example 10.3 (Ellipse described by a non-uniform rational B-spline). In Fig. 10.8 we use again a right-angled polygon. The vector of weights is [1 1/4 1]. We obtain the non-uniform rational B-spline drawn as a black, dash-dash line. We expect this curve to be an ellipse. Assuming the parametric equations

$$x = a\cos\theta$$
$$y = y_0 + b\sin\theta \tag{10.7}$$

let us find the coordinate y_0 of the centre, and the semi-axes a and b. We observe that the curve is tangent to the control polygon at its first point. Let the coordinates of this point be x_1, y_1, and the parameter corresponding to this point be θ_0. The slope of the tangent to an ellipse is

$$\frac{dy}{dx} = -\frac{b\cos\theta}{a\sin\theta} = -\frac{b}{a}\frac{1}{\tan\theta} \tag{10.8}$$

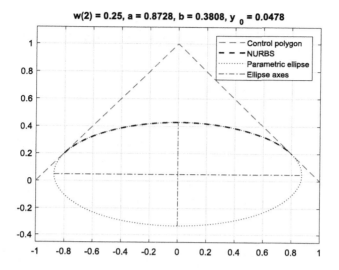

w(2) = 0.25, a = 0.8728, b = 0.3808, y$_0$ = 0.0478

Figure 10.8 Ellipse as non-uniform rational B-spline

The angle between the left tangent to the ellipse and the x-axis is $\pi/4$. Then, at the point of tangency $dy/dx = 1$ and Eq. (10.8) yields

$$\tan\theta_0 = -\frac{b}{a} \tag{10.9}$$

On the other hand, dividing side by side the parametric equations of the ellipse and substituting the values corresponding to x_1, y_1 we get

$$\frac{y_1 - y_0}{x_1} = \frac{b}{a}\tan\theta_0 \tag{10.10}$$

Noting by y_m the maximum y-coordinate of the ellipse we have

$$y_m - y_0 = b \tag{10.11}$$

Extracting y_0 and substituting it into Eq. (10.10) we obtain

$$\frac{y_1 - y_m + b}{x_1} = \frac{b}{a}\tan\theta_0$$

This equation can be rewritten as

$$\frac{y_1 - y_m}{x_1} = b\left(\frac{\tan\theta_0}{a} - \frac{1}{x_1}\right)$$

The coordinates appearing in the left-hand side can be obtained from the statements that calculate the non-uniform rational B-spline. We substitute the value of x_1 given by

$$x_1 = a\cos\theta_0$$

and use for the left-hand side the notation

$$\frac{y_m - y_1}{x_1} = c$$

After some algebraical manipulations we obtain

$$-c = -\frac{\tan\theta_0}{\cos\theta_0}(1 - \sin\theta_0)$$

From here we get the result that interests us

$$\sin\theta_0 = -\frac{c}{1 + c} \tag{10.12}$$

Now, knowing the value of the parameter θ at the first point of the NURBS, and the coordinates of this point, Eqs (10.7) yield the values of the semi-axes, and Eq. (10.11) the coordinate y_0 of the centre of the ellipse. The equations developed above are used in the function `FunNurbEllip` shown below. This function draws as a black dash-dash line the NURBS corresponding to the input value $w(2)$ and completes it with the arc of ellipse calculated from parametric equations. This arc is drawn as a dot-dot line, blue in the electronic version of the book.

```
function FunNurbEllip(w2)


%FUNNURBELLIP For the right-angled control polygon defined below,
%    and the weight of the second point as an input argument, this
%    function calculates the corresponding quadratic NURBS and the
%    semi-axes of the resulting ellipse.
%    Input: w2, should be smaller than one.

P1 = [ -1; 0 ]; P2 = [ 0; 1 ]; P3 = [ 1; 0 ];
w  = [ 1; w2; 1 ];
if w2 >= 1
    warndlg('Input weight should be less than 1', 'Input error', 'modal')
else
    % plot control polygon
```

```
pline([ P1 P2 P3 ], 1, 'r', '--')
axis equal, grid
hold on
% convert control points to homogeneous coordinates
Ph1 = [ w(1)*P1; w(1) ]; Ph2 = [ w(2)*P2; w(2) ];
Ph3 = [ w(3)*P3; w(3) ]; Ph = [ Ph1 Ph2 Ph3 ];
% calculate rational spline
t1 = (0: 0.02: 1)'; t2 = t1.^2; t0 = ones(size(t1));
M2 = [ 1   1 0
      -2   2 0
       1  -2 1 ]/2;
B3 = [ t0 t1 t2 ]*M2*Ph';
B2 = [ B3(:, 1)./B3(:, 3) B3(:, 2)./B3(:, 3) ];
pline(B2', 1.5, 'k', '--')
% calculate ellipse data
x1   = B2(1, 1); y1   = B2(1, 2); ym   = max(B2(:, 2))
% find angle theta0
c    = (ym - y1)/x1;
t0   = asind(-c/(1 + c));
% find axes and centre
a    = -x1/cosd(t0); b    = a*tand(t0);
y0   = ym - b;
% plot whole ellipse
t    = 0: 2.5: 360; % curve parameter, degrees
xe   = a*cosd(t); ye   = y0 + b*sind(t);
plot(xe, ye, 'b:', 'LineWidth', 1)
plot([ -a a ], [ y0 y0 ], 'k-.')
plot([ 0 0 ], [ (y0 - b) (y0 + b) ], 'k-.')
legend('Control polygon', 'NURBS', 'Parametric ellipse',...
       'Ellipse axes')
Tt   = [ 'w(2) = ' num2str(w(2), 4) ', a = ', num2str(a, 4) ];
Tt   = [ Tt ', b = ' num2str(b, 4) ', y_0 = ' num2str(y0, 3) ];
title(Tt)
end   % of conditional statement
hold off
end % of function
```

10.9 SUMMARY

A B-spline is composed of one or more polynomial segments. If K denotes the *order* of the spline, the polynomials are of the degree $K - 1$, possibly part of them lower. The segments join at points t_i, $i \in [0\ m]$ called *knots*.

The set of all knots is known as the *knot vector*. Let the control points be $\mathbf{P}_i$, $i \in [0 \ n]$. Then, the B–spline is defined by

$$\mathbf{P}(u) = \sum_{i=0}^{n} \mathbf{P}_i N_{i,K}(u)$$

The *basis functions*, $N_{i,k}$, are defined recursively by the *Cox–De Boor* relations

$$N_{i,1}(u) = \begin{cases} 1 & \text{if } t_i \leq u \leq t_{i+1} \\ 0 & \text{otherwise} \end{cases}$$

and

$$N_{i,k}(u) = \frac{(u - t_i)N_{i,k-1}(u)}{t_{i+k-1} - t_i} + \frac{(t_{i+k} - u)N_{i+1,k-1}(u)}{t_{i+k} - t_{i+1}}$$

where k assumes integer values ranging from 2 to K. Please make a distinction between lower case k and upper case K. In general a B–spline does not pass through the first and last control points. For example, a second–degree, or *quadratic* B–spline begins and ends at the points

$$\mathbf{P}_{start} = \frac{\mathbf{P}_0 + \mathbf{P}_1}{2}, \ \mathbf{P}_{end} = \frac{\mathbf{P}_{n-1} + \mathbf{P}_n}{2} \tag{10.13}$$

We list the methods of forcing the spline to *interpolate* the first and the last control points.

Multiple points — The first and the last point are repeated $K - 1$ times. This method produces straight–line segments at the ends.

Phantom points — The first and the last control points are replaced by points calculated from equations like (10.13). Such points are also called *pseudovertices*; in general they are not displayed and therefore they are not accessible to the user.

Clamped spline — The first and the last knot in the knot vector are repeated K times. This is the only B–spline dealt with in some textbooks, and the one available in some software, for example MultiSurf.

The points of a B–spline are affine combinations of the control points and they enjoy the same properties as the Bézier curves.

The *non-uniform rational B-splines*, shortly *NURBS*, are defined by

$$\mathbf{P}(u) = \frac{\sum_{i=1}^{n+1} \mathbf{P}_i w_i N_{i,k}(u)}{\sum_{i=1}^{n+1} w_i N_{i,k}(u)}$$

where the numbers w_i are called *weights*. The NURBS inherit the properties of the B-splines and, in addition, are invariant also under projective transformations. This means that the projection of a rational B-spline can be obtained by projecting the control polygon and using this projection to generate the curve. Like the Bézier curves, the B-splines can represent parabolas, but no other conic sections. Like the rational Bézier curves, the NURBS can produce any conic section. As the conics are described by second-degree curves, using a quadratic NURBS with three control points it is possible to generate a conic arc. A common practice is to set the weights of the first and third point to 1, and the weight of the second point to

- $w(2) < 1$ for an ellipse;
- $w(2) = 1$ for a parabola;
- $w(2) > 1$ for a hyperbola.

To calculate NURBS the coordinates of the control points are first converted to homogeneous coordinates using the weights as the additional coordinate. Next, calculate in homogeneous coordinates the corresponding B-spline. Finally, project the spline back into the initial space or plane.

10.10 EXERCISES

Exercise 10.1 (Quadratic B-spline, end points and tangents). We consider a quadratic B-spline based on the control polygon $\mathbf{P}_0, \mathbf{P}_1, \ldots, \mathbf{P}_n$. Using Eq. (10.4) show that the curve starts at the point $(\mathbf{P}_0 + \mathbf{P}_1)/2$, ends at the point $(\mathbf{P}_{n-1} + \mathbf{P}_n)/2$, and is tangent to the control-polygon segments

$$\overline{\mathbf{P}_0, \mathbf{P}_1}, \ \overline{\mathbf{P}_{n-i}, \mathbf{P}_n}$$

Exercise 10.2 (Quadratic B-spline reduced to a Bézier curve). Write a MATLAB script that produces Fig. 10.5.

Exercise 10.3 (Local influence of control points). Eq. (10.4) shows that a segment of a quadratic B-spline is influenced by three control points, and Eq. (10.5) shows that a segment of a cubic B-spline is influenced by four control points. The question in this exercise is how many segments are influenced by a control point if the B-spline is
a. quadratic;
b. cubic.

Exercise 10.4 (Cubic B-spline, end points). We consider a cubic B-spline based on the control polygon $\mathbf{P}_0, \mathbf{P}_1, \ldots, \mathbf{P}_n..$

1) Using Eq. (10.5) show that the curve starts at the point

$$Q_1 = \frac{P_0 + 4P_1 + P_2}{6} \tag{10.14}$$

and ends at the point

$$Q_2 = \frac{P_{n-2} + 4P_{n-1} + P_n}{6} \tag{10.15}$$

2) Let the control points be

$$P_0 = \begin{vmatrix} 0 \\ 0 \end{vmatrix}, \ P_1 = \begin{vmatrix} 3 \\ -1 \end{vmatrix}, \ P_2 = \begin{vmatrix} 5 \\ 2 \end{vmatrix}, \ P_3 = \begin{vmatrix} 3 \\ 4 \end{vmatrix}, \ P_4 = \begin{vmatrix} 1 \\ 3 \end{vmatrix}$$

1. Plot the corresponding B-spline using the function BsplineCubC. Calculate and plot the first and the last point of the spline using Eqs (10.14) and (10.15).
2. Use multiple control points to force the spline to pass through the first and last control points.
3. Use phantom points to force the spline to pass through the first and last control points.

Exercise 10.5 (Quadratic B-spline with phantom end points). Modify the function BsplineQuadC, Section 10.3, to plot a quadratic B-spline that interpolates the first and the last control point by using phantom points (pseudovertices). Experiment with this function.

Exercise 10.6 (Cubic B-spline with phantom end points). Modify the function BsplineCubC, Section 10.5, to plot a quadratic B-spline that interpolates the first and the last control point by using phantom points (pseudovertices). Experiment with this function.

Exercise 10.7 (Sum of basis functions). Read the first paragraph of Section 10.7. Next check that the sums of the polynomials that multiply the coordinates of the control points in Eqs (10.4) and (10.5) equal 1 for all values of the parameter u.

Exercise 10.8 (Compare quadratic and cubic B-splines). Given the control polygon

```
P = [ -20 -10 0 10 20; 0 10 0 10 0 ];
```

plot in the same figure the control polygon and the corresponding quadratic and cubic splines.

Exercise 10.9 (Multiple points vs. phantom points — Quadratic spline). Given the initial control polygon

```
P = [ -20 -10 0 10 20; 0 10 0 10 0 ];
```

plot in the same figure the control polygon and two corresponding quadratic B-splines, one with multiple end points, the other with phantom end points.

Exercise 10.10 (Multiple points vs. phantom points — Cubic spline). Given the initial control polygon

```
P = [ -20 -10 0 10 20; 0 10 0 10 0 ];
```

plot in the same figure the control polygon and two corresponding cubic B-splines, one with multiple end points, the other with phantom end points.

Exercise 10.11 (Multiple control points — Quadratic spline). Given the initial control polygon

```
P = [ -20 -10 0 10 20; 0 10 0 10 0 ];
```

plot the quadratic B-spline defined by this polygon. Change the control polygon by repeating the middle control point

```
P1 = [ -20 -10 0 0 10 20; 0 0 10 0 10 0 ];
```

Calculate the corresponding quadratic B-spline and plot it in the same figure.

Exercise 10.12 (Multiple control points — Cubic spline). Given the initial control polygon

```
P = [ -20 -10 0 10 20; 0 10 0 10 0 ];
```

plot the cubic B-spline defined by this polygon. Change the control polygon by repeating the middle control point

```
P1 = [ -20 -10 0 0 0 10 20; 0 0 0 10 0 10 0 ];
```

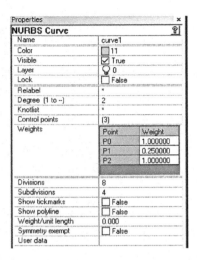

Figure 10.9 The Properties Manager for NURBS

Calculate the corresponding cubic B-spline and plot it in the same figure.

Exercise 10.13 (Rational B-spline reduced to an integral B-spline).
1. Show that when all weights are equal to one a rational B-spline reduces to a non-rational B-spline.
2. As a consequence, a non-uniform rational B-spline that produces a parabola can be reduced to a non-rational B-spline. Modify the script MyNurbsParab described in Example 10.2 so as to use a non-rational B-spline instead of the rational one.
3. Check that you obtain the same curve as in the example.

Exercise 10.14 (NURBS drawing a circle as special case of the ellipse). The circle is the special case of an ellipse in which the semi-axes are equal, that is $a = b$. Experiment with the function shown in Example 10.2 until you obtain a circle. **Hint:** play with $w(2)$.

APPENDIX 10.A   A NOTE ON B-SPLINES AND NURBS IN MULTISURF

To define a B-spline in MultiSurf select the control points in the correct order. Next, use

$$\text{Insert} \to \text{Curve} \to \text{B-spline Curve}$$

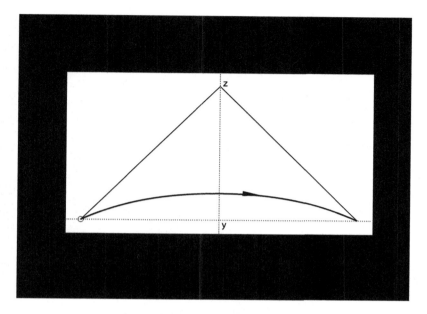

Figure 10.10 A non-uniform rational B-spline

The resulting curve passes through the end points and is identical to the corresponding curve obtained in MATLAB using phantom end points. The MultiSurf B-spline is of the *clamped* type. To define a non-uniform rational B-spline begin again by selecting the control points in the correct order. Next, use

Insert → Curve → NURBS Curve

In the Properties Manager fill in the weights. As an example, we used the same right-angled polygon as in Example 10.3. The Properties Manager of the curve, including the weights, is shown in Fig. 10.9. The resulting curve appears in Fig. 10.10. For readability we show the negative of the MultiSurf screen. The result differs from that shown in Fig. 10.8. This time the curve passes through the first and last control points. The spline is of the clamped type.

CHAPTER 11

Computer Representation of Surfaces

Contents

11.1 INTRODUCTION

In Chapters 7 and 10 we learned that a complex curve can be described 'piecewise' by curve segments that fulfill, if required, continuity conditions at the joining points. In an analogous way surfaces can be described by **patches** that fulfill, if required, certain continuity conditions along the joining *edges*. In Chapters 9 and 10 we have seen that the shape of curves is defined and can be modified by control polygons. In this chapter we are going to see that the shape of patches is determined by **control nets**. The shape of a patch can be modified by moving the control points of the control net. In Chapter 6 we have shown that the parametric representation of surfaces requires two parameters; we noted them by u and v. In computer aided geometrical design (shortly CAGD) the *isoparametric curves* of a surface, that is the constant-u and constant-v curves, can be of one of the types described in Chapters 9 and 10, but there are more possibilities not discussed in this book. The coordinates of a point on a patch are calculated as sums of products of a polynomial in u by a polynomial in v and by the coordinates of control points. The qualifier used for such surfaces is *tensor product*.

Geometry for Naval Architects
https://doi.org/10.1016/B978-0-08-100328-2.00022-5

Copyright © 2019 Elsevier Ltd.
All rights reserved.
411

Frequently surfaces are created starting with one or more curve segments that can be open or closed, in most cases plane, and define sections of the surface. This process is called *cross-sectional design*; it includes *sweeping*, *skinning* and *lofting*. The latter term comes from Naval Architecture and refers to a method similar to that has been used in shipyards during several centuries.

In this chapter we deal only with Bézier tensor-product patches; their mathematical description, geometrical interpretation, and implementation in MATLAB are simple. It may be easy to figure out how to generalize to rational Bézier, B-spline, or NURBS patches. The reader can study this subject in one of the textbooks cited in the previous chapter. The computer implementation of the more complex surfaces requires much more efforts and we suppose that the interested reader will use for this one of the specialized software packages available on the market.

11.2 BÉZIER PATCHES

A **Bézier patch** is described by

$$S(u, v) = \sum_{i=0}^{n} \sum_{j=0}^{m} P_{i,j} B_i^n(u) B_j^m(v) \tag{11.1}$$

where

u, v are the parameters of the surface;

i, j, $i \in \{0, 1, \ldots, n\}, j \in \{0, 1, \ldots, m\}$, are indices;

$\mathbf{P}_{i,j}$ are the control points;

B_i^n are the Bernstein polynomials of the n-th degree Bézier curves that run in the u direction (constant-v curves, see Section 9.5);

B_j^m are the Bernstein polynomials of the m-th degree Bézier curves that run in the v direction (constant-u curves, see Section 9.5).

Note that the indices are defined over a rectangular, $(n+1) \times (m+1)$ domain. In most textbooks such a surface is called **tensor product surface**, but Rogers (1977) writes about 'Cartesian-product' surfaces, and Rogers (2001) introduces this subject as 'A Cartesian or tensor product Bézier surface'. While it would be more difficult to explain the term *tensor product*, it is easy to define a *Cartesian product*. We are going to do this starting with the simple case of the first degree Bézier patch.

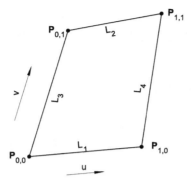

Figure 11.1 The control net of a bilinear patch

11.2.1 A Bilinear Patch

We consider here a patch generated by first degree Bézier curves, that is by straight lines. This is a **bilinear** patch for which

$$n = m = 1$$

To explain our notations we show an example in Fig. 11.1. For constant v curves the set of Bernstein polynomials is $U = \{(1 - u), u\}$, and for constant u it is $V = \{(1 - v), v\}$. Expanding Eq. (11.1) we obtain

$$S(u, v) = \sum_{i=0}^{1} \sum_{j=0}^{1} \mathbf{P}_{i,j} B_i^1(u) B_j^1(v)$$

$$= \sum_{i=0}^{1} B_i^1(u)[\mathbf{P}_{i,0}(1 - v) + \mathbf{P}_{i,1}v]$$

$$= (1 - u)(1 - v)\mathbf{P}_{0,0} + (1 - u)v\mathbf{P}_{0,1}$$

$$+ u(1 - v)\mathbf{P}_{1,0} + uv\mathbf{P}_{1,1} \tag{11.2}$$

Now we can explain the term *Cartesian product surface*. Given two sets, A and B, their Cartesian product, $A \times B$, is the set of all possible ordered pairs

$$a_i, b_j : a_i \in A, \ b_j \in B$$

For example, to obtain the Cartesian product of the sets U and V defined above we use Table 11.1. An element of this table multiplies the control point in the homologous position in Table 11.2. This is another

Table 11.1 Cartesian product of 1st-degree Bernstein polynomials

	$1 - u$	u
v	$(1 - u)v$	uv
$1 - v$	$(1 - u)(1 - v)$	$u(1 - v)$

Table 11.2 The control points of the bilinear patch

$\mathbf{P}_{0,1}$	$\mathbf{P}_{1,1}$
$\mathbf{P}_{0,0}$	$\mathbf{P}_{1,0}$

way to get Eq. (11.2). The products that multiply the control points in this equation are commutative, for example

$$(1 - u)v = v(1 - u)$$

Then, was it, indeed, necessary to specify 'ordered pairs'? Our answer is *yes* because it helps to prevent entering, for example, both $(1 - u)v$ and $v(1 - u)$.

We can find Eq. (11.2) by using Tables 11.1 and 11.2. There is a third way to obtain Eq. (11.2). Referring to Fig. 11.1 the equation of the line L_1 is

$$(1 - u)\mathbf{P}_{0,0} + u\mathbf{P}_{1,0}$$

and the equation of the line L_2

$$(1 - u)\mathbf{P}_{0,1} + u\mathbf{P}_{1,1}$$

For the given control net, the first order Bézier surface is generated by the straight lines that connect the points on the line L_1, with the points on the line L_2 that correspond to the same u-values. The equation of the resulting surface is

$$\begin{aligned}
S &= (1 - v)L_1 + vL_2 \\
&= (1 - u)(1 - v)\mathbf{P}_{0,0} + u(1 - v)\mathbf{P}_{1,0} \\
&\quad + (1 - u)v\mathbf{P}_{0,1} + uv\mathbf{P}_{1,1}
\end{aligned} \tag{11.3}$$

In Exercise 11.1 we ask the reader to check that Eq. (11.3) can also be obtained by deriving the equations of the lines L_3 and L_4 and in contin-

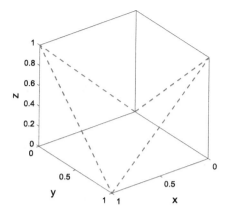

Figure 11.2 The control net that defines a patch of a hyperbolic paraboloid

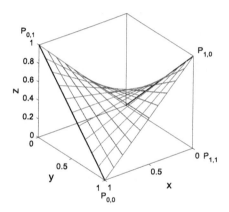

Figure 11.3 A bilinear patch, wireframe representation

uation the equation of the lines that join corresponding points on these lines.

Example 11.1 (A patch of the hyperbolic paraboloid). As shown in Fig. 11.2, in this example we use as control points four vertices of a unit cube. The plot of the patch is done by the script shown below. It begins with the definition of the control points. In continuation, the script plots several *isoparametric lines*, that is constant-*u* and constant-*v* lines. The resulting wireframe representation is shown in Fig. 11.3. Without giving a formal proof we observe that this patch is part of a hyperbolic paraboloid.

```
%HYPERBPARABPATCH Draws a bilinear patch that belongs to a
%    hyperbolic paraboloid.

% define axes and control net
P00 = [ 1; 1; 0 ];  P01 = [ 1; 0; 1 ];
P10 = [ 0; 1; 1 ]; P11 = [ 0; 0; 0 ];
pline([ P00 P01 ], 1.5, 'k', '-')
ax = gca; ax.XDir = 'reverse'; ax.YDir = 'reverse';
axis equal, box
hl = xlabel('x'); set(hl, 'FontSize', 14)
hl = ylabel('y'); set(hl, 'FontSize', 14)
hl = zlabel('z'); set(hl, 'FontSize', 14)
hold on
text(1.2, 1.2, 0, 'P_{0,0}'), text(1.2, 0, 1.2, 'P_{0,1}')
text(0, 1.1, 1.1, 'P_{1,0}'), text(0, 1.2, 0, 'P_{1,1}')
pline([ P10 P11 ], 1.5, 'k', '-');
% loops for drawing isoparametric curves
d = 10;
for v = 0: 1/d: 1          % draw constant-v lines
    Pl = (1-v)*P00 + v*P01; % left end
    Pr = (1-v)*P10 + v*P11; % right end
    pline([ Pl Pr ], 1, 'r', '-')
end
for u = 0: 1/d: 1          % draw constant-u lines
    Pl = (1-u)*P00 + u*P10; % left end
    Pr = (1-u)*P01 + u*P11; % right end
    pline([ Pl Pr ], 1, 'b', '-')
end
hold off
```

The coordinates of points belonging to the bilinear patch can be calculated in matrix form using the equation

$$S(u, v) = \begin{bmatrix} (1 - u) & u \end{bmatrix} \begin{bmatrix} \mathbf{P}_{00} & \mathbf{P}_{01} \\ \mathbf{P}_{10} & \mathbf{P}_{11} \end{bmatrix} \begin{bmatrix} (1 - v) \\ v \end{bmatrix} \quad (11.4)$$

To implement this equation we can use, for example, the script HyperbparabSurf shown below. The resulting plots appear in Figs 11.4, 11.5, and 11.6.

```
%HYPERBPARABSURF Plots the axonometric view and the
% contours of a bilinear patch.
```

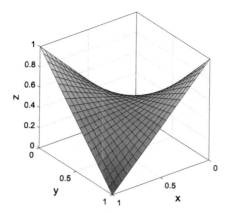

Figure 11.4 The bilinear patch drawn by the MATLAB function `surf`

```
% define axes and control net
P00 = [ 1; 1; 0 ];  P01 = [ 1; 0; 1 ];
P10 = [ 0; 1; 1 ]; P11 = [ 0; 0; 0 ];
d   = 20; u   = 0: 1/d: 1; v = u;
U   = [ (1-u') u' ]; V = [ (1- v); v ];
Px  = [ P00(1) P01(1); P10(1) P11(1) ];
X   = U*Px*V;
Py  = [ P00(2) P01(2); P10(2) P11(2) ];
Y   = U*Py*V;
Pz  = [ P00(3) P01(3); P10(3) P11(3) ];
Z   = U*Pz*V;
surf(X, Y, Z)
ax = gca; ax.XDir = 'reverse'; ax.YDir = 'reverse';
axis equal, box
hl = xlabel('x'); set(hl, 'FontSize', 14)
hl = ylabel('y'); set(hl, 'FontSize', 14)
hl = zlabel('z');set(hl, 'FontSize', 14)
pause
contour(X, Y, Z,[ 0.15: 0.15: 0.9 ], 'ShowText', 'On')
ax = gca; ax.XDir = 'reverse'; ax.YDir = 'reverse';
axis equal, box, grid
hl = xlabel('x'); set(hl, 'FontSize', 14)
hl = ylabel('y'); set(hl, 'FontSize', 14)
pause
contour3(X, Y, Z,[ 0.15: 0.15: 0.9 ], 'ShowText', 'On')
ax = gca; ax.XDir = 'reverse'; ax.YDir = 'reverse';
```

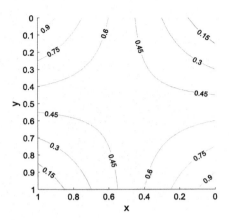

Figure 11.5 Sections of the bilinear patch drawn by the MATLAB function `contour`

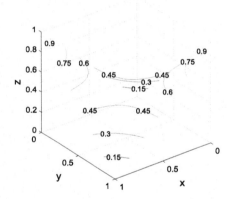

Figure 11.6 Sections of the bilinear patch drawn by the MATLAB function `contour3`

```
axis equal, box
hl = xlabel('x'); set(hl, 'FontSize', 14)
hl = ylabel('y'); set(hl, 'FontSize', 14)
hl = zlabel('z'); set(hl, 'FontSize', 14)
```

Figs 11.5 and 11.6 show horizontal sections of the bilinear surface obtained with the MATLAB functions `contour` and `contour3`. These curves are arcs of hyperbolas. Sections by planes that contain the z-axis are parabolas. We can check this visually by clicking on the `Rotate 3D` button in the toolbar of the figure and using the mouse to rotate the figure. Another possibility, which requires more work, is to plot adequate curves that lie on the surface.

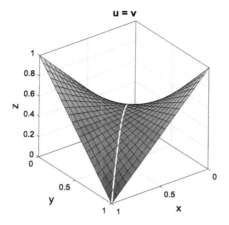

u = v

Figure 11.7 The curve $u = v$ on the bilinear patch

11.2.2 Curve on Surface

In Section 6.3 we have shown that if we define a relationship between the parameters u and v of a surface the equation of that surface becomes the equation of a curve lying on the given surface. Let, for example, $u = v = t$ in Eq. (11.3). We get

$$C = (1 - t)^2 \mathbf{P}_{0,0} + (1 - t)t\mathbf{P}_{0,1} + (1 - t)t\mathbf{P}_{1,0} + t^2 \mathbf{P}_{1,1}$$
$$= (1 - t)^2 \mathbf{P}_{0,0} + 2(1 - t)t\frac{\mathbf{P}_{0,1} + \mathbf{P}_{1,0}}{2} + t^2 \mathbf{P}_{1,1} \qquad (11.5)$$

This is the equation of a quadratic Bézier curve. To plot the curve add the following lines to the script `HyperbParabPatch`.

```
hold on
CP = ([ P00 (P01+ P10)/2 P11])
% build array of parameter values
t  = 0: 0.05: 1;
% build array of Bernstein-polynomial values
t1 = (1 - t);
A  = [ t1.^2; 2*t1.*t; t.^2 ];
% calculate and plot curve points
C  = CP*A;
pline(C, 1.5, 'r', '-')
hold off
```

Above we developed the explicit equation of the curve because it has an interesting form and a clear geometric meaning. In general, it is possible

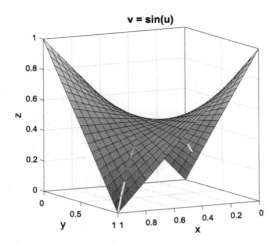

Figure 11.8 The curve $v = \sin u$ on the bilinear patch

to use the equation of the surface after imposing the relationship between the two parameters u and v. For example, we can add to our script the lines

```
hold on
% Calculate and plot curve on surface
v = sin(u);
C = P00*(1-u).*(1-v) + P01*(1-u).*v + P10*u.*(1-v) + P11*u.*v;
pline(C, 2, 'w', '-')
title('v = sin(u)', 'FontSize',14)
hold off
```

The result is shown in Fig. 11.8. We used the rotation button in the toolbar of the figure to show a more interesting view of the curve.

11.3 BICUBIC BÉZIER PATCH

In Mortenson (1997) we find the following matrix equation of a bicubic Bézier patch

$$S(u, v) = \begin{bmatrix} (1-u)^3 & 3u(1-u)^2 & 3u^2(1-u) & u^3 \end{bmatrix} \mathbf{P} \begin{bmatrix} (1-v)^2 \\ 3v(1-v)^2 \\ 3v^2(1-v) \\ v^3 \end{bmatrix}$$

$$(11.6)$$

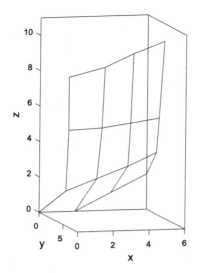

Figure 11.9 The control net of the bicubic Bézier patch

where **P** is the array of control points

$$
\mathbf{P} = \begin{bmatrix} \mathbf{P}_{1,1} & \mathbf{P}_{1,2} & \mathbf{P}_{1,3} & \mathbf{P}_{1,4} \\ \mathbf{P}_{2,1} & \mathbf{P}_{2,2} & \mathbf{P}_{2,3} & \mathbf{P}_{2,4} \\ \mathbf{P}_{3,1} & \mathbf{P}_{3,2} & \mathbf{P}_{3,3} & \mathbf{P}_{3,4} \\ \mathbf{P}_{4,1} & \mathbf{P}_{4,2} & \mathbf{P}_{4,3} & \mathbf{P}_{4,4} \end{bmatrix}
\tag{11.7}
$$

The control net is defined by **P**, a 4×4 array of elements that are themselves 3×1 arrays. We chose to store **P** as a MATLAB $4 \times 4 \times 3$ array. For an example, which could be a patch of a ship hull, we create the array in the following MATLAB script that calls the function `cat`

```
%BICUBICNET Creates a 4x4x3 array of control points

X = [ 0 0   0 0; 2 2   2   2; 4 4   4 4 ; 6 6 6   6 ];
Y = [ 0 5.5 6 6; 0 4.5 5.5 6; 0 3.5 4 4.5; 0 2 2.5 3.5 ];
C = cat(3, X, Y);
Z = [ 0 2 5.5 8.5; 0 2.5 5.5 9; 1 3 5.5 9.5; 2 3.5 5.5 10 ];
Pcubic = cat(3, C, Z)
```

This control net is shown in Fig. 11.9. Next, we write the following function

```
function  CubicBezSurf(P)
%CUBICBEZSURF Given the control net P, calculates
%      and plots a bicubic Bezier patch.
%      Input argument P, 4x4 array of 3D points

m = 20;      % divisions of the unit interval
t = 0: 1/m: 1; u = t'; v = t;
U = [ (1 - u).^3 3*u.*(1 - u).^2 3*u.^2.*(1 - u) u.^3 ];
V = [ (1 - v).^3; 3*v.*(1 - v).^2; 3*v.^2.*(1 - v); v.^3 ];
X = U*P(:,:,1)*V; Y = U*P(:,:,2)*V; Z = U*P(:,:,3)*V;
surf(X, Y, Z)
ax = gca; ax.YDir = 'reverse';
axis equal, box
xlabel('x', 'FontSize', 14), ylabel('y', 'FontSize', 14)
zlabel('z', 'FontSize', 14)
pause
contour(X, Y, Z, 'ShowText', 'On')
ax = gca; ax.YDir = 'reverse';
axis equal, box
xlabel('x', 'FontSize', 14), ylabel('y', 'FontSize', 14)
pause
contour3(X, Y, Z)
ax = gca; ax.YDir = 'reverse';
axis equal, box
xlabel('x', 'FontSize', 14), ylabel('y', 'FontSize', 14)
zlabel('z', 'FontSize', 14)
end
```

We call this function with the commands

```
>> BicubicNet
>> CubicBezSurf(Pcubic)
```

The resulting plots are shown in Figs 11.10, 11.11 and 11.12.

Example 11.2 (A ruled patch between two cubic Bézier curves). It is possible to build a Cartesian-product Bézier patch with isoparametric curves of different degrees in the two directions. We give here an example that could be part of the hull surface of a dinghy. The lower and the upper borders of the surface are cubic Bézier curves, where the former may be the intersection with the bottom, and the latter the deck at side. Let the

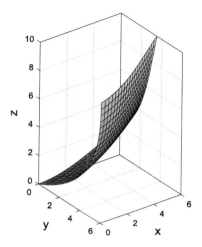

Figure 11.10 A bicubic Bézier patch drawn by the MATLAB function `surf`

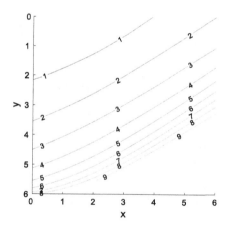

Figure 11.11 Horizontal sections of the bicubic Bézier patch drawn by the MATLAB function `contour`

control net have the coordinates

$$x = \begin{bmatrix} 0 & 5 & 10 & 15 \\ 0 & 5 & 10 & 16 \end{bmatrix}, \quad y = \begin{bmatrix} 3.0 & 4.0 & 4.0 & 0 \\ 3.0 & 4.5 & 4.5 & 0 \end{bmatrix},$$
$$z = \begin{bmatrix} 0 & 0 & 0 & 0 \\ 3 & 3 & 3 & 4 \end{bmatrix}$$

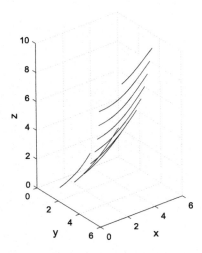

Figure 11.12 Horizontal sections of the bicubic Bézier patch drawn by the MATLAB function `contour3`

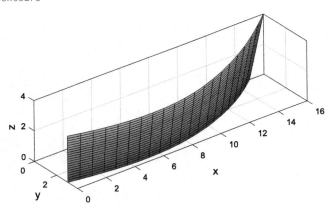

Figure 11.13 A ruled patch between cubic Bézier curves

We store these coordinates in a $2 \times 4 \times 3$ array that we call **P**. The equation of the surface is

$$S(u, v) = \left[\ [(1 - u)^3 \quad 3u(1 - u)^2 \quad 3u^2(1 - u) \quad u^3] \ \right] \mathbf{P}' \begin{bmatrix} 1 - v \\ v \end{bmatrix} \quad (11.8)$$

We use the MATLAB function `surf` to show the surface as in Fig. 11.13. The function *contour* yields the waterlines that appear in Figs 11.14 and 11.15.

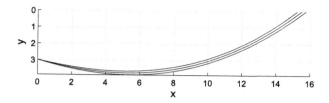

Figure 11.14 The waterlines of the dinghy

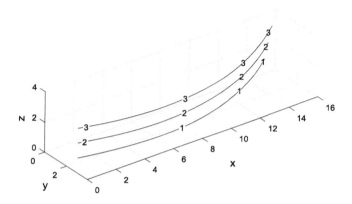

Figure 11.15 The waterlines of the dinghy — axonometric view

11.4 JOINING TWO BÉZIER PATCHES

In Section 9.9 we learned how to join two Bézier segments so as to achieve the continuity of the composed line and of the tangent. In this section we give an example of joining two Bézier patches while obtaining the continuity of the surface and of the tangents to it along the joining edge. Each patch is defined by a control net based on 4 × 4 points. Foley and Van Dam (1984, p. 530) write

> 'Continuity across patch edges is obtained by making the four control points on the edges equal. Continuity of tangent vector ... is obtained by additionally making the two sets of control points on either side of the edge collinear with the points of the edge... the ratios of the lengths of the collinear segments must be constant.'

Leiceaga et al. (2007) formulate these conditions by stating that
1. the points of the edges to be joined must be identical;
2. the points of the common edge must be affine combinations of corresponding points on either side of the common edge.

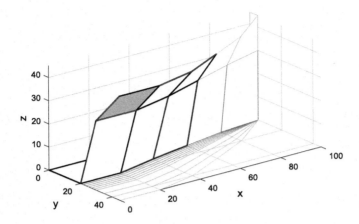

Figure 11.16 Surface composed of two cubic patches — Control net

In our example the control points of the first patch are stored in the arrays $X1$, $Y1$, $Z1$, and those of the second patch in the arrays $X2$, $Y2$, $Z2$ defined below.

```
dx  = 20;
X1  = [ 0   0   0   0;  1  1   1   1;  2  2   2   2;  3 3   3   3 ]*dx;
Y1  = [ 0  20  30  45;  0  20  30  45;  0  15  25  40;  0  11  18  30 ];
Z1  = [ 0  0  30  45;  0  0  30  45;  0  0  30  45;  0  0  30  45 ];
X2  = [ 3 3   3   3;  4  4   4   4;  5  5  5  5;  5  5  5  5 ]*dx;
Y2  = [ 0  11  18  30;  0  7  11  20;  0  5  5  5;  0  0  0  0 ];
Z2  = [ 0  0  30  45;  0  0  30  45;  0  0  30  45;  0  0  30  45 ];
```

The conditions stated by Foley and Van Dam (1984) are satisfied. We can verify this using the formulation of Leiceaga et al. (2007). In MATLAB notation we have

```
Y1(4, :) = (1 - 0.5)*Y1(3, :) + 0.5*Y2(2, :)
```

that is

$$\begin{vmatrix} 0 & 11 & 18 & 30 \end{vmatrix} = (1 - 0.5)\begin{vmatrix} 0 & 15 & 25 & 40 \end{vmatrix} \\ + 0.5\begin{vmatrix} 0 & 7 & 11 & 20 \end{vmatrix}$$

The corresponding control net is shown in Fig. 11.16.

The fact is visible also in Fig. 11.17 obtained with the MATLAB function `meshc`. Below the axonometric view of the control net we see a set

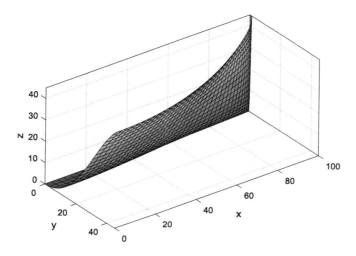

Figure 11.17 Surface composed of two bicubic patches — Axonometric view

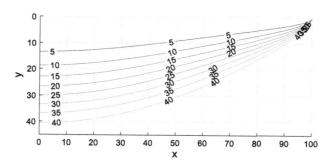

Figure 11.18 Surface composed of two cubic patches — Horizontal contours

of contours (horizontal lines) lying on the surface of the net. The horizontal contours are shown in Fig. 11.18, the longitudinal contours in Fig. 11.19. The composed surface is continuous and resembles the forward part a ship hull. To obtain this figure we stored the control nets in a file called TwoPatches2 and used the following MATLAB commands

```
TwoPatches2
m = 20;      % divisions of the unit interval
t = 0: 1/m: 1; u = t'; v = t;
U = [ (1 - u).^3 3*u.*(1 - u).^2 3*u.^2.*(1 - u) u.^3 ];
V = [ (1 - v).^3; 3*v.*(1 - v).^2; 3*v.^2.*(1 - v); v.^3 ];
X1 = U*Pc1(:,:,1)*V; Y1 = U*Pc1(:,:,2)*V; Z1 = U*Pc1(:,:,3)*V;
```

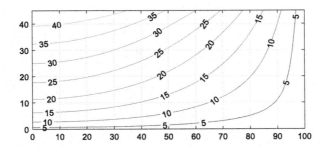

Figure 11.19 Surface composed of two bicubic patches — Longitudinal contours

```
surf(X1, Y1, Z1)
axis equal, axis([ 0 100 0 45 0 45 ]), box
ax = gca; ax.YDir = 'reverse';
xlabel('x', 'FontSize', 14), ylabel('y', 'FontSize', 14)
zlabel('z', 'FontSize', 14)
hold on
X2 = U*Pc2(:,:,1)*V; Y2 = U*Pc2(:,:,2)*V; Z2 = U*Pc2(:,:,3)*V;
surf(X2, Y2, Z2)
hold off
```

11.5 SWEPT SURFACES

Sweeping is a way of creating a surface by moving a given plane curve along a *trajectory* that is called also *spine curve*. In some languages the corresponding term is *translation surface*. For example, in Italian we would talk about 'superficie de traslazione'. During the translation the given curve remains perpendicular to the spine; it is the cross-section of the surface. The trajectory may be a straight line or a curve. In this section we use the following MATLAB commands to generate a swept surface by moving along a straight line the curve shown in Fig. 11.20. Using homogeneous coordinates we can perform the translation by matrix multiplication and we allow the reader to combine the translation with another motion, for example rotation. Fig. 11.21 shows the sweep surface.

```
% control points of cross-section curve
P0 = [ 0; 0; 0 ]; P1 = [ 0; 1.5; 1 ];
P2 = [ 0; 1.5; 3 ]; P3 = [ 0; 0; 4 ];
P  = Bezier3([ P0 P1 P2 P3 ]); % cubic Bezier curve
[ m, n ] = size(P);
Ph = [ P; ones(1, n) ];        % homogeneous coordinates
```

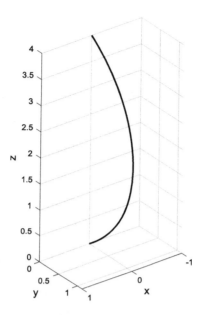

Figure 11.20 A cubic Bézier cross-section

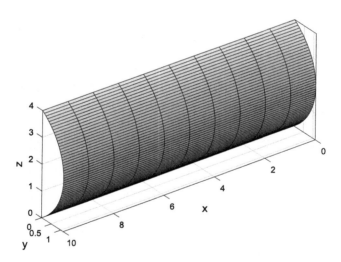

Figure 11.21 Sweep surface based on a given Bézier curve

```
l   = 10; v  = 0: 1/l: 1;          % trajectory parameter
dx = 10;                          % translation vector
X = zeros((l+1), n); Y = zeros(size(X)); Z = zeros(size(X));
```

```
for k = 1:(1 + 1)
        T = [ 1 0 0 v(k)*dx % translation matrix
              0 1 0 0
              0 0 1 0
              0 0 0 1 ];
        B = T*Ph;
        x = B(1, :); X(k, :) = x;
        y = B(2, :); Y(k, :) = y;
        z = B(3, :); Z(k, :) = z;
end
surf(X, Y, Z)
ax = gca; ax.XDir = 'reverse'; ax.YDir = 'reverse';
axis equal, box
xlabel('x', 'FontSize', 14), ylabel('y', 'FontSize', 14)
zlabel('z', 'FontSize', 14)
```

11.6 LOFTED SURFACES

Given several curves that fulfill certain conditions, it is possible to build a surface that interpolates them. The process is called **lofting**, a term that comes from Naval Architecture. This is the method that has been used during several hundred years to build hull surfaces based on given frames, sometimes also diagonals. Relevant, historical references can be found in Chapter 2. As Nowacki et al. (1995) write, 'The skin curves must be of equivalent segmentation, parametrization and polynomial degrees in order to achieve the fundamental tensor product form ...'. Often these requirements are not met; therefore, the above-mentioned authors indicate a procedure for adequately transforming the curves. In this section we show how to create a bicubic patch that resembles the forward end of a ship hull. The given curves are shown in Fig. 11.22; in the terminology of MULTISURF they are called *master curves*. The first three curves are constant-x curves, that is transverse hull sections. The fourth curve lies entirely in the centreplane $y = 0$; it may be the stem line that continues forward the keel. We obtained the coordinates of the control net by concatenating the control polygons of the desired cross-sections

```
X = [ 0 0    0 0; 3.9 3.9 3.9 3.9; 7.8 7.8 7.8 7.8; 12.5 16.5 17.5 22 ];
Y = [ 0 5.5 6 6; 0 4.5 5.5 5.5; 0 2.5 3.5 4.5; 0 0 0 0 ];
Z = [ 0 2 5.5 8.5; 0 2.4 5.5 9; 1 2.5 5.0 9.5; 2 3.5 5.5 10 ];
```

We create the surface with the help of the function CubicBezSurf described in Section 11.3 and obtain Fig. 11.23. To show constant-z sections

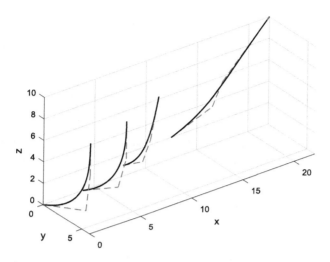

Figure 11.22 The master curves of a hull forward end

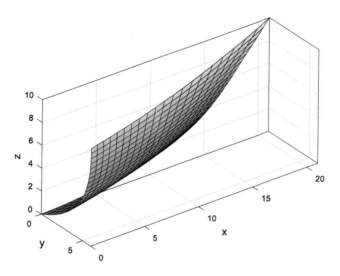

Figure 11.23 Lofted surface based on Bézier curves

we use the MATLAB function `contour` as explained in the documentation of this software:

```
contour(X, Y, Z, [1 2 4 6 8 ], 'ShowText', 'On')
ax = gca; ax.YDir = 'reverse';
axis equal, grid, box
```

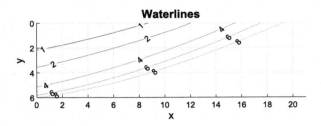

Figure 11.24 Lofted surface based on Bézier curve — Waterlines

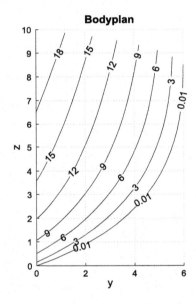

Figure 11.25 Lofted surface based on Bézier curves — Body plan

```
xlabel('x', 'FontSize', 14), ylabel('y', 'FontSize', 14)
title('Waterlines', 'FontSize', 14)
```

The contours are displayed in Fig. 11.24. Can we plot also constant-x sections? The answer is yes. To do this we observe that the MATLAB function surf used, as recommended, with the command surf(X, Y, Z), plots z as a function of x and y. Therefore, calling the function contour with the input arguments X, Y, Z in the same order as for surf plots constant z curves. Changing the commands as shown below we get Fig. 11.25.

```
contour(Y, Z, X, [ 0.01 3 6 9 12 15 18 ], 'k', 'ShowText', 'On')
axis equal, grid, box
```

```
xlabel('y', 'FontSize', 14), ylabel('z', 'FontSize', 14)
title('Bodyplan', 'FontSize', 14)
```

11.7 COMPUTER-AIDED DESIGN OF HULL SURFACES

In Section 2.10 we have written that the computer does not shorten the time needed to design a ship hull surface, but once the model is ready it can be used for transformations and calculations and this is the real advantage of a computer hull model. In fact, an experienced designer needs no more time to draw ship lines manually than with the aid of a computer. We like to mention that a computer can do only what it 'was taught' to do, while the designer is not so limited. To quote a developer of hull-design software (Hollister, 2002), 'There is nothing more flexible than a piece of paper, a pencil, and the human brain. Once you begin using a computer for hull shape design, you are forced into the limitations and idiosyncrasies of the program and its underlying hull geometry.' In continuation this author lists the advantages of the model stored in the computer. Krueger (2001) also writes that, when working with a spline to produce a line drawing, an experienced designer is 'exactly' as fast as computer program. The advantages of a computer model appear only when transferring automatically the data of the ship lines to other programs.

The first ideas about using mathematically defined curves for describing the ship lines appeared by the end of the 17th century. It took a long time and many proposals to reach the present state of computer-aided design of hull surfaces. A comprehensive historical review can be found in Alaez (1991). Many papers have been published on this subject and bibliographical lists are attached to the above mentioned report and in other books and papers cited in our book. We would like to include here only a few additional citations. Söding (1963) reviews the mathematics that represented the state of the art at the time of his paper. They deal mainly with polynomial interpolation and modelling of the 'elastic' line assumed by the drawing instrument called spline. At about the same time Kantorowitz (1967) describes the work done at the Danish Sip Research Institute. A book entirely dedicated to computer-aided modelling of ship lines is Kuo (1971). Kouh (1985) develops a system of representing ship lines by rational cubic splines. Krueger (2001) advocates the use of transverse lines (constant-x curves) and longitudinal lines for describing a hull surface. A point of the surface is defined as the intersection of a transverse line with a longitudinal one. The description must enable the automatic transfer of the definition to other calculations. In the traditional definition of ship lines at least three projections are used, the body plan, the waterlines plan and the sheer plan. This means redundant information that cannot be introduced in computer algorithms. A presentation that includes historical and bibliographic data as well as computer-aided examples is that of Ventura (no year indicated).

The software available today uses tensor-product patches to model the ship hull. Certain features of the ship lines do not allow the representation by a single patch. For example, a keel must be represented as another surface than the main hull. So it may happen for the bow, specially for a bulbous bow, sonar domes, or at the ship aft. Obviously, different patches are needed to account for tangent discontinuities due to a hard chine or a knuckle. The various patches that compose a hull must be joined with surface continuities (C^0) and in most cases tangent continuity (C^1). In Section 11.4 we have given the continuity conditions for bicubic Bézier patches. Leiceaga et al. (2007) describe algorithms for smoothly joining B-spline surfaces in two cases:

- two patches with a common border;
- four patches with a common vertex.

As written above, one of the advantages of a computer hull model is that data obtained from it can be transferred automatically to programs for hydrostatic calculations, plate development, strength assessment, and seakeeping analyses. However, Ko (2010) explains that the information obtained from the ship lines stored in the computer does not suit always the requirements of some programs. This author describes two methods for overcoming the problems.

In Section 2.6 we explained the fairing of ship lines and the difficulties of achieving it. Various authors devised methods for carrying on the fairing on computers. We cite only a few references an early one being Kantorowitz (1966). The use of curvature as criterion for fairing is treated by Narli and Sariöz (2004). Pérez-Arribas et al. (2006) discuss the modelling of a ship hull as a NURBS surface while taking care of fairness. Pérez et al. (2008) show how to model a hull starting from sectional area curve and the waterline.

Specific notes and tutorials about hull modelling with the aid of software available on the market can be downloaded freely from the web. As we have given examples in MultiSurf, we refer in the Appendix to the literature of this software.

11.8 SUMMARY

The prevailing way of representing free-form surfaces is as *tensor product surfaces*. An alternative term is *Cartesian product*. The general equation of a *patch* of such a surface is

$$S(u, v) = \sum_{i=0}^{n} \sum_{j=0}^{m} P_{i,j} B_i^n(u) B_j^m(v) \tag{11.9}$$

where

u, v are the parameters of the surface;

i, j, $i \in \{0, 1, \ldots, n\}, j \in \{0, 1, \ldots, m\}$, are indices;

$\mathbf{P}_{i,j}$ are the control points;

B_i^n are the blending functions of the n-th degree curves that run in the u direction (constant-v curves, see Section 9.5);

B_j^m are the blending functions of the m-th degree curves that run in the v direction (constant-u curves, see Section 9.5).

The blending functions, also known as *basis functions*, are the Bernstein polynomials for Bézier patches. Other polynomials are used for B-spline, rational B-spline, or NURBS surfaces. There are more possibilities not covered in this book.

Constant-u and constant-v curves on the surface are known as *isoparametric curves*. Using first-degree Bézier lines for both directions results in a *bilinear surface*, which is a patch of a hyperbolic paraboloid. Points on such a surface can be calculated in matrix form using the equation

$$S(u, v) = \begin{bmatrix} (1 - u) & u \end{bmatrix} \begin{bmatrix} \mathbf{P}_{00} & \mathbf{P}_{01} \\ \mathbf{P}_{10} & \mathbf{P}_{11} \end{bmatrix} \begin{bmatrix} (1 - v) \\ v \end{bmatrix} \tag{11.10}$$

where $\mathbf{P}_{00}, \mathbf{P}_{01}, \mathbf{P}_{10}, \mathbf{P}_{11}$ are the control points, each one of them stored as a 3×1 array.

Points on a bicubic Bézier patch can be calculated with the matrix equation

$$S(u, v) = \begin{bmatrix} (1 - u)^3 & 3u(1 - u)^2 & 3u^2(1 - u) & u^3 \end{bmatrix} \mathbf{P} \begin{bmatrix} (1 - v)^2 \\ 3v(1 - v)^2 \\ 3v^2(1 - v) \\ v^3 \end{bmatrix} \tag{11.11}$$

where **P** is an array of 16 control points, each one of them stored as a 3×1 array.

If a relationship between the parameters u and v is imposed, the equation of the patch becomes the equation of a curve on the given surface. In Section 9.9 we learned that two Bézier curves can be joined with line and tangent continuity if

1. the last point of the first curve and the first point of the second curve are identical;
2. the point before the last of the first curve, the common point, and the second point of the second curve are collinear.

In an analogous way, two Bézier patches can be joined with surface and tangent continuities if

1. the control points on the edges to be joined are identical;
2. the control points on either side of the common edge are collinear with the points on the common edge;
3. the ratios of the collinear segments are constant.

Swept, lofted and skinned surfaces are based on curves that are cross-sections of the surface. A *swept* surface is generated by translating a curve on a given *trajectory* also called *spine curve*. The trajectory can be a straight line or a curve and the given cross-section remains perpendicular to it during the process. *Lofted* surfaces are defined by curves of the same polynomial degree and with identical parametrization and segmentation. This is the process that has been used during centuries for generating ship hulls and the term comes from this activity. In skinned surfaces the given curves are orthogonal to a spine curve. In the terminology of MultiSurf the given cross-sections are called *master curves*.

11.9 EXERCISES

Exercise 11.1 (A bilinear patch). Return to Fig. 11.1 and find the expression of the bilinear patch (first-degree Bézier surface) by considering the straight lines that join corresponding points on the edges L_3 and L_4. The result should be identical to Eq. (11.3).

Exercise 11.2 (The equation of a bilinear patch). Derive Eq. (11.3) from Eq. (11.4).

Exercise 11.3 (The curve $u = v$ on a bilinear patch). Plot in Fig. 11.7 the control polygon of the curve $u = v$.

Exercise 11.4 (A curve on a bilinear patch). Consider the bilinear patch treated in Example 11.1. You are asked to

1. derive the equation of the curve $u + v = 1$ lying on this patch. **Hint:** see Section 11.2.2;
2. draw the curve;
3. find a geometric interpretation of the equation derived at 1.

Exercise 11.5 (Another curve on the bilinear patch). On the patch treated in Example 11.1 plot the curve defined by

$$u = 0.075 \sin(2\pi t) + 0.3, \quad v = 0.075 \cos(2\pi t) + 0.3, \quad t \in [0, \ 1]$$

Use the rotation button in the toolbar of the figure to check that the curve lies entirely on the given surface.

Exercise 11.6 (Again a curve on the bilinear patch). On the patch treated in Example 11.1 plot the curve defined by

$$t \in [0, \ 1], \quad u = 1 - t^2, \quad v = \frac{1}{20t}$$

Use the rotation button in the toolbar of the figure to better see the curve.

Exercise 11.7 (A plane curve on the bilinear patch). **1.** On the patch treated in Example 11.1 plot the curve defined by

$$t \in [0, \ 1], \quad u = 1 - t^2, \quad v = \frac{1/2 + t^2}{2}$$

2. Use the Rotate 3D button in the toolbar of the figure to check that the curve lies entirely in a vertical plane.
3. Use the coordinates x, y to plot the horizontal projection of the curve and confirm the same thing.

Exercise 11.8 (A ruled patch between two cubic Bézier curves). This exercise continues Example 11.2. Your tasks are:

1. write a MATLAB script that implements the example;
2. add the plot obtained with the MATLAB function contour3.

Exercise 11.9 (A swept surface based on a cubic Bézier curve). In the example of swept surface given in Section 11.5 change the translation vector to $dx = 10$, $dy = 2$, $dz = 5$.

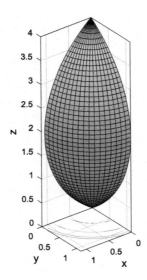

Figure 11.26 Surface of revolution as example of swept surface

Exercise 11.10 (Surface of revolution). Consider the cubic Bézier curve described in Fig. 11.20. Build a swept surface by rotating the curve as shown in Fig. 11.26. **Hint:** see Sections 8.3 and 11.5.

Exercise 11.11 (Lofted surface — Buttocks). We refer to the lofted surface shown in Fig. 11.23. Plot the sections (buttocks) corresponding to $y = 0.01$, $y = 2.5$, and $y = 5$.

APPENDIX 11.A A NOTE ON SURFACES IN MULTISURF

CAGD packages for Naval Architecture provide interactive facilities for designing surfaces. Fully worked examples created in MultiSurf by Reinhard Siegel can be downloaded freely. To do this look on the web for AeroHydro and follow the sequence

Aero Hydro Products → Support → Advanced Tutorial Topics

Especially interesting are the possibilities of achieving 'accurate, durable joins between surfaces'. This includes C^1 continuity.

PART 4

Applications in Naval Architecture

CHAPTER 12

Hull Transformations by Computer Software

Contents

12.1 INTRODUCTION

The creation from scratch of new ship lines is a difficult and costly task. Therefore, a common practice is to produce a new hull surface by modifying an existing, proven hull. More recently, several researchers created new ship lines by combining the lines of two or more parent ships. Methods have been developed for calculating the transformations that change hydrostatic properties in a desired way. When it is necessary to change hydrodynamic properties, such as seakeeping performances, a set of new hulls is obtained by systematic changes and their hydrodynamic behaviour is assessed with the help of *CFD* (computational fluid dynamics) programs.

The simplest transformations are *scalings* such as described in Section 8.2.7. In naval architecture the resulting surfaces are known as *affine hulls*. The hydrostatic properties of the new hulls can be derived from those of the parent hull by simple algebraic expressions. When the scales along the three coordinate axes are equal, the new hull is *geometrically similar* to the parent hull, shortly **geosim**. This is the relationship between the lines of an actual ship and those of her basin model.

More elaborate transformations are due to Lackenby. As explained in detail in Section 3.9, certain properties can be modified by translating the stations in a calculated way, while keeping constant the waterline length. Thus, the x-coordinates are changed while the y- and z-coordinates remain constant.

Geometry for Naval Architects
https://doi.org/10.1016/B978-0-08-100328-2.00024-9

Copyright © 2019 Elsevier Ltd.
All rights reserved.

A third way of generating new hulls is to combine the lines of two or more parent hulls. Schneekluth and Bertram (1998) propose to interpolate by hand between the lines of two hulls. Several researchers developed computer procedures that define points on the new hull surface as affine combinations of points on the surfaces of parent hulls (for the term 'affine combinations' see Section 8.5). Some authors use the term **morphing** for their transformations.

All methods mentioned above perform linear transformations. German researchers developed non-linear transformations and software for their implementation. The German term for such transformations is 'Verzerrung', which literally means 'distortion'.

12.2 AFFINE HULLS

In Section 2.10 we have mentioned the possibility of obtaining a new hull surface by scaling with the factor r_x all x-coordinates, by the factor r_y all y-coordinates, and by the factor r_z all z-coordinates. Using the subscript 'o' for the coordinates of the parent ship, and 'N' for those of the new ship, we have

$$\begin{vmatrix} x_N \\ y_N \\ z_N \end{vmatrix} = \begin{vmatrix} r_x x_o \\ r_y y_o \\ r_z z_o \end{vmatrix} \tag{12.1}$$

We derive immediately

$$\begin{vmatrix} dx_N \\ dy_N \\ dz_N \end{vmatrix} = \begin{vmatrix} r_x dx_o \\ r_y dy_o \\ r_z dz_o \end{vmatrix} \tag{12.2}$$

The subject is briefly discussed in Biran and López-Pulido (2014), Section 4.6, where we show that the displacement volume, ∇_{new}, of the new hull obtained in this way is related to the displacement volume of the parent ship by

$$\nabla_N = r_x r_y r_z \nabla_o$$

In the cited book it is mentioned that the affine hull has the same form coefficients as the parent hull. For example, the block coefficient of the new hull is

$$C_{bN} = \frac{r_x r_y r_z \nabla_o}{r_x L_o r_y B_o r_z T_o} = C_{bo}$$

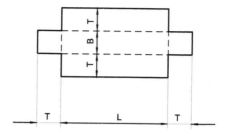

Figure 12.1 The developed wetted surface of a parallelepipedic hull

In Exercise 12.1 we ask the reader to check that other coefficients of form are also invariant.

When the scales are equal, $r_x = r_y = r_z = r$, the parent and the new hull are geometrically similar, shortly *Geosim*. Then, we can write

$$
\begin{vmatrix} x_N \\ y_N \\ z_N \end{vmatrix} = r \begin{vmatrix} x_o \\ y_o \\ z_o \end{vmatrix}, \quad \begin{vmatrix} dx_N \\ dy_N \\ dz_N \end{vmatrix} = r \begin{vmatrix} dx_o \\ dy_o \\ dz_o \end{vmatrix} \tag{12.3}
$$

An example of consequence is $\nabla_N = r^3 \nabla_o$. Schneekluth and Bertram (1998) remark that when $r_y \neq r_z$ a round bilge transforms into an elliptic one. These authors discuss also the influence of transformations on details like propeller clearance.

Example 12.1 (Affine hulls — Wetted surface coefficient). Doctors (1995a) uses the definition of the wetted–surface coefficient

$$
C_s = S / \nabla^{2/3}
$$

where S is the wetted-surface area, and mentions that this coefficient is invariant under scaling only if the scaling factors are equal, that is $r_x = r_y = r_z$. In this example we give a proof for the simple case of a parallelepipedic hull with length L, breadth B, and draught T. The developed wetted surface is shown in Fig. 12.1. In naval architectural terms this is a *shell expansion*. First, we calculate the wetted-surface coefficient of the initial hull

$$
C_{S0} = \frac{2LT + 2BT + LB}{(LBT)^{2/3}} = 2\frac{L^{1/3} T^{1/3}}{B^{2/3}} + 2\frac{B^{1/3} T^{1/3}}{L^{2/3}} + \frac{L^{1/3} B^{1/3}}{T^{2/3}} \tag{12.4}
$$

The wetted-surface coefficient of the affine hull is

$$
\begin{aligned}
C_{sN} &= \frac{2r_x r_z LT + 2r_y r_z BT + r_x r_y LB}{(r_x r_y r_z LBT)^{2/3}} \\
&= 2\frac{r_x^{1/3} r_z^{1/3}}{r_y^{2/3}} \cdot \frac{L^{1/3} T^{1/3}}{B^{2/3}} + 2\frac{r_y^{1/3} r_z^{1/3}}{r_x^{2/3}} \cdot \frac{B^{1/3} T^{1/3}}{L^{2/3}} \\
&\quad + \frac{r_x^{1/3} r_y^{1/3}}{r_z^{2/3}} \cdot \frac{L^{1/3} B^{1/3}}{T^{2/3}}
\end{aligned}
\tag{12.5}
$$

The coefficients C_{so} and C_{sN} are equal if the following three conditions are fulfilled

$$
\frac{r_x^{1/3} r_z^{1/3}}{r_y^{2/3}} = 1
\tag{12.6}
$$

$$
\frac{r_y^{1/3} r_z^{1/3}}{r_x^{2/3}} = 1
\tag{12.7}
$$

$$
\frac{r_x^{1/3} r_y^{1/3}}{r_z^{2/3}} = 1
\tag{12.8}
$$

We can see immediately that

$$
r_x = r_y = r_z = r
\tag{12.9}
$$

is a sufficient condition for the invariance of the wetted-surface coefficient under scaling transformations. This condition is also necessary. Subtracting Eq. (12.6) from Eq. (12.7) and simplifying we obtain $r_y = r_z$, and subtracting Eq. (12.7) from Eq. (12.8) and simplifying we get $r_y = r_z$. In other words, the wetted-surface coefficient is an invariant only for geometrically similar hulls.

Example 12.2 (Affine hulls — Metacentric height). For completeness we repeat a few definitions. The transverse metacentric height of the parent ship is given by

$$
\overline{GM}_o = \overline{KB}_o + \overline{BM}_o - \overline{KG}_o
$$

(see, for example Biran and López-Pulido, 2014) where $\overline{KB}$ is also called 'vertical centre of buoyancy', $\overline{BM}$ 'transverse metacentric radius', and $\overline{KG}$ 'vertical centre of gravity'. We invite the reader to prove that for the new, affine ship we have

$$
\overline{KB}_N = r_z \overline{KB}_z
$$

The transverse metacentric radius of the parent ship is calculated as

$$\overline{BM}_o = \frac{I_{xxo}}{\nabla_o}$$

where I_{xxo} is the second moment of the waterplane area of the parent ship about the axis x. Noting x_A the aftermost x-coordinate of the waterplane, and x_F the foremost one, we have (see Biran and López-Pulido, 2014, pp. 98–99)

$$I_{xxo} = \frac{2}{3} \int_{x_A}^{x_F} y_o^3 \, dx$$

For the new ship we have

$$I_{xxN} = \frac{2}{3} \int_{x_A}^{x_F} r_y^3 y_0^3 r_x \, dx = r_x r_y^3 I_{xxo}$$

In consequence, we write for the new ship

$$\overline{BM}_N = \frac{r_x r_y^3 I_{xxo}}{r_z r_y r_z \nabla_o} = \frac{r_y^2}{r_x} \overline{BM}_o$$

Schneekluth and Bertram (1998), pp. 150–1, show that the vertical centre of gravity can be assumed proportional to the depth, D, or to a corrected depth

$$D_A = D + \frac{\nabla_A + \nabla_{DH}}{L_{pp} B}$$

where ∇_A is the volume of superstructures, and ∇_{DH} that of deckhouses. For affine hulls $T_N = r_z T$, while D_N is not necessarily equal to $r_z D_o$. The freeboard, $D_N - T_N$, is based on special considerations and regulations, such as *The international convention on load lines*. Therefore, for a good calculation it is necessary to evaluate the freeboard and take into account the drawings of superstructures and deckhouses. If in the initial stages of the ship design we ignore this fact and assume the approximation

$$D_{AN} \simeq r_z D_{Ao}$$

we can write $\overline{KG}_N \simeq r_z \overline{KG}_o$ and obtain

$$\overline{GM}_N \simeq r_z \overline{KB}_o + \frac{r_y^2}{r_z} \overline{BM}_o - r_z \overline{KG}_o$$

For geometrically similar hulls this relationship reduces to

$$\overline{GM}_N \simeq r_z \overline{GM}_o$$

12.3 A NOTE ON LACKENBY'S TRANSFORMATION

Versluis (1977) describes a program that performs one of Lackenby's trans-formations. The title of the article is 'Computer aided design of shipform by affine transformation'. Although not mentioned, the transformations dealt with perform the 'swinging' of stations described by Lackenby (1950). Are the transformations affine as noted in the title? In fact, the swinging is equivalent to a shearing of the sectional-area curve. Returning to Eq. (3.91) we write

$$\delta x = \frac{\delta LCB}{\gamma_B} A_x = S_x A_x$$

where LCB is the notation of the Longitudinal Centre of Buoyancy, and A_x the sectional area at the coordinate x. In matrix form the swinging transformation is

$$\begin{vmatrix} x_N \\ A_{xN} \end{vmatrix} = \begin{vmatrix} 1 & S_x \\ 0 & 1 \end{vmatrix} \begin{vmatrix} x_o \\ A_{xo} \end{vmatrix}$$

This is, indeed, an affine transformation. The method is applied to the normalized sectional-area curve so that x varies from 0 to 1. Let us check what happens in the 'one minus prismatic' method not treated by Versluis (1977). The shift of stations calculated as in Section 3.9 is

$$\delta x = \frac{\delta C_p}{1 - C_p}(1 - x) = S_p(1 - x)$$

The matrix form of the transformation is

$$\begin{vmatrix} x_N \\ y_N \\ z_N \end{vmatrix} = \begin{vmatrix} 1 - S_p & 0 & 0 \\ 0 & 1 & 0 \\ 0 & 0 & 1 \end{vmatrix} \begin{vmatrix} x_o \\ y_o \\ z_o \end{vmatrix} + \begin{vmatrix} S_p \\ 0 \\ 0 \end{vmatrix} \qquad (12.10)$$

This is a scaling followed by a translation, two affine transformations.

12.4 AFFINE COMBINATIONS OF OFFSETS

Doctors (1995a) describes an interesting method of generating a new hull from a number of parent hulls. The method assumes the existence of a 'library of practical hull forms'. After choosing N hulls from the library, a computer program combines the forms. The author uses the terms 'merges' and 'blends'. In fact, the coordinates of points on the new hull are linear combinations of the coordinates of points on the parent hulls. First, the parent hulls must be scaled to the same main dimensions. Next, using our notations we write for the new coordinates

$$x_N = \sum_{i=1}^{N} r_x(i)x(i)$$

$$y_N = \sum_{i=1}^{N} r_y(i)y(i)$$

$$z_N = \sum_{i=1}^{N} r_z(i)z(i) \tag{12.11}$$

where $r_x(i)$, $r_y(i)$, $r_z(i)$ are scaling factors in the directions of the coordinate axes, and $x(i)$, $y(i)$, $z(i)$ the coordinates of points of the i-th parent hull. As recommended also by Krueger (2001), the points are defined by the intersection of longitudinal lines with 'girthwise lines', which can be transverse contours, such as ship stations. The paper exemplifies the method by showing how to generate 'demihulls suitable for a catamaran' and compares the seakeeping performances of the new hulls. An application to a 'proboscidean' bow is described in Doctors (1995b), a 'platypus bow' is studied in Doctors (1996), and semi–SWATH models in Doctors et al. (1996). Some sets of scaling factors in Doctors (1995a) are

```
{ -1 2 }, { -1 -6 8 }, { -1 -2 4 }, { -1 -2 4 }
```

In all sets the factors sum up to 1. Thus, the points of the new hulls are affine combinations of the points of the parent hulls. In Exercise 12.4 we show what happens when the new points are linear, but not affine combinations of the parent points. We discussed in Chapter 8 the properties of collinearity and coplanarity, and the invariance of affine combinations under affine transformations.

12.5 MORPHING

Some authors use the term **morphing** instead of 'transformation'. Marsh (2000) gives the following definition

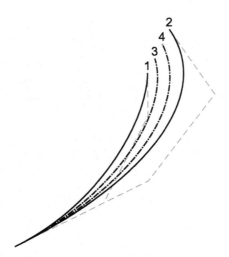

Figure 12.2 Morphing Bézier curves

'Morphing is a technique used in computer graphics in which a shape is gradually deformed over a period of time. Morphing has been used in animation sequences of feature films.'

Similar definitions can be found in dictionaries of the English language, some of which mention that the term comes from 'metamorphosis'. The Webster's New World College Dictionary, for example, adds that the process should result in a 'smoothly continuous series of images'. Marsh gives an example of morphing a B-spline. Given the control polygon, P_1, of the initial spline, and the control polygon, P_2, of the final spline, intermediate splines are defined by the polygons

$$P_\lambda = (1 - \lambda_1)P_1 + \lambda P_2, \ \ 0 \le \lambda \le 1 \tag{12.12}$$

Using the function `MorphBez` described below, we give an example of morphing a cubic Bézier curve. Calling the function with the input arguments *1/3* and *2/3* we obtain Fig. 12.2. The initial and final control polygons are shown in dash–dash lines, red in the electronic edition of the book. The corresponding curves are noted *1* and *2*. In a first phase the control polygons are transformed using the values 1/3 and 2/3 for the parameter λ. The curves defined by the new control polygons are shown as black dash–dash lines and are marked *3* and *4*. In the second phase two new curves are obtained by applying the transformation directly to the curves *1* and *2*. The resulting curves are plotted as dot–dot lines. They coincide with the curves *3* and *4* plotted in the first phase.

```
function MorphBez(lambda1, lambda2)
%MURPHBEZ Exemplifies the morphing of a Bezier curve.

% initial control polygon and curve
P1= [ 0 2 3 3; 0 1 3 4 ]; pline(P1, 1, 'r', '--')
axis equal, axis off
hold on
B1 = Bezier3(P1); pline(B1, 1.5, 'k', '-')
ht = text(B1(1, end), B1(2, end), '1');
set(ht, 'FontSize', 14, 'HorizontalAlignment', 'center',...
        'VerticalAlignment', 'bottom')
% final control polygon and curve
P2 = [ 0 3 4.5 3.5; 0 1.5 3.5 5 ]; pline(P2, 1, 'r', '--')
B2 = Bezier3(P2); pline(B2, 1.5, 'k', '-')
ht = text(B2(1, end), B2(2, end), '2');
set(ht, 'FontSize', 14, 'HorizontalAlignment', 'center',...
        'VerticalAlignment', 'bottom')
% morph polygons, next calculate and plot curves
P3 = (1 - lambda1)*P1 + lambda1*P2;
B3 = Bezier3(P3); pline(B3, 1.5, 'k', '--')
ht = text(B3(1, end), B3(2, end), '3');
set(ht, 'FontSize', 14, 'HorizontalAlignment', 'center',...
        'VerticalAlignment', 'bottom')
P4 = (1 - lambda2)*P1 + lambda2*P2;
B4 = Bezier3(P4); pline(B4, 1.5, 'k', '--')
ht = text(B4(1, end), B4(2, end), '4');
set(ht, 'FontSize', 14, 'HorizontalAlignment', 'center',...
        'VerticalAlignment', 'bottom')
% interpolate directly between the initial and the final curves
B5 = (1 - lambda1)*B1 + lambda1*B2; pline(B5, 1.5, 'k', ':')
B6 = (1 - lambda2)*B1 + lambda2*B2; pline(B6, 1.5, 'k', ':')
hold off
end
```

Ang et al. (2016) study the morphing of two-dimensional ship lines. First, the parent hulls are scaled to the same main dimensions. Next, cubic-spline interpolations are used to obtain the same number of data points on the two parent curves. Finally, the morphing is carried on as in Eq. (12.12). The authors use sometimes different parent hulls for the afterbody and the forebody. In some examples the parameter λ is varied from station to station.

The hull description by transverse sections and longitudinal curves does not suit all calculations. In CFD programs the surface is modelled by meshes. The morphing of meshes is a difficult problem. See, for example, Kang and Lee (2012).

12.6 NON-LINEAR TRANSFORMATIONS

Söding and Rabien (1977) and Rabien (1977) describe software that performs non-linear transformations of ship lines. More applications of the corresponding mathematics can be found in Krueger (2002) and Koechert (2010). In our notations, the equations used by Krueger are

$$x_N = x_o + f_1 x_o f_2 y_o f_3 z_o \qquad (12.13)$$
$$y_N = y_o + f_4 x_o f_5 y_o f_6 z_o$$
$$z_N = z_o + f_7 x_o f_8 y_o f_9 z_o$$

If the points are given on stations, that is transversal planes for which the x-coordinate is constant, the functions f_2, f_3 should be equal to 1. The functions $f_4, \ldots f_9$ are chosen so as to achieve desired results.

12.7 SUMMARY

New hulls are usually developed from parent hulls. There are several ways for doing this.

1. An *affine hull* is obtained by scaling the coordinates with factors constant along the three axes. Let these factors be r_x in parallel with the x-axis, r_y in parallel with the y-axis, and r_z for the z-axis

$$\begin{vmatrix} x_N \\ y_N \\ z_N \end{vmatrix} = \begin{vmatrix} r_x x_o \\ r_y y_o \\ r_z z_o \end{vmatrix}$$

When

$$r_x = r_y = r_z = r$$

the hulls are *geometrically similar*, shortly *Geosim*. This is the relationship between an actual ship and her basin model. These transformations do not change the coefficients of form. Other geometric properties can be easily derived from those of the parent hull.

2. Lackenby's transformations change some geometric properties by translating stations in calculated ways and keeping the waterline length constant.

3. A new hull surface can be defined by points that are affine combinations of points on the two or more parent-hull surfaces. The same new hull can be created by control polygons, P_λ, that are affine combinations of the control polygons, P_1, P_2, of parent-surface curves

$$P_\lambda = (1 - \lambda_1)P_1 + \lambda P_2, \; 0 \leq \lambda \leq 1$$

The extension to three parent hulls is

$$P = \lambda P_1 + \mu P_2 + \nu P_3, \; \lambda + \mu + \nu = 1$$

Some authors use the term *morphing* for such transformations.

4. German researchers developed a system of transformations in which the coordinates of the new hull are related to those of the parent hull by

$$x_N = x_o + f_1(x_o)f_2(y_o)f_3(z_o)$$
$$y_N = y_o + f_4(x_o)f_5(y_o)f_6(z_o)$$
$$x_N = z_o + f_7(x_o)f_8(y_o)f_9(z_o)$$

Transverse sections remain plane if $f_2 = f_3 = 1$. The functions are determined by conditions such as points through which they have to pass and slopes.

12.8 EXERCISES

Exercise 12.1 (Affine hulls — Coefficients of form). **1.** Prove that the scaling transformations that generate affine hulls leave unchanged the coefficients of form C_M, C_{WL}, C_P, C_{VP}, and the ratio D/T.
2. What are the conditions for invariance of the ratios L/B, B/T, and ∇/L^3, under scalings that produce an affine hull?

Exercise 12.2 (Affine hulls — Wetted-surface coefficient). Consider a floating body that has the form of a right cylinder of rotation, with the axis vertical. Let the radius of the cylinder be R, and the draught T. Find the necessary and sufficient condition for invariance under scaling.

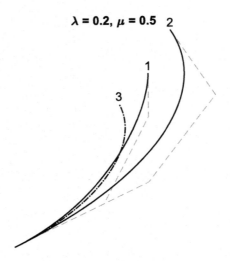

$\lambda = 0.2,\ \mu = 0.5$

Figure 12.3 Non-affine combination of points of Bézier curves

Exercise 12.3 (Geosim hulls — Longitudinal metacentric height). The longitudinal metacentric height is given by

$$\overline{GM}_L = \overline{KB} + \overline{BM}_L - \overline{KG}$$

where $\overline{BM}_L$ is the 'longitudinal metacentric radius', and the other terms have the meanings explained in Example 12.2. For real ships $\overline{KB} - \overline{KG} \ll \overline{BM}_L$. Taking this into account derive an approximate relationship between the longitudinal metacentric height of a new hull and that of the Geosim parent ship.

Exercise 12.4 (Morphing Bézier curves — Non-affine combinations of points). Starting with the curves B1 and B2 used in the morphing example of Section 12.5, show that a curve

$$B = \lambda B_1 + \mu B_2, \quad \lambda + \mu \neq 1$$

does not interpolate between the given curves. Fig. 12.3 shows an example in which, given the control polygons, P_1 and P_2, of two Bézier curves, the intermediate curve is defined by

$$P_3 = 0.2 P_1 + 0.5 P_2$$

To solve the exercise play with two sets of factors $\lambda + \mu < 1$ and $\lambda + \mu > 1$.

CHAPTER 13

Conformal Mapping

Contents

13.1 INTRODUCTION

English	French	German	Italian
complex number	nombre complexe	komplexe Zahl	numero complesso
conformal mapping	transformation conforme	konforme Abbildung	trasformazione conforme
imaginary part	partie imaginaire	Imaginärteil	parte immaginaria
imaginary unit	unité imaginaire	imaginäre Einheit	unità immaginaria
modulus, absolute value	module	Betrag	modulo, valore assoluto
real part	partie réelle	Realteil	parte reale

The calculations of motions and loads in waves require the solution of difficult mathematical problems and considerable computational resources. To overcome these problems various simplifying assumptions were researched and used in practical calculations. Thus, the *slender body* assumption supposes that the length of the underwater hull is much larger than the draught and the beam. Next, the slender hull is divided into *strips*, that is short volumes limited by two transverse cross-sections. The *strip theory* assumes that the 'flow field at any cross-section of the ship may be approximated by the assumed two-dimensional flow in that strip' (Beck et al., 1989). Certain values calculated in this way are integrated along the hull length to yield results for the whole ship. More details can be found in the literature of specialty, for example McCormick (2010). The first step is the calculation of the flow around the ship section and the derivation of the quantities required in seakeeping calculations. One possibility is to use the results known

Geometry for Naval Architects
https://doi.org/10.1016/B978-0-08-100328-2.00025-0

Copyright © 2019 Elsevier Ltd.
All rights reserved.

for a circular cylinder and to transform them for a form that approximates sufficiently well the given ship section. The transformation is carried on with the aid of functions of complex variables and is known as **conformal mapping**. F.M. Lewis introduced in 1929 a three-term *mapping function* and the ship sections produced by it are known as **Lewis forms**. The range of section forms that can be approximated in this way is limited. To approximate more sections, nearly thirty years later Landweber and Macagno added a term to the mapping function. More authors followed introducing other *extended Lewis forms* and this activity continues to these days.

In this chapter we introduce the subjects of conformal mapping and Lewis forms. We hope that after reading this introduction the reader will be able to understand the more recent developments. We give examples in MATLAB, the computer environment we used throughout the book. For an easy-to-read treatment of the theory of complex variables we like the classic book of Churchill (1960). There are newer editions of this book with coauthor J.W. Brown. An example of textbook dedicated to functions of complex variables is Saff and Snider (2003). Chapters on this theory are included in texts of engineering mathematics, such as Kreyszig (2011). Examples of working with complex variables in MATLAB can be found, for example, in Biran and Breiner (2002), and Biran (2011).

13.2 WORKING WITH COMPLEX VARIABLES

For completeness, we introduce in this section a few elementary concepts of the theory of complex numbers and exemplify their implementation in MATLAB. The reader familiar with these matters may skip the section and go directly to the next one. To keep the input values and the intermediary results, the operations described below should be carried on in a single MATLAB section. In accordance with the general practice we use the notation i for the square root of -1

$$i = \sqrt{-1}$$

MATLAB allows also the use of the symbol j. This notation can be found in electrical engineering and in many German books. Try the commands

```
>> i^2, j^2
ans =
    -1
```

```
ans =
    -1
```

A *complex number* has the form

$$z = x + iy$$

where x and y are real numbers. Here x is called the *real part* of z, y the *imaginary part*. The *imaginary unit* is

```
>> i
ans =
    0.0000 + 1.0000i
```

Let us define in MATLAB two complex numbers

```
>> z1 = 3 + 7*i
z1 =
    3.0000 + 7.0000i
>> z2 = 2 + 3*i
z2 =
    2.0000 + 3.0000i
```

In MATLAB we can work with them as easily as with real numbers, for example

```
>> z3 = z1 + z2
z3 =
    5.0000 +10.0000i
>> z4 = z1*z2
z4 =
  -15.0000 +23.0000i
>> z5 = z1/z2
z5 =
    2.0769 + 0.3846i
```

We retrieve the real and the imaginary parts of a complex number with the aid of the MATLAB functions real and imag

```
>> real(z5)
ans =
    2.0769
>> imag(z5)
ans =
    0.3846
```

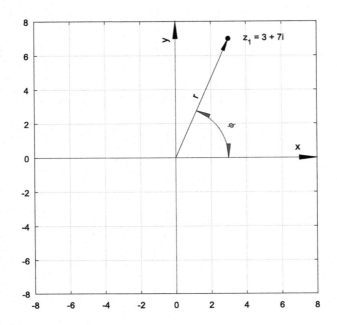

Figure 13.1 Geometrical representation of the complex number z_1

A graphic representation of complex numbers was devised first by Caspar Wessel (Norwegian–Danish, 1745–1818) and soon after him by Jean Robert Argand (Swiss, 1768–1822). Fig. 13.1 shows this representation for the point z_1 calculated above. In this cartesian plot the real part of the complex number is measured along the horizontal axis, and the imaginary part along the vertical axis. This representation is known in German literature as *Gaußsche Zahlenebene*, and in Italian as *diagramma di Argand-Gauss*. The length r of the vector from the origin of coordinates to a point z is called **modulus,** or *absolute value* of the complex number and can be retrieved in MATLAB with the aid of the function abs

```
>> r = abs(z1)
r =
    7.6158
```

The angle ϕ between the above vector and the real axis is called **argument,** or *phase* of the complex number and can be retrieved in MATLAB with the function angle

```
>> phi = angle(z1)
```

```
phi =
    1.1659
```

This angle is measured in radians. As can be easily seen in Fig. 13.1, a complex number having the module r and the argument ϕ can be expressed in **polar form** as

$$z = r(\cos\phi + i\sin\phi)$$

In continuation of the above calculations try

```
>> r*cos(phi) + i*r*sin(phi)
ans =
    3.0000 + 7.0000i
```

An equivalent representation is yielded by *Euler's equation* (Leonhard, Swiss, 1707–1783)

$$e^{i\phi} = \cos\phi + i\sin\phi$$

Try in MATLAB

```
>> r*exp(i*phi)
ans =
    3.0000 + 7.0000i
```

We recover the number z_1 defined at the beginning of this section.

13.3 CONFORMAL MAPPING

Let us consider a plane in which we define points $z = x + iy$, and a second plane in which we define points $w = u + iv$. If there exists a function f such that to each point z corresponds one point $w = f(z)$, we say that the function f is a *mapping* or *transformation* of the plane z into the plane w. If the function $f(z)$ is complex differentiable at every point in a region R of the complex plane we say that the function is **analytic** in the region R. Another term for this notion is *holomorphic function*. Mappings by an analytic function preserve angles; therefore, such mapping are qualified as **conformal**. In other words, if in the plane z the angle between two curves is α, in the plane w the angle between the conformal mappings of the given curves is also α.

There is another invariant of conformal mappings and the application discussed in this chapter is based on it. If we study a two-dimensional flow and assume that it is irrotational and inviscid, there is a **potential** function, ϕ, such that the components of the fluid velocity are the partial derivatives of this function with respect to coordinates. The potential must be a solution of the *Laplace equation*

$$\frac{\partial^2 \phi_1}{\partial x^2} + \frac{\partial^2 \phi_1}{\partial y^2} = 0 \tag{13.1}$$

and fulfill adequate boundary conditions. If in the plane z we are given a potential Φ_1, and apply to it a conformal transformation, in the plane w we obtain a potential Φ_2 that is a solution of the Laplace equation

$$\frac{\partial^2 \phi_2}{\partial u^2} + \frac{\partial^2 \phi_2}{\partial v^2} = 0 \tag{13.2}$$

A proof of this theorem is given in Wylie (1966). Now, let us assume that we want to study the flow around a given ship section. We have a solution for the flow around a circle. We must find a function that maps the circle into a shape that approximates well the given section and then we use that function to map the potential flow.

Example 13.1 (The Joukovski transformation). Zhukovski (Nikolai Egorovich, Russian, 1847–1921) published in 1910 a paper that describes what is known as the famous *Joukovski transformation*. This conformal mapping transforms a circle into an airfoil and its main use is in aeronautical engineering. Airfoil shapes, however, appear also in naval architecture. Ruders have often the form of a symmetrical airfoil, while cambered shapes are used for hydrofoil craft or roll stabilizers. To exemplify the Joukovski transformation we represent the circle in the z-plane by the polar equations

$$x = x_0 + r_0 \cos\theta$$
$$y = y_0 + r_0 \sin\theta \tag{13.3}$$

and define the mapping function

$$w = \left(z + \frac{b^2}{z} \right) / 2 \tag{13.4}$$

A MATLAB function that carries on the mapping is

```
function Joukovski(x0, y0, b)

%JOUKOVSKI Joukowski transformation of circle with centre
```

```
%    at x0, y0, and parameter b.

% define and plot circle in z plane
theta = 0: pi/60: 2*pi;          % circle parameter
r0    = 1.5;                      % circle radius
x     = x0 + r0*cos(theta); y = y0 + r0*sin(theta);
z     = x + i*y;
subplot(1, 2, 1)           % z-plane
    plot(x, y, 'r-'), grid, axis equal
    tt = [ 'x_0 = ', num2str(x0), ', y_0 = ', num2str(y0) ];
    tt = [ tt, ', b = ', num2str(b) ];
    title({'z-plane'; tt})
    xlabel('\Re'), ylabel('\Im')
    hold on
    point([ x0; y0 ], x0/8)
    hold off
% Joukowski transformation
w = (z + b^2./z)/2;
subplot(1, 2, 2)           % w plane
    plot(real(w), imag(w), 'r-'), grid
    axis equal
    title({'w-plane'; 'w = (z + b^2/z)/2'})
    xlabel('\Re'), ylabel('\Im')
```

Three examples appear in Figs 13.2 to 13.4; they show how the centre of the circle in the z-plane determines the shape of the foil.

13.4 LEWIS FORMS

Assume that in the z-plane we define a circle by an expression $\zeta(z)$. A general mapping function can have the form

$$w = a_0\zeta + a_1\zeta^{-1} + a_2\zeta^{-2} + a_3\zeta^{-3} + \ldots + a_i\zeta^{-i} + \ldots$$

Ship sections are symmetric about the centreplane. To simplify the mathematics we consider that the shape is also symmetric about the waterplane, an assumption explained in the literature on motions in waves. Wilson (1987) shows that these symmetries are achieved when the coefficients of even powers are zero and those of odd powers are real. Therefore, the series used in naval architecture have the form

$$w = a_0\zeta + a_1\zeta^{-1} + a_3\zeta^{-3} + \ldots + a_{2i+1}\zeta^{-(2i+1)} + \ldots \tag{13.5}$$

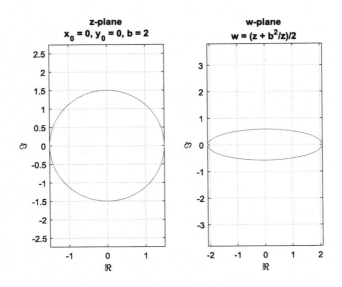

Figure 13.2 The Joukovski transformation, circle to ellipse

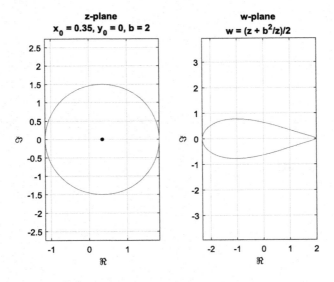

Figure 13.3 The Joukovski transformation, symmetrical airfoil

Lewis used only the first three terms of this series. To explain this mapping we follow mainly a paper of Bishop et al. (1978). In both complex planes we choose systems of coordinates with the real axis positive towards right, and the imaginary axis positive down. The circle to be transformed

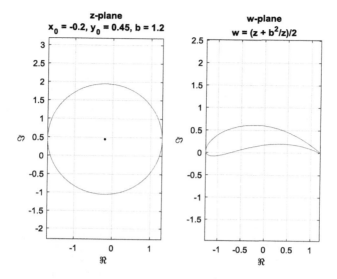

Figure 13.4 The Joukovski transformation, cambered airfoil

is described in the ζ-plane by the equation

$$\zeta = ie^{-i\phi} \qquad (13.6)$$

where the angle ϕ is measured as in Fig. 13.5. The mapping function is

$$w = a(\zeta + a_1 \zeta^{-1} + a_3 \zeta^{-3}) \qquad (13.7)$$

The coefficient a is a scaling factor, the coefficients a_1, a_3 are the parameters of the transformation. Therefore, this mapping is known as a *two-parameter transformation*. In the w-plane the origin of coordinates is the intersection of the local waterline with the centreplane of the ship section. During the derivation of the equations that define the coefficients of the mapping function we use the identities

$$i^{-1} = \frac{1}{i} = \frac{i}{i \cdot i} = -i, \quad i^{-3} = \frac{i^{-1}}{i^2} = \frac{-i}{(-1)} = i$$

and De Moivre's theorem

$$[r(\cos\theta + i\sin\theta)]^n = r^n(\cos n\theta + i\sin n\theta)$$

in the form

$$(re^{i\theta})^n = r^n e^{in\theta}$$

Expanding Eq. (13.7) we obtain

$$
\begin{aligned}
w &= a[ie^{-i\theta} - a_1(ie^{-i\theta})^{-1} + a_3(ie^{-i\theta})^3] \\
&= ai(e^{-i\theta} - a_1 e^{i\theta} + a_3 e^{3i\theta})
\end{aligned}
\tag{13.8}
$$

The angle θ is measured as in Fig. 13.5. Expanding the exponential form of the various terms we obtain their polar forms, for example

$$e^{i\theta} = \cos\theta + i\sin\theta$$

Naming u the real component, and v the imaginary one, we have

$$
\begin{aligned}
u &= -a[\sin(-\theta) - a_1 \sin\theta + a_3 \sin 3\theta] \\
v &= a[\cos(-\theta) - a_1 \cos\theta + a_3 \cos 3\theta]
\end{aligned}
\tag{13.9}
$$

There are three numbers that must be defined, namely a, a_1 and a_3. Therefore, we choose three constants of the given ship section: the section breadth, B, the section draught, T, and the sectional area, A. For $\theta = 0$ Eqs (13.9) yield

$$
\begin{aligned}
u &= 0 \\
v &= a(1 - a_1 + a_3) = T
\end{aligned}
\tag{13.10}
$$

and for $\theta = \pi/2$

$$
\begin{aligned}
u &= -a(-1 - a_1 - a_3) = a(1 + a_1 + a_3) = B/2 \\
v &= a(0 - 0 + 0) = 0
\end{aligned}
\tag{13.11}
$$

The area of the ship section is given by

$$A = \int_{u=0}^{B/2} v\, du \tag{13.12}$$

The solution of this integral yields a third equation in a, a_1, a_3 that must be solved together with the second Eq. (13.10) and the first Eq. (13.11). The calculations are rather tedious and involve the solution of a second-degree equation. A full derivation can be found in Ciobanu et al. (2006).

With the notations

$$\sigma = \frac{A}{BT}, \quad H = \frac{B}{2T} \tag{13.13}$$

and

$$C = \left(3 + \frac{4\sigma}{\pi}\right) + \left(1 - \frac{4\sigma}{\pi}\right)\left(\frac{H-1}{H+1}\right)^2 \tag{13.14}$$

the final results are

$$a = \frac{B}{2(1 + a_1 + a_3)}$$

$$a_1 = (1 + a_3)\frac{H-1}{H+1}$$

$$a_3 = \frac{3 - C + \sqrt{9 - 2C}}{c} \tag{13.15}$$

A MATLAB script that implements these results for a particular example is

```
%LEWIS Plots a quarter of a circle in the zeta plane
%    and the corresponding Lewis form in the w plane.

% input
B = 10;              % section breadth
T = 8;               % section draught
A = B*T*0.625;       % section area
phi   = 0: pi/90: pi/2;
zeta  = i*exp(-i*phi);
nphi  = 25;          % defines point to be displayed
subplot(1, 2, 1)     % zeta plane
    plot(real(zeta), imag(zeta), 'k-'), grid, axis equal
    axis ij      % reverts direction of vertical axis
    O = [ real(zeta(nphi)); imag(zeta(nphi)) ];
    point(O, B/500)
    ht1 = text(1.05*O(1), O(2), '\zeta (\phi)');
    set(ht1, 'FontSize', 14, 'HorizontalAlignment', 'left')
    title('\zeta plane')
    xlabel('\xi'), ylabel('\eta')
    P = [ real(zeta(nphi)); imag(zeta(nphi)) ];
    hold on
    plot([ 0 P(1) ], [ 0 P(2) ], 'r-')
    ht = text(0.14, 0.4, '\phi');
```

```
    set(ht, 'FontSize', 14, 'Color', 'r')
    % mark angle phi
    phi1 = 0: pi/90: phi(nphi);
    plot(0.4*sin(phi1), 0.4*cos(phi1), 'r-')
    hold off
% calculate mapping coefficients
sigma = A/(B*T);    % section-area coefficient
H     = B/(2*T);    % aspect ratio
HH    = (H - 1)/(H + 1);
C     = (3 + 4*sigma/pi) + (1 - 4*sigma/pi)*HH^2;
a3    = (3 - C + (9 - 2*C)^(1/2))/C;
a1    = (1 + a3)*HH;
a     = B/(2*(1 + a1 + a3));
% conformal mapping
w     = a*(zeta + a1*zeta.^(-1) + a3*zeta.^(-3));
y     = real(w); z = imag(w);
subplot(1, 2, 2)    % w plane
    plot(y, z, 'k-'), grid, axis equal
    axis ij % reverts direction of vertical axis
    t = [ 'B = ', num2str(B), ', T = ', num2str(T) ];
    t = [ t ', \sigma = ', num2str(sigma) ];
    title({'w plane'; t})
    xlabel('y'), ylabel('z')
    Ow = [ 0; 0 ];        % origin of axes
    hold on
    Pw = [ y(nphi); z(nphi) ];
    point(Pw, 0.1)
    ht1 = text(1.05*Pw(1), Pw(2), 'w (\theta)');
    set(ht1, 'FontSize', 14, 'HorizontalAlignment', 'left')
    plot([ 0 Pw(1) ], [0 Pw(2) ], 'r-')
    hold off
```

Three examples of valid Lewis forms obtained with the above script are shown in Figs. 13.5 to 13.7. Fig. 13.10 shows a ship section and the corresponding Lewis form. In this case the approximation is very good. One can also appreciate that the given ship section and the Lewis form have the same area. The appearance of a square root in the equation that defines a_3 limits the domain of H and σ pairs for which usable Lewis forms can be obtained. In Figs 13.8 and 13.9 we see two examples of what may happen in some cases. To improve the possibilities of approximating the shape of ship sections Landweber and Macagno added in 1959 a further term to the mapping section. An additional characteristic of the section was necessary

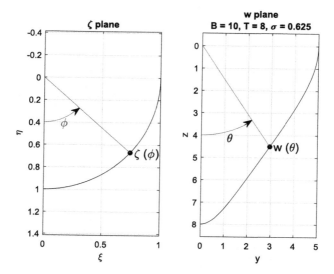

Figure 13.5 A first example of Lewis transformation

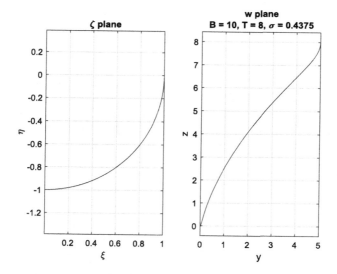

Figure 13.6 A second example of Lewis transformation

to determine all the coefficients. The authors chose the moment of inertia of the section. An iterative process was used to solve the resulting system of equations and the process is concisely described in Hoffman and Zielinski (1977). These authors describe also a development published by Kerczek

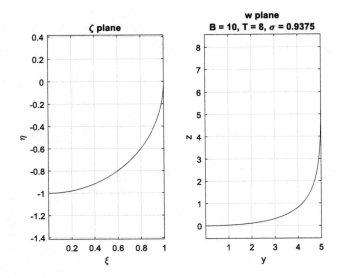

Figure 13.7 Lewis form, a third example

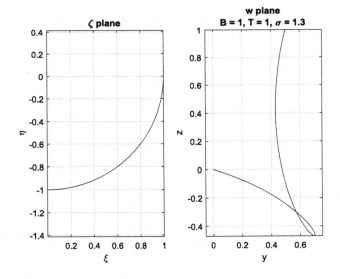

Figure 13.8 Lewis form, unacceptable example

and Tuck in 1969. Athanassoulis and Loukakis (1983) use a mapping function of the form

$$w = c_1\zeta + c_2\zeta^{-1} + c_3\zeta^{-3} + c_4\zeta^{-5}$$

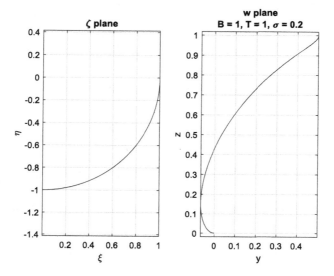

Figure 13.9 Lewis form, another unacceptable example

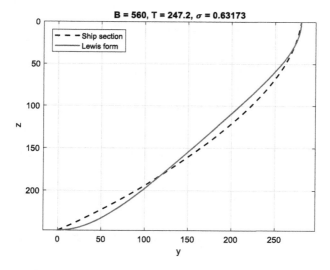

Figure 13.10 A ship section and the corresponding Lewis form. Dimensions measured in mm from a lines plan

The coefficients are determined as functions of the draught, T, used as a scaling factor, and three non-dimensional geometrical parameters: the breadth-to-draught ratio, $B/(2T)$, the sectional area coefficient, $A/(2BT)$, and the vertical centre of buoyancy to draught ratio, $\overline{KB}/T$ ratio. Simulations proved that the latter ratio has a significative influence on seakeeping performances. References to additional, newer extended forms can be found on the web.

13.5 SUMMARY

Conformal mappings are used in seakeeping calculations under the assumption of the *strip theory* that the flow in a transverse section of the hull is not influenced by the flow in the adjacent sections. Then, as the potential flow around a circle is known, the circle is mapped into a form that approximates the ship section. Once the mapping function is known, it is used also for mapping the flow. The transformation is carried on in complex numbers. The circle is represented in the complex plane z, for example with coordinates x and y, its mapping in the complex plane w, with coordinates u and v. The mapping function, $w = f(z)$, is an *analytic function*, that is a function of z that has a derivative in the region of interest. Angles between curves are invariants of such mappings that are called *conformal*. Under the assumptions of irrotational and inviscid flow, the components of the fluid velocity can be calculated as the partial derivatives of a *potential* function, Φ, that is the solution of the *Laplace equation*

$$\frac{\partial^2 \phi_1}{\partial x^2} + \frac{\partial^2 \phi_1}{\partial y^2} = 0$$

and fulfill adequate boundary conditions. If in the plane z we are given a potential Φ_1, and apply to it a conformal transformation, in the plane w we obtain a potential Φ_2 that is a solution of the Laplace equation

$$\frac{\partial^2 \phi_2}{\partial u^2} + \frac{\partial^2 \phi_2}{\partial v^2} = 0$$

Lewis introduced in 1929 a three–term mapping function of the form

$$w = a(z + a_1 z^{-1} + a_2 z^{-3})$$

The resulting shapes are known as *Lewis forms*. The range of shapes that can be approximated by this mapping is limited. To enhance the possibilities other authors added more terms to the mapping functions. For reasons of symmetry the series employed contain only odd powers of z and have real coefficients.

13.6 EXERCISES

Exercise 13.1 (Complex number — Elementary operations). This exercise refers to Section 13.2.
1. Plot in the same figure the points z_1, z_2, z_3, z_4.
2. Give a geometric interpretation of the addition that yields the number z_3.
3. Give a geometric interpretation of the multiplication that yields the number z_4. **Hint:** perform the multiplication using the exponential representation and check the results in the graphic representation.

Exercise 13.2 (Symmetrical Lewis mapping). In Section 13.4 we mentioned that, given the ζ function that defines the circle, when the mapping function includes only odd negative powers of ζ the resulting conformal mapping is symmetric about the imaginary axis and reflected in the real axis. To exemplify this return to one of the examples given in Section 13.4 and define the whole circle (360°).

Exercise 13.3 (Asymmetric conformal mapping). To show that when the mapping function contains even, negative powers of ζ the conformal mapping is asymmetric about the imaginary axis define the semicircle

```
zeta  = i*exp(-i*phi);
```

and use the mapping function

```
w     = zeta + 2*zeta.^(-1) + 4*zeta.^(-2) + 5*zeta.^(-3);
```

Exercise 13.4 (Aft section, Lewis form). Below are the offsets of an aft section of a patrol boat. Plot the section and the corresponding Lewis form analogously to Fig. 13.10.

```
y  = [ 0.0 29.0 43.0 48.0 49.0 50.0 51.0 51.5 ];
z0 = [ 3.0  9.0 18.0 27.0 36.0 45.0 54.0 63.0 ];
```

Exercise 13.5 (Forward section, Lewis form). Below are the offsets of a forward section of a patrol boat. Plot the section and the corresponding Lewis form analogously to Fig. 13.10.

```
y  = [  0.0  2.0  4.0  7.0 11.0 15.0 17.0 ];
z0 = [ 27.0 36.0 45.0 54.0 63.0 72.0 81.0 ];
```

BIBLIOGRAPHY

Abate, M., 1996. Geometria. McGraw-Hill Libri Italia, Milano.

Abbot, E.A., 1992. Flatland — A Romance of Many Dimensions. Dover Publications, New York.

Alaez, J.A., 1991. Computer Hull Form Design — A Historical Review. Publication 125. Canal de Experencias Hidrodinamicas, Madrid.

Ang, J.H., Goh, C., Li, Y., 2016. Hybrid evolutionary shape manipulation for efficient hull form design optimisation. In: Conference on Computer Applications and Information Technology in the Maritime Industries.

Anonymous, 2014. 2D in a 3D world — Mark Waldie, Darren Larkis and Denis Morais. In: SSI, Canada Explain How 3D Technologies Can Enhance 2D Approved Workshop Drawings. The Naval Architect Oct., 51–52.

Athanassoulis, G.A., Loukakis, T.A., 1983. An Extended-Lewis Form Family of Ship-Sections and Its Applications to Seakeeping Calculations. Report No. 12-1983. National Technical University of Athens, Dept. of Naval Architecture and Marine Engineering, Athens.

Barbarin, P., 1880. Note sur le planimètre polaire. Nouvelles annales de mathématiques, 2e série 19, 212–215.

Barker, R., 2003. Whole-moulding: a preliminary study of early English and other sources. In: Nowacki, H., Valleriani, M. (Eds.), Shipbuilding Practice and Ship Design Methods From the Renaissance to the 18th Century. Max-Planck Institut für Wissenschaftsgeschichte, pp. 33–66. Preprint 245.

Barker, R., 2006. Two architectures — a view of sources and issues. In: Nowacki, H., Lefèvre, W. (Eds.), Creating Shapes in Civil and Naval Architecture: A Cross-Disciplinary Comparison, Workshop December 2006. Max Plank Institute, Berlin, pp. 41–133. Preprint 338.

Bäschlin, G.F., 1947. Einführung in die Kurven- und Flächenthorie. Orell Füssli Verlag, Zührich.

Beck, R.F., Cummins, W.E., Dalzell, J.F., et al., 1989. Motion in waves. In: Lewis, E.V. (Ed.), Principles of Naval Architecture, 2nd revision. SNAME, Jersey City, NJ.

Bertoline, G.R., Wiebe, E.N., 2003. Technical Graphics Communications, 3rd edition. McGraw-Hill, Boston.

Bertram, V., 2004. Considerations of design for production principles in ship hull design. Journal of Marine Engineering 1 (1), 1–14.

Bieri, H.P., Prautsch, H., 1999. Preface. Computer Aided Geometric Design 16, 579–581.

Biran, A., 2003. Ship Hydrostatics and Stability, 1st edition. Butterworth-Heinemann, Oxford, UK.

Biran, A., 2005. Ship Hydrostatics and Stability. Butterworth-Heinemann, Oxford, UK. Enlarged reprint with contributions by López-Pulido.

Biran, A., 2006. A MATLAB digitizer-integrator. https://www.mathworks.com/matlabcentral/fileexchange/43744-ship-hydrostatics-and-stability-by-adrian-biran-and-ruben-lopez-pulido-isbn-9780080982878?s_tid=srchtitle.

Biran, A., 2011. What Every Engineer Should Know About MATLAB and Simulink. CRC Press, Taylor and Francis, Boca Raton, FL.

Biran, A., Breiner, M., 2002. MATLAB 6 for Engineers, 3rd edition. Prentice-Hall, Harlow, England.

Biran, A., López-Pulido, R., 2014. Ship Hydrostatics and Stability, 2nd edition. Butterworth-Heinemann, Oxford, UK.

Bishop, R.E.D., Price, W.G., Tam, P.K.Y., 1978. Hydrodynamic coefficients of some heaving cylinders. International Journal for Numerical Methods 17, 17–33.

Böge, A., 1959. Blechköper — Abwicklungs- und Fertigungsverfahren. Fachbuchverlag Dr. Pfanneberg & Co., Giessen.

Branco, J.R., Gordo, J.M., 2008. Tecnologia de estaleiros navais — Lições de traçagem. Instituto Superior Técnico, Lisbon.

Branco, J.N.R., Guedes Soares, C., 2005. Mapping of shell plates of double curvature into plane surfaces. Journal of Ship Production 21 (4), 248–257.

Brannan, D.A., Esplen, M.F., Gray, J.J., 1999. Geometry. Cambridge University Press, Cambridge.

Bugrov, Ya.S., Nikolsky, S.M., 1982. Fundamentals of Linear Algebra and Analytical Geometry. Mir Publishers, Moscow.

Bundjulov, V.St., Dimovski, Iv.Il., Petrov, D.N., 1964. Desfășuratele pieselor din tablă (The Development of Plate Elements). Editura Tehnică, Bucharest (in Romanian). The original was published in Bulgarian, in 1961.

Castro, F., 2007. Rising and narrowing: 16th century geometric algorithms used to design the bottom of ships in Portugal. International Journal of Naval Archeology 36, 148–154.

Chirone, E., Tornincasa, S., 2005. Disegno tecnico industriale, vol. 1. Il Capitello, Torino.

Churchill, R.V., 1960. Complex Variables and Applications, 2nd edition. McGraw-Hill Book Company, New York.

Ciobanu, C., Cață, C., Anghel, A., 2006. Conformal mapping in hydrodynamics. Bulletin of the Transylvania University of Brașov 13 (48), 85–95.

Clements, J.C., 1981. A computer system to derive developable hull surfaces and tables of offsets. Marine Technology 18 (3), 227–233.

Clements, J.C., 1984. Developed plate expansion using geodetics. Marine Technology 21 (4), 384–386.

Coolidge, J.L., 1963. A History of Geometrical Methods. Dover Publications, New York.

Courant, R., Robbins, H., 1996. What Is Mathematics? 2nd edition. Oxford University Press, New York.

Davies, A., Sammuels, Ph., 1996. An Introduction to Computational Geometry for Curves and Surfaces. Clarendon Press, Oxford.

De Casteljau, P., 1999. De Casteljau's autobiography. My time at Citroën. Computer Aided Geometric Design 16, 583–586.

Delachet, A., 1964. La géométrie projective. Que sais-je? 1103. Presses Universitaires de France, Paris.

Devlin, Keith, 2001. The Maths Gene — Why Everyone Has It, But Most People Don't Use It. Phoenix, London.

DIN 5 Teil 1, 1970. Zeichnungen — Axonometrische Projektionen, Isometrische Projektion. Beuth Verlag, Berlin. Replaced by DIN ISO 54563 (4.1998).

DIN 5 Teil 2, 1970. Zeichnungen — Axonometrische Projektionen, Dimetrische Projektion. Beuth Verlag, Berlin. Replaced by DIN ISO 5456-3 (4.1998).

DNV-GL, 2015. Rules for Classification, Ships, Part 3 Hull, Chapter 5, Hull Girder Strength. DNV GL AS.

Doctors, L.J., 1995a. A versatile hull-generation program. In: Proceedings Twenty-First Century Shipping Symposium. Sydney, pp. 140–158.

Doctors, L.J., 1995b. The influence of the proboscidean bow on ship motions. In: Proceedings Twelfth Australasian Fluid Mechanics Conference. Sydney, pp. 263–266.

Doctors, L.J., 1996. The influence of a duckbilled platypus bow on ship motions. In: Proceedings Small Craft Marine Engineering, Resistance and Propulsion Symposium. Ann Arbor. 7_1–16.

Doctors, L.J., Holloway, D., Davis, M.R., 1996. The effect of hull-form on ship motions. In: Proceedings The Twenty-Sixth Israel Conference on Mechanical Engineering. Haifa, pp. 5–9.

Dormidontow, W.K., 1954. Technologie des Schiffbaues. Fachbuchverlag Leipzig. Translation of the Russian original: Tekhnologhya sudostroyenya, Sudpromghis, Leningrad, 1949.

Eggerton, P.A., Hall, W.S., 1999. Computer Graphics — Mathematical First Steps. Prentice Hall, Harlow, England.

Epstein, M.P., 1976. On the influence of parametrization in parametric interpolation. SIAM Journal on Numerical Analysis 13, 261–268.

Farin, G., 1999. NURBS from Projective Geometry to Practical Use, 2nd edition. A.K. Peters, Natick, MA.

Farin, G., 2002. Curves and Surfaces for Computer-Aided Geometric Design — A Practical Guide, 5th edition. MK (Academic Press), San Diego.

Farin, G., Hansford, 2005. Practical Linear Algebra — A Geometry Toolbox. A.K. Peters, Wellesley, MA.

Flocon, A., Taton, R., 1984. La perspective. Que sais-je? 1050. 4th edition. Presses Universitaires de France, Paris.

Foley, J.D., Van Dam, A., 1984. Fundamentals of Interactive Computer Graphics. Addison-Wesley Publishing Company, Reading, MA.

Fornero, E., 2011. Calcolo del raggio di curvatura di una curva regolare di equazione $F = f(x)$. http://www.superzeko.net.

Fuller, T., Tarwater, D., 1992. Analytic Geometry. Addison-Wesley Publishing Company, Reading, MA.

Gallier, J., 2011. Geometric Methods and Applications for Computer Science and Engineering, 2nd edition. Springer, New York.

Gauss, C.F., 1828. Disquisitiones generales circa superficies curvas. Typis Diederichianis, Göttingen.

Gotman, A.Sh., no year indicated. Use of developable surfaces for designing well-streamlined ship shapes. shipdesign.ru/Gotman/Streamlines_Shape_Hull.pdf.

Hahn, H.G., 1992. Technische Mechanik, 2nd edition. Carl Hanser Verlag, München.

Hamlin, N.A., 1988. Ship geometry. In: Lewis, E.V. (Ed.), Principles of Naval Architecture, vol. 1. SNAME, Jersey City, NJ.

Henschke, W. (Ed.), 1957. Schiffbautechnisches Handbuch, vol. 1, 2nd edition. VEB Verlag Technik, Berlin.

Hervieu, R., 1985. Statique du navire. Masson, Paris.

Hilbert, D., Cohn-Vossen, S., 1944. Anschauliche Geometrie. New York Dover Publications. The book has an English translation entitled Geometry and the Imagination.

Hoffman, D., Zielinski, T., 1977. The use of conformal mapping techniques for hull surface definition. In: SCAHD77 — First International Symposium on Computer-Aided Hull Surface Definition. Annapolis, Md, pp. 159–174.

Hollister, S.M., 2002. The dirty little secrets of hull design by computer. www.newavesys. com/secrets.htm.

Hoschek, J., Lasser, D., 1993. Fundamentals of Computer Aided Geometric Design. A.K. Peters, Wellesley, MA. Translation of Grundlagen der geometrischen Datenverarbeitung, Teubner, Stuttgart, 1989.

Ilie, D., 1974. Teoria generală a plutitorillor (The General Theory of Floating Bodies). Editura Academiei Republicii Socialiste România, Bucharest (in Romanian).

INSEAN, 1962. Carene di pescherecci. Vasca Navale, Roma.

INSEAN, 1963. Carene di petroliere. Vasca Navale, Roma.

INSEAN, 1965. Carene di navi passeggeri. Vasca Navale, Roma.

ISO 5456-1:1996(E), 1996. Technical Drawings — Projection Methods — Part 1: Synopsis. ISO.

Kang, J.Y., Lee, B.S., 2012. Application of morphing technique with mesh merging in rapid hull form generation. International Journal of Naval Architecture and Ocean Engineering 4, 228–240.

Kantorowitz, E., 1966. Fairing and mathematical definition of ship surface. Shipbuilding and Shipping Record 108, 348–351.

Kantorowitz, E., 1967. Mathematical Definition of Ship Surfaces. Danish Ship Research Institute Report, No. DSF-14. Danish Technical Press, Copenhagen.

Kim, M.-S., Hoschek, J., Farin, G.E., 2002. Handbook of Computer Aided Geometric Design. North Holland, Amsterdam.

Ko, K.H., 2010. A survey: application of geometric modeling techniques to ship modeling and design. International Journal of Naval Architecture and Ocean Engineering 2, 177–184.

Koechert, C., 2010. Entwicklung eines Verfahrens zur semiautomatischen Optimierung von Schiffsrümpfen bezüglich minimalen Schiffswiderstandes. Report 651. TUHH, Hamburg-Harburg.

Konesky, B., 2005. Newer and more robust algorithms for computer-aided design of developable surfaces. Marine Technology and SNAME News 42 (2), 71–79.

Kouh, J.S., 1985. Rechnergestützte Darstellung von Schiffsformen mit rationalen kubischen Splines. Schriftenreihe Schiffbau No. 459. TUHH.h.

Kreyszig, E., 2011. Advanced Engineering Mathematics, 10th edition. John Wiley and Sons, Hoboken, NJ.

Krieger, E. (Ed.), 2010. Johows Hilfsbuch für den Schiffbau, vol. 1. Salzwasser Verlag, Bremen (1910).

Krueger, S., 2001. Rechnergestützte Beschreibung der Schiffsform. TUHH, Harburg. http://www.ssi.tu-harburg.de/doc/webseiten_dokumente/ssi/vorlesungsunterlagen/cadform.pdf.

Krueger, S., 2002. Verzerren von Schiffslinien. TU Harburg. http://www.ssi.tu-harburg.de/doc/webseiten_dokumente/ssi/vorlesungsunterlagen/verzerr.pdf.

Kuo, Ch., 1971. Computer Methods for Ship Surface Design. Longman, London.

Lackenby, M., 1950. On the systematic geometrical variation of ship forms. Transactions INA 92, 289–316.

Leiceaga, X.A., Soto, E., Ruiz, O.E., et al., 2007. Surface fairing for ship hull design. In: Graphica 2007 — VII International Conference on Graphics Engineering for Arts and Design XVIII Simpósio Nacional de Geometria Descritiva e Desenho Técnico.

Lessenich, J., 2009. Draughting curves used in ship design. In: Nowacki, H., Lefèvre, W. (Eds.), Creating Shapes in Civil and Naval Architecture. Brill, Leiden.

Letcher, J.S., 1993. Lofting and fabrication of compound-curved plates. Journal of Ship Research 37 (2), 166–175.

Letcher, J.S., 2009. The Geometry of Ships. The Principles of Naval Architecture Series. SNAME, Jersey City.

Letcher, J.S., Shook, D.M., Shepherd, S.G., 1995. Relational geometric synthesis — Part 1. Computer Aided Design 27 (11), 821–832.

Letuppe, J., 2010. Les origines de la construction "à carvel" au Levant et en Atlantique au travers des sources archéologiques. Chronique d'Histoire Maritime 69, 121–139.

Lloyd's Register, 2016. Rules and Regulations for the Classification of Ships. Lloyd's Register Group, London.

Loureiro, V., 2006. O Padre Fernando Oliveira e o Liuro da Fabrica das Naos. Revista Portugues de Archeologia 2, 353–367.

Lourenço, R., 2010. Automatic Method to Develop Plates, Suitable to the Production Processes. Instituto Superior Técnico, Lisbon. https://fenix.tecnico.ulisboa.pt/downloadFile/395142233596/resumo.pdf.

Marsh, D., 2000. Applied Geometry for Computer Graphics and CAD, 2nd printing. Springer, London.

McCleary, J., 1997. Geometry From a Differentiable Viewpoint. Cambridge University Press, Cambridge, UK.

McCormick, M.E., 2010. Ocean Engineering Mechanics With Applications. Cambridge University Press, Cambridge.

McGee, D., 1998. The Amsler Integrator and the Burden of Calculation. Material History Review (Revue d'histoire de la culture matérielle) 48, 57–74.

McGrail, S., 2001. Portuguese-derived ship design methods in Southern India? In: Proc. International Symposium on Archeology of Medieval Ships of Iberian-Atlantic Tradition. Lisbon, 1998. Alves, Francisco J.S., Lisbon.

Meriam, J.L., Kraige, L.G., 2006. Engineering Mechanics — Dynamics, 2nd edition. John Wiley & Sons, Hoboken, NJ.

Merker, J., 2014. Courbure des surfaces dans l'espace: le Theorema Egregium de Gauss. ArXiv preprint arXiv:1402.1018.

Miranda, S., 2012. Appunti Geometria della Nave. Università degli Studi di Napoli Federico II.

Mitamura, T., 1961. A new method of shell development — Geodesic line method. In: Kihara, H., Otani, M., Fujita, Y. (Eds.), Recent Developments in Shipbuilding Practice in Japan. The Society of Naval Architects of Japan, Tokyo, pp. 25–39.

Molland, A.F. (Ed.), 2008. The Maritime Engineering Reference Book. Butterworth-Heinemann, Amsterdam.

Mortenson, M.E., 1997. Geometric Modeling, 2nd edition. John Wiley and Sons, New York.

Narli, E., Sariöz, K., 2004. The automated fairing of ship hull lines using formal optimisation methods. Turkish Journal of Environmental Sciences 28, 157–166.

Nowacki, H., 2002. Splines im Schiffbau. In: Proc. 21st Duisburg Colloquium on Ship and Ocean Technology: The Ship Out of the Computer. Gerhard Mercator University, Duisburg, pp. 27–53.

Nowacki, H., Bloor, M.I.G., Oleksiewicz, B., 1995. Computational Geometry for Ships. World Scientific, Singapore.

Nowacki, Horst, Ferreiro, L.D., 2003. Historical roots of the theory of hydrostatic stability of ships. In: Rojas, Luis Perez (Ed.), Proceedings of the 8th International Conference on the Stability of Ships and Ocean Vehicles. ETSIN, Madrid, pp. 1–30. Reprinted as Preprint 237. Max Planck Institute for the History of Science (MPIWG), Berlin, 2003.

Nowacki, H., Valleriani, M., 2003. Shipbuilding Practice and Ship Design Methods From the Renaissance to the 18th Century. Max-Planck Institut für Wissenschaftsgeschichte. Preprint 245.

Pavanelli, F., Miliani, M., Marchesini, I., 2003. Lezioni di disegno, vol. 3, Assonometria. Hoepli, Milano.

Penna, M.A., Patterson, R.R., 1986. Projective Geometry and Its Applications to Computer Graphics. Prentice-Hall, Englewood Cliffs, NJ.

Pérez, F.L., Clemente, J.A., Suárez, J.A., et al., 2008. Parametric generation, modeling, and fairing of simple hull lines with the use of nonuniform rational B-spline surfaces. Journal of Ship Research 52 (1), 1–15.

Pérez-Arribas, F., Suárez, J.A., Fernández-Jambrina, L., 2006. Automatic surface modelling of a ship hull. Computer Aided Design 38, 584–594.

Piegl, L., 1991. On NURBS: a survey. IEEE Computer Graphics & Applications, 54–71.

Piegl, L., Tiller, W., 1997. The Nurbs Book, 2nd edition. Springer, Berlin.

Pierrotet, E., 1942. Archittetura Navale. Istituto per le Edizioni scientifiche ed academiche, Milano.

Piskunov, N., 1965. Differential and Integral Calculus. Peace Publishers, Moscow.

Rabelo, A.B., Manso, F.F., 2004. O Planímetro e o Teorema de Green. http://www.mat.ufmg.br/comed/2004/e2004/planimetro.pdf.

Rabien, U., 1977. Transformation von Schiffsformen. Report No. 24. Technische Universität Hannover, Lehrstuhl und Institut für Entwerfen von Schiffen und Schiffstheorie, pp. 8–27.

Reutter, F., 1972. Darstellende Geometrie, vol. 1. G. Braun, Karslruhe.

Rieth, E., 2001. Le cas de la France à la fin du XVIIe siècle: une même méthode de conception des navires en Ponant et au Levant. In: Proc. International Symposium on Archeology of Medieval Ships of Iberian-Atlantic Tradition. Lisbon, 1998. Alves, Francisco J.S., Lisbon.

Rieth, E., 2003a. First archaeological evidence of the Mediterranean whole moulding ship design method: the example of the Culip VI wreck, Spain (XIIIth–XIVth c.). In: Nowacki, H., Valleriani, M. (Eds.), Shipbuilding Practice and Ship Design Methods From the Renaissance to the 18th Century. Max-Planck Institut für Wissenschaftsgeschichte, pp. 9–16. Preprint 245.

Rieth, E., 2003b. La méthode moderne de conception des carènes du "whole-moulding": une mémoire des chantiers navals méditerranéens du Moyen Age. In: Nowacki, H., Valleriani, M. (Eds.), Shipbuilding Practice and Ship Design Methods From the Renaissance to the 18th Century. Max-Planck Institut für Wissenschaftsgeschichte, pp. 17–32. Preprint 245. This paper has been published first in Neptunia 220, pp. 10–21.

Rising, J.S., Almfeldt, M.W., 1964. Engineering Graphics: An Integration of Engineering Drawing, 3rd edition. Wm. C. Brown Co..

Rogers, D.F., 1977. B-spline curves and surfaces for ship hull definition. In: SCAHD77 — First International Symposium on Computer-Aided Hull Surface Definition. Annapolis, Md, pp. 71–96.

Rogers, D.F., 2001. An Introduction to NURBS With Historical Perspective. Morgan Kaufman Publishers, San Francisco.

Rogers, D.F., Adams, A.J., 1990. Mathematical Elements for Computer Graphics, 2nd edition. McGraw-Hill Publishing Company, New York.

Rondeleux, M., 1911. Stabilité du navire. Augustin Challamel, Paris.

Rupp, K.-H., 1981. Zeichen von Schiffslinien. Universität Hannover, Institut für Schiffen und Schiffstheorie. Nr. 41.

Saff, E.B., Snider, A.D., 2003. Fundamentals of Complex Analysis, 3rd edition. Pearson Education, Upper Saddle River.

Scheltema de Heere, R.F., 1970. Buoyancy and Stability of Ships. George G. Harrap & Co., London. With a supplement by Bakker, A.R.

Schmidt, R., 1977. Darstellende Geometrie mit Stereo-Bilder. Bauverlag, Wiesbaden & Berlin. Spanish translation: Geometria descriptiva con figuras estereoscopicas, Editorial Reverté, Barcelona, 1993.

Schneekluth, H., 1980. Entwerfen von Schiffen, 2nd edition. Koehler, Herford.

Schneekluth, H., Bertram, V., 1998. Ship Design for Efficiency & Economy, 2nd edition. Butterworth-Heinemann, Oxford, UK.

Selkirk, O., Nieremair, J.C., 1977. Geometry of the ship. In: Comstock, J.P. (Ed.), Principles of Naval Architecture, 4th reprint. SNAME, New York.

Semyonov-Tyan-Shansky, V., 2004. Statics and Dynamics of the Ship. Peace Publishers, Moscow.

Sessa, C.T., 2008. Rappresentazione e trasformazione delle carene navali. Doctoral research thesis. Federico II University, Napoli.

Shin, J.Z.G., Ryu, C.H., 2000. Nonlinear kinematic analysis of the deformation of plates for ship hull fabrication. Journal of Ship Research 44 (4), 270–277.

Simmons, C., Maguire, D., Phelps, N., 2012. Manual of Engineering Drawing, 4th edition. Butterworth-Heinemann.

SNAME - D-9, 1966. Model Resistance Data Sheets, Set of 9 Trawlers. SNAME, New York.

Söding, H., 1963. Das Straken von Schiffslinien mit Digitalrechnern. HANSA 104 (1), 1387–1394.

Söding, H., Rabien, U., 1977. Hull surface design by modifying hull. In: SCAHD77 — First International Symposium on Computer-Aided Hull Surface Definition. Annapolis, Md, pp. 19–29.

Struik, D.J., 1961. Lectures on Classical Differential Geometry. Addison-Wesley Publishing Company, Reading, MA.

Thuillier, P., Belloc, J.C., de Villèle, A., 1991. Mathématiques — Géométrie différentielle. Masson, Paris.

Townsend, A., 2014. History of the Spline Computational Curve and Design — AlaTown. www.alatown.com/spline/ originally published in the International Journal of Interior Architecture + Spatial Design, Applied Geometries (Jonathon Anderson & Meg Jackson, 2014).

Valenti, V.N., 2009. Transitions in Medieval Mediterranean Shipbuilding — A Reconstruction of the "Nave Quadra" of the Michael of Rhodes Manuscript. MA thesis. Texas A&M University.

Ventura, M., no year indicated. Geometric Modeling of the Hull Form. IST, Lisbon. http://www.mar.ist.utl.pt/mventura/Projecto-Navios-I/EN/SD-1.5.3-HullVol. 24. pp. 147–160.

Versluis, 1977. Computer aided design of shipform by affine transformation. International Shipbuilding Progress 24 (274), 147–160.

Vrijlandt, W., 1948. Theoretische Scheepsbouwkunde – Part I, 2nd reprint. H. Stam, Haarlem. With Molenaar, J., Fabery de Jonge, A.M.

Vuibert, H., 1912. Les anaglyphes géométriques. Librairie Vuibert, Paris.

Walser, H., 2003. Kurvenkrümmung. ETH, Zürich. http://e-collection.library.ethz.ch/eserv/eth:25629/eth-25629-12.pdf.

Watson, D.G.M., 1998. Practical Ship Design. Elsevier, Amsterdam.

Weinblum, G., 1953. Systematische Entwicklung von Schiffsformen. Schriftenreihe Schiffbau, Technische Univesität Hamburg-Harburg.

Weisstein, E.W., no year indicated. Green's Theorem from MathWorlds — A Wolfram Web Resource. http://mathworld.wolfram.com/GreensTheorem.html.

Wilson, P.A., 1987. The Use of Conformal Mapping in Seakeeping Calculations. Ship Science Report No. 32. University of Southampton.

Wylie Jr., C.R., 1966. Advanced Engineering Mathematics. McGraw-Hill Book Company, New York.

Zhao, H., Wang, G., 2008. A new method for designing a developable surface utilizing the surface pencil through a given curve. Progress in Natural Science 18, 105–110.

ANSWERS TO SELECTED EXERCISES

Exercise 1.6. See Fig. 1.

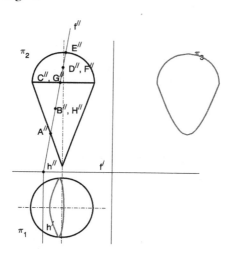

Figure 1 An ice-cream cone

Exercise 1.7. See Fig. 2.

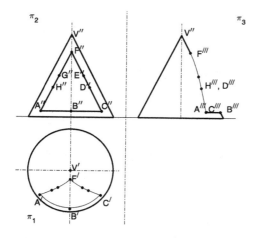

Figure 2 Cone with hole

Exercise 3.2. 6 m^2.

Exercise 3.3.

$$H = \frac{ah}{a-b}, \quad A = \frac{a+b}{2}h, \quad y_c = \frac{h}{3} \cdot \frac{a+2b}{a+b}$$

Exercise 4.5. See Fig. 3.

Figure 3 Trochoid compared to sine

Exercise 5.2. $\rho = 2p$.

Exercise 6.1. See Fig. 4.

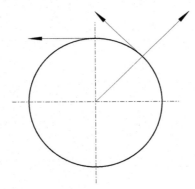

Figure 4 Cylinder — Tangents and normal to a constant-v line

Exercise 6.2.

$$\mathbf{P} = \begin{vmatrix} \frac{(1-v)d}{2}\cos 2\pi u \\ \frac{(1-v)d}{2}\sin 2\pi u \\ hv \end{vmatrix}$$

$$\mathbf{P}_u = \begin{vmatrix} -\pi(1-v)d\sin 2\pi u \\ \pi(1-v)d\cos 2\pi u \\ 0 \end{vmatrix}, \quad \mathbf{P}_v = \begin{vmatrix} \frac{-d}{2}\cos 2\pi u \\ \frac{-d}{2}\sin 2\pi u \\ h \end{vmatrix}$$

Let $d/2 = h\tan\alpha$, where α is the half-angle at the cone vertex. Then,

$$\frac{\mathbf{P}_u \times \mathbf{P}_v}{\|\mathbf{P}_u \times \mathbf{P}_v\|} = \begin{vmatrix} \cos\alpha \cos 2\pi u \\ \cos\alpha \sin 2\pi u \\ \sin\alpha \end{vmatrix}$$

Find the geometrical interpretation of this result.

Exercise 6.6.

$$E = 4\pi^2 r^2, \ F = 0, \ G = h^2, \ A = 2\pi rh$$

Exercise 6.8.

$$z = H\left(1 - \sqrt{\frac{x^2}{a^2} + \frac{y^2}{b^2}}\right)$$

Exercise 7.1. See Fig. 5.

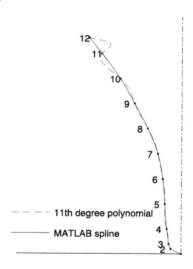

Figure 5 Station 19 of INSEAN hull 1189

Exercise 7.4.

Station	$y = f(x)$	Parametric spline
3	0.2318	0.2293
7	0.3874	0.3887
17	0.2878	0.2942

Exercise 7.8. See Fig. 6.

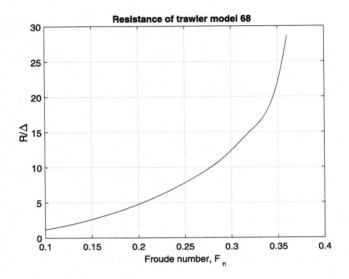

Figure 6 Resistance curve of trawler model

Exercise 8.8. See Fig. 7.

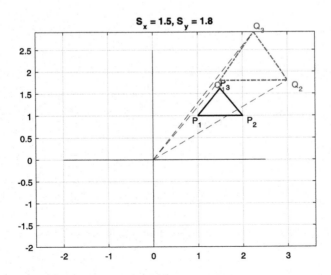

Figure 7 Scaling a triangle with different scales

Exercise 8.9. See Fig. 8.

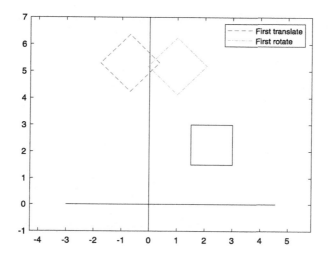

Figure 8 Red dash line: first translate. Magenta point-dash line: first rotate

$$
T_{TR} = \begin{vmatrix} 0.7071 & -0.7071 & -0.7071 \\ 0.7071 & 0.7071 & 2.1213 \\ 0 & 0 & 1.0000 \end{vmatrix},
$$

$$
T_{RT} = \begin{vmatrix} 0.7071 & -0.7071 & 1.0000 \\ 0.7071 & 0.7071 & 2.0000 \\ 0 & 0 & 1.0000 \end{vmatrix}
$$

Exercise 8.10.

$$
\begin{vmatrix} 1 & 0 \\ Sh_y & 1 \end{vmatrix}
$$

Exercise 8.12. Only the distances between points on a horizontal line are preserved.

Exercise 8.13.

$$
\begin{vmatrix} S_x & 0 & 0 \\ 0 & S_y & 0 \\ 0 & 0 & 1 \end{vmatrix} \begin{vmatrix} x \\ y \\ 1 \end{vmatrix} = \begin{vmatrix} xS_x \\ yS_y \\ 1 \end{vmatrix}
$$

Exercise 8.19. Solution in Fig. 9.

Point	Mass	x	x-moment	y	y-moment	z	z-moment
P_1	1.00	1.00	1.00	1.00	1.00	0.00	0.00
P_2	1.30	2.30	2.99	3.00	3.90	0.50	5.07
P_3	2.10	2.00	4.20	4.50	9.45	1.70	3.57
P	4.40	1.86	8.19	3.26	14.35	1.96	8.64
Q_1	3.00	0.40	1.20	0.70	2.10	3.10	9.30
Q_2	3.50	0.40	1.40	0.80	4.90	3.70	12.95
Q_3	4.00	0.80	3.20	1.20	4.80	3.90	15.60
Q	10.50	0.55	5.80	1.12	11.80	3.60	37.85
Total	14.90	0.94	13.99	1.76	26.15	3.12	46.49

Check							
Subsystem	Mass	x	x-moment	y	y-moment	z	z-moment
P	4.40	1.86	8.19	3.26	14.35	1.96	8.64
Q	10.50	0.55	5.80	1.12	11.80	3.60	37.85
Total	14.90	0.94	13.99	1.76	26.15	3.12	46.49

Figure 9 A system of mass points

Exercise 9.1. See Fig. 10.

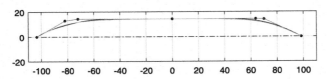

Figure 10 Bézier-curve approximation of a waterline

Exercise 9.2. See Figs 11 and 12.

Exercise 9.4.

$$B_0^3 = \begin{pmatrix} 3 \\ 0 \end{pmatrix} t^0 (1-t)^{3-0} = 1 \cdot 1 \cdot (1-t)^3$$

$$B_1^3 = \begin{pmatrix} 3 \\ 1 \end{pmatrix} t^1 (1-t)^{3-1} = 3 \cdot t \cdot (1-t)^2$$

$$B_2^3 = \begin{pmatrix} 3 \\ 2 \end{pmatrix} t^2 (1-t)^{3-2} = 3 \cdot t^2 \cdot (1-t)$$

$$B_3^3 = \begin{pmatrix} 3 \\ 3 \end{pmatrix} t^3 (1-t)^{3-3} = 1 \cdot t^3 \cdot 1$$

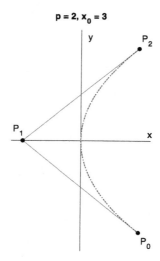

p = 2, x₀ = 3

Figure 11 A parabola drawn as a Bézier curve

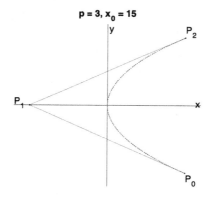

p = 3, x₀ = 15

Figure 12 Another parabola drawn as a Bézier curve

Exercise 9.5.

```
%TRANSLBEZ Exercise on invariance of Bezier curves under translation.

% Initial Bezier curve
% Define control points
P0 = [ 1;    0.5 ];P1 = [ 2.75; 1.5 ];
P2 = [ 4;   3 ]; P3 = [ 2;   4 ];
Pp = [ P0 P1 P2 P3 ];              % control polygon
plot(Pp(1, :), Pp(2, :), 'r-'), grid
xlabel('x'), ylabel('y')
```

```
axis equal
hold on
point(P0, 0.025), point(P1, 0.025)
point(P2, 0.025), point(P3, 0.025)
P  = Bezier3(Pp);
hp = plot(P(1, :), P(2,:), 'k-');
set(hp, 'LineWidth', 1.5)
%%%%%%%%%%%%%%%%%%%% FIRST BEZIER , NEXT TRANSLATION  %%%%%%%%%%%%%%%%%%%%
% Shear Bezier curve
Tt = [  1.8 ;                     % vector of translation
       -1.5 ];
Pr = P + Tt;
hp = plot(Pr(1, :), Pr(2, :), 'b--');
set(hp, 'LineWidth', 1.5)
%%%%%%%%%%%%%%%%%%%% FIRST TRANSLATION , NEXT BEZIER  %%%%%%%%%%%%%%%%%%%%
% Reflect control polygon
P0r = Pp + Tt;
plot(P0r(1, :), P0r(2, :), 'r-')
point(P0r(:,1), 0.03), point(P0r(:, 2), 0.03)
point(P0r(:, 3), 0.03), point(P0r(:, 4), 0.03)
% Calculate and plot Bezier curve
Q  = Bezier3(P0r);
plot(Q(1, :), Q(2,:), 'm:');
hold off
```

The result is shown in Fig. 13.

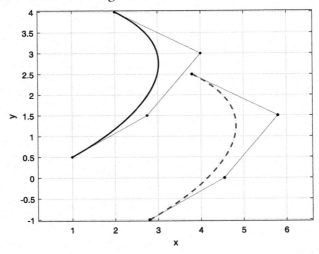

Figure 13 Translation of a Bézier curve

Exercise 9.6. See Fig. 14.

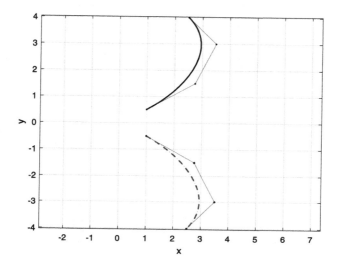

Figure 14 Reflection of a Bézier curve in the *x*-axis

Exercise 9.9. See Fig. 15.

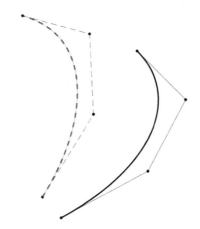

Figure 15 Rotating a Bézier curve

Exercise 9.10. See Fig. 16.

Exercise 9.11. See Fig. 17.

Exercise 9.12. See Fig. 18.

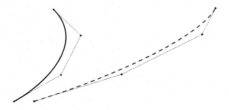

Figure 16 Transforming a Bézier curve by shearing

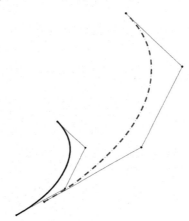

Figure 17 Scaling a Bézier curve

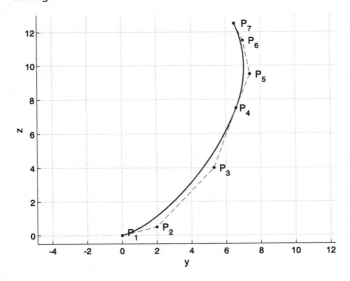

Figure 18 A section composed of two Bézier curves

Exercise 10.4. See Fig. 19.

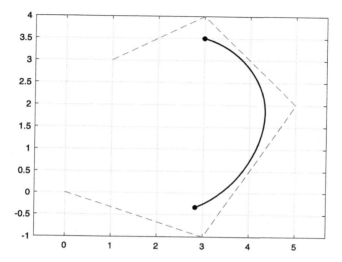

Figure 19 The first and the last points of cubic B-spline

Exercise 10.8. See Fig. 20.

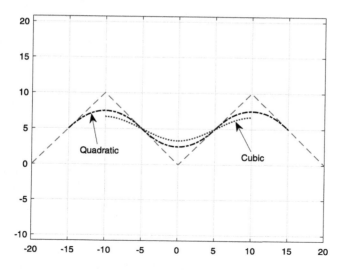

Figure 20 Comparing quadratic and cubic B-spline

Exercise 10.9. See Fig. 21.

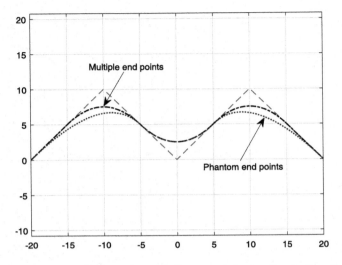

Figure 21 Comparing multiple points with phantom points — Quadratic spline

Exercise 10.10. See Fig. 22.

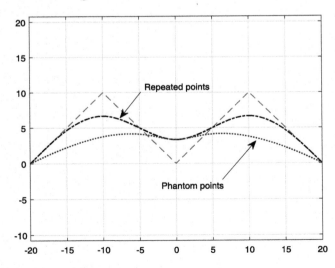

Figure 22 Comparing multiple points with phantom points — Cubic spline

Exercise 10.12. See Fig. 23.

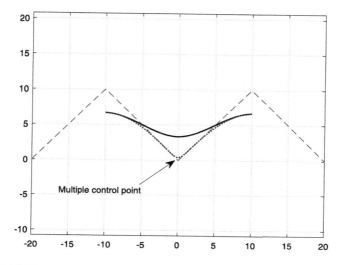

Figure 23 Middle control point repeated — Cubic spline

Exercise 10.14. See Fig. 24.

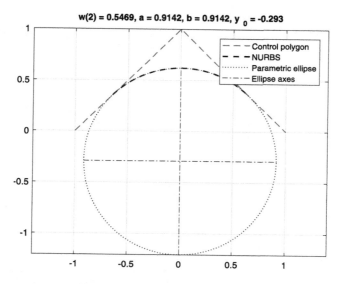

Figure 24 NURBS representing a circle as a special ellipse

Exercise 11.4.

$$C = (1 - t)^2 \mathbf{P}_{0,0} + 2(1 - t)t \frac{\mathbf{P}_{0,0} + \mathbf{P}_{1,1}}{2} + t^2 \mathbf{P}_{1,0}$$

See Fig. 25.

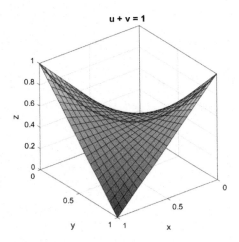

Figure 25 The curve $u + v = 1$ on a patch of a hyperbolic paraboloid

Exercise 11.5. See Fig. 26.

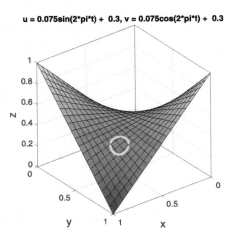

Figure 26 A circle on a patch of a hyperbolic paraboloid

Exercise 13.3. See Fig. 27.

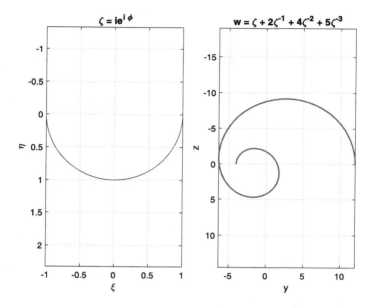

Figure 27 An asymmetrical conformal transformation

Exercise 13.4. See Fig. 28.

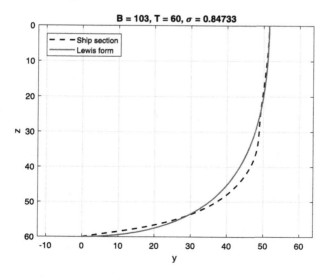

Figure 28 Aft section and its Lewis form

INDEX